Studies in Fuzziness and Soft Computing

Volume 358

About this Series

The series "Studies in Fuzziness and Soft Computing" contains publications on various topics in the area of soft computing, which include fuzzy sets, rough sets, neural networks, evolutionary computation, probabilistic and evidential reasoning, multi-valued logic, and related fields. The publications within "Studies in Fuzziness and Soft Computing" are primarily monographs and edited volumes. They cover significant recent developments in the field, both of a foundational and applicable character. An important feature of the series is its short publication time and world-wide distribution. This permits a rapid and broad dissemination of research results.

More information about this series at http://www.springer.com/series/2941

Carlos Cruz Corona
Editor

Soft Computing for Sustainability Science

Editor
Carlos Cruz Corona
Department of Computer Science
and Artificial Intelligence
University of Granada
Granada
Spain

ISSN 1434-9922 ISSN 1860-0808 (electronic)
Studies in Fuzziness and Soft Computing
ISBN 978-3-319-87300-8 ISBN 978-3-319-62359-7 (eBook)
DOI 10.1007/978-3-319-62359-7

Softcover reprint of the hardcover 1st edition 2017

Printed on acid-free paper

This Springer imprint is published by Springer Nature
The registered company is Springer International Publishing AG
The registered company address is: Gewerbestrasse 11, 6330 Cham, Switzerland

To my sons, Carlos and Frank

Foreword

Finding a single definition of Sustainability Science is a hard task. The next three paragraphs may help the reader to get an idea.

In (Sustainability Science: A room of its own, PNAS 2007 104(6), 1737–1738), one can read *Like "agricultural science" and health science, sustainability science is a field defined by the problems it addresses rather than by the disciplines it employs*.

In the Proceedings of the National Academy of Sciences of the USA (http://sustainability.pnas.org), it is stated that Sustainability Science is *an emerging field of research dealing with the interactions between natural and social systems, and with how those interactions affect the challenge of sustainability: meeting the needs of present and future generations while substantially reducing poverty and conserving the planet's life support systems*.

Bettencourt and Kaur (Evolution and structure of sustainability science, PNAS 2011 108 (49) 19540–19545), after a wide analysis of related scientific publications, concluded: *These developments demonstrate the existence of a growing scientific field of sustainability science as an unusual, inclusive and ubiquitous scientific practice and bode well for its continued impact and longevity*.

Nowadays our planet faces several challenges related to the efficient and careful use of scarce resources. Fighting against climate change, taking actions towards sustainable agriculture and forestry; promoting research on secure, clean and efficient energies and smart, green, integrated transportation modes; developing policies on smart cities and so on are in clear connection with Sustainability Science.

As a consequence, there is a clear consensus on the need of promoting, supporting and developing the field of Sustainability Science as the quality of life of our future generations depends on what we do today.

In order to cope with the challenges and high degree of complexity of the problems arising in this context, there is a clear need for the integration of many sources of knowledge, new methodological approaches, tools and techniques. One key aspect of these problems is the presence of fuzziness, vagueness and imprecision in one or more problems' characteristics. Allowing a proper modeling and management of these types of uncertainties are clearly "must-have" features that, in turn, will enable decision-makers and stakeholders to better understand and solve the problems at hand.

It is here where the application of Soft Computing methodologies can play an essential role. Soft Computing has a long history of successful developments that I would not pretend to summarize here.

In my opinion, the idea of addressing sustainability science's problems using Soft Computing techniques is up-to-date, relevant and well aligned with the research challenges proposed by the most prominent scientific agencies and governments.

This book collects a showcase on the application of different Soft Computing techniques in the context of Sustainability Science. Fuzzy optimization, machine learning, decision making, prediction and so on, are applied to location problems, agriculture, topography, multimodal transport, Vehicular Ad-Hoc Networks, vessels fuel optimization, etc. The chapters are scientifically sound, quite easy to read and follow and the authors are worldwide distributed: Italy, Spain, México, Cuba, Portugal, Brazil, Belgium, Singapore and Sweden.

To conclude, I would like to say some words about the Editor, Dr. Carlos Cruz. I know him since 2004, when I co-advised his Ph.D. thesis in Soft Computing based, parallel coordinated strategies to solve optimization problems. Since then, Dr. Cruz developed an excellent research profile, together with the constant participation in national and international projects. The use of Soft Computing techniques for solving relevant problems (for example, transportation problems in Cuba) is being constant along his career as a quick check of his publication record can reveal.

Overall, I consider this book a great piece of work, not only for well-established researchers but also for Ph.D. students trying to find good research opportunities.

March 2017

David Pelta
Department of Computer Science
and Artificial Intelligence, University of Granada,
Granada, Spain

Acknowledgements

This work has been partially funded by the Spanish Ministry of Economy and Competitiveness with the support of the project TIN2014-55024-P, and by the Regional Government of Andalusia—Spain with the support of the project P11-TIC-8001 (both including funds from the European Regional Development Fund, ERDF).

Contents

Contributors

Pavel Anselmo Álvarez Carrillo Department of Economic and Management Sciences, University of Occident, Culiacan, Sinaloa, Mexico

Davide Anguita DIBRIS—University of Genoa, Genoa, Italy

Leticia Arco Department of Computer Science, Central University of Las Villas, Santa Clara, Cuba

Alfonso Bahillo DeustoTech-Fundacion Deusto, Deusto Foundation, Bilbao, Spain; Faculty of Engineering, University of Deusto, Bilbao, Spain

Francesco Baldi Department of Shipping and Marine Technology, Chalmers University of Technology, Gothenburg, Sweden

Guillermo Bárcena-Gonzalez Department of Computer Science and Engineering, University of Cádiz, Cádiz, Spain

Andrea Coraddu R&D Department, DAMEN Shipyard Singapore, Singapore, Singapore

Carlos Cruz Corona Department of Computer Science and A.I. University of Granada, Granada, Spain

Luis Enrique Diez DeustoTech-Fundacion Deusto, Deusto Foundation, Bilbao, Spain; Faculty of Engineering, University of Deusto, Bilbao, Spain

Rayner Domínguez García Departamento de Automática y Computación, Instituto Superior Politécnico José A. Echevarría CUJAE, La Habana, Cuba

Christopher Expósito-Izquierdo University of La Laguna, San Cristóbal de La Laguna, Spain

Airam Expósito-Márquez University of La Laguna, San Cristóbal de La Laguna, Spain

Armando Fernandes INOV—INESC Inovação, Lisboa, Portugal

Pedro L. Galindo Department of Computer Science and Engineering, University of Cádiz, Cádiz, Spain

Véronique Gomes CITAB-Centre for the Research and Technology of Agro-Environmental and Biological Sciences, Universidade de Trás-os-Montes e Alto Douro, Vila Real, Portugal

Elisa Guerrero Department of Computer Science and Engineering, University of Cádiz, Cádiz, Spain

Maria P. Guerrero-Lebrero Department of Computer Science and Engineering, University of Cádiz, Cádiz, Spain

Alejandro Gómez-Boix Department of Computer Science, Central University of Las Villas, Santa Clara, Cuba

Virgilio C. Guzmán Unidad Académica de Ciencias y Tecnologías de la Información, Universidad Autónoma de Guerrero, Mexico, Mexico

Unai Hernandez-Jayo DeustoTech-Fundacion Deusto, Deusto Foundation, Bilbao, Spain; Faculty of Engineering, University of Deusto, Bilbao, Spain

Idoia de la Iglesia DeustoTech-Fundacion Deusto, Deusto Foundation, Bilbao, Spain; Faculty of Engineering, University of Deusto, Bilbao, Spain

Vicente Liern Universitat de València, Valencia, Spain

Orestes Llanes Santiago Departamento de Automática y Computación, Universidad Tecnológica de La Habana José A. Echevarría CUJAE, La Habana, Cuba

Juan Carlos Leyva López Department of Economic and Management Sciences, University of Occident, Culiacan, Sinaloa, Mexico; Autonomous University of Sinaloa, Culiacan, Sinaloa, Mexico

Manuel López-Coello Department of Computer Science and Engineering, University of Cadiz, Puerto Real (Cádiz), Spain

Jaime Martel Itelligent Information Technologies. Parque Tecnológico CEEI, El Puerto de Sta Maria (Cádiz), Cádiz, Spain

Luis Martínez University of Jaen, Campus Las Lagunillas, s/n, Jaen, Spain

Antonio D. Masegosa DeustoTech-Fundacion Deusto, Deusto Foundation, Bilbao, Spain; Faculty of Engineering, University of Deusto, Bilbao, Spain; IKERBASQUE, Basque Foundation for Science, Maria Diaz de Haro, 3, Bilbao, Spain

Belén Melián-Batista University of La Laguna, San Cristóbal de La Laguna, Spain

Pedro Melo-Pinto CITAB-Centre for the Research and Technology of Agro-Environmental and Biological Sciences, Departamento de Engenharias, Escola de Ciências e Tecnologia, Universidade de Trás-os-Montes e Alto Douro, Vila Real, Portugal

J. Marcos Moreno-Vega University of La Laguna, San Cristóbal de La Laguna, Spain

Ann Nowé COMO Lab, Vrije Universiteit Brussel, Brussel, Belgium

Luca Oneto DIBRIS—University of Genoa, Genoa, Italy

Enrique Onieva DeustoTech-Fundacion Deusto, Deusto Foundation, Bilbao, Spain; Faculty of Engineering, University of Deusto, Bilbao, Spain

David A. Pelta Universidad de Granada, Granada, Spain

Blanca Pérez-Gladish Universidad de Oviedo, Asturias, Spain

Rafael Pino-Mejías Department of Statistics and Operational Research, University of Seville, Seville, Spain

J.F. Reinoso-Gordo Department of Architectonic and Engineering Graphic Expression, University of Granada, Granada, Spain

Mario Rivas-Sánchez Itelligent Information Technologies. Parque Tecnológico CEEI, El Puerto de Sta Maria (Cádiz), Cádiz, Spain

Adrián Rodríguez Ramos Departamento de Automática y Computación, Universidad Tecnológica de La Habana José A. Echevarría CUJAE, La Habana, Cuba

Rosa M. Rodríguez Department of Computer Science and Artificial Intelligence, University of Granada, Granada, Spain

R. Romero-Zaliz Department of Computer Science and Artificial Intelligence, University of Granada, Granada, Spain

Ricardo C. Silva Institute of Science and Technology, Federal University of São Paulo, São José dos Campos, SP, Brazil

Esther-Lydia Silva-Ramírez Department of Computer Science and Engineering, University of Cadiz, Puerto Real (Cádiz), Spain

Omar Ahumada Valenzuela CONACYT Research Fellow, Management Sciences Doctorate Program, University of Occident, Culiacan, Sinaloa, Mexico

José Luis Verdegay Galdeano Department of Computer Science and Artificial Intelligence, University of Granada, Granada, Spain

Juliana Verga Department of Telematics, School of Electrical and Computer Engineering, University of Campinas, Campinas, SP, Brazil

Akebo Yamakami Department of Telematics, School of Electrical and Computer Engineering, University of Campinas, Campinas, SP, Brazil

Yeleny Zulueta University of Informatics Science, Havana, Cuba

Soft Computing Techniques and Sustainability Science, an Introduction

Carlos Cruz Corona

Abstract Sustainability Science is a research field that seeks to understand the fundamental character of interactions between Nature and Society. Because of the high degree of complexity of the problems and challenges it faces, this field requires new methodological approaches, tools and techniques that enable decision-makers and stakeholders can evaluate and make decisions based on a wide range of uncertainty and little information. It is here where Soft Computing methodologies can play an important role in addressing many of these challenges. The inherent tolerance of uncertainty and imprecision, and the robustness of the techniques that make up this paradigm can help solve or reduce the impact of these problems. This chapter introduces this book as a a catalogue of the most successful Soft Computing methodologies applied to Sustainability Science.

According to the National Centers for Environmental Information, NCEI, the global annual temperature has increased at an average rate of 0.07 °C (0.13 °F) per decade since 1880 and at an average rate of 0.17 °C (0.31 °F) per decade since 1970 [7]. It is an undeniable fact that the climate of our planet is warming at an alarming and unprecedented rate compared to past events.

The change in global surface temperature relative to 1880–2016 average temperatures is illustrated in Fig. 1. The temperature average is calculated for five-year variation. We can see that the 10 warmest years have all occurred since 2000 (with the exception of 1998), and 2015 ranks as the warmest year on record.

The dimension of the consequences of this situation is uncertain and is already dramatic. The frequency of extreme weather events is increasing today, such as a decrease in cold temperature extremes, an increase in warm temperature extremes, an increase in extreme high sea levels and an increase in the number of heavy precipitation events in a number of regions, an abrupt and possible irreversible changes in some physical systems and ecosystems are occurring such as the threat of disap-

C.C. Corona (✉)
Department of Computer Science and A.I. University of Granada,
Granada, Spain
e-mail: carloscruz@decsai.ugr.es

C. Cruz Corona (ed.), *Soft Computing for Sustainability Science*,
Studies in Fuzziness and Soft Computing 358, DOI 10.1007/978-3-319-62359-7_1

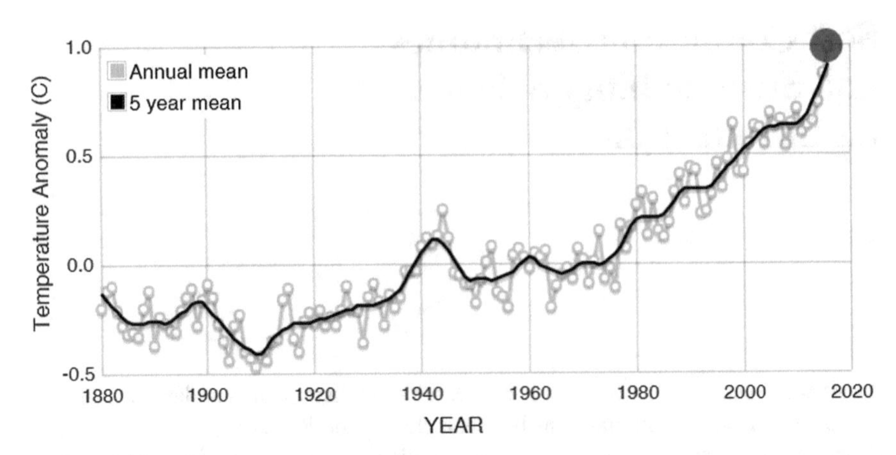

Fig. 1 Change in global surface temperature. *Source* NASA's Goddard Institute for Space Studies (GISS)

pearing coral reefs by the process of acidification due to an increased uptake of CO_2 by the oceans.

Then, how does this affect us as a biological species living on this planet? It depends on a multitude of factors that can be grouped within the concept of social vulnerability [2, 8], which refers to the lack of capability of individuals, groups, or communities to cope with and adapt to external stresses placed on their livelihoods and wellbeing. The 2030 Agenda for Sustainable Development, Member States worries that "its adverse impacts undermine the ability of all countries to achieve sustainable development... The survival of many societies, and of the biological support systems of the planet, is at risk" [13].

Climate change and sustainable development are very closely related. Poor and developing countries, particularly the Least Developed Countries, will be among those most adversely affected and where populations are most vulnerable and least likely to easily adapt to climate change [14]. Therefore, the effects will not be the same for one who lives on a small low-lying island in the Caribbean Sea as it will for one who lives in an industrialised and developed nation.

A typical example of the effects of sea-level rise on Caribbean nations is Haiti, which is projected to experience 0.61 m (low estimate: 0.41 m, high estimate: 1.04 m) of sea-level rise in a 4 °C world [12]. Another example is Cuba, where the predictions say that by 2050 over 2 percent of the country could be submerged and the rise in average sea level could be 34 mm/year [1]. In nations like these, protecting the coasts is a matter of national security. Nevertheless, irrespective of its size or strength, no country is immune from the impact of climate change, and no country can afford to tackle the climate challenge alone.

The cost of adapting to the impact of climate change will affect the cost of living if people, through serious policies of their respective countries do not commit themselves to achieving the goal of limiting temperature rise to well below 2 °C. The economist Nicholas Stern in his famous review for the Government of the United Kingdom in 2006, has already made due reference to this subject: "the evidence gathered by the Review leads to a simple conclusion: the benefits of strong, early action considerably outweigh the costs" [10].

Having arrived at this point, the question is: What can we all do? Human influence on the climate system is clear, but we have the means to limit climate change and build a more prosperous, sustainable future. These are some of the key findings of the Synthesis Report launched by the United Nation's Intergovernmental Panel on Climate Change [4]. Policies, integrated responses and measures at all levels: international, regional, national and sub-national are needed.

Resources, science, technology, education and training, capacity development, partnerships, and data and information are all essential to empowering countries to take action in all climate-related sectors. The United Nations Climate Change Conference (COP 22) in Marrakech marked a new era of implementation and action on climate and sustainable development. In the Marrakech Action Proclamation, State Parties to the United Nations Framework Convention on Climate Change affirmed their "commitment" to the "full implementation" of the Paris Agreement [6].

Acting on climate change is everyone's responsibility. All the people on this planet should be involved in actions for creating a safer climate. The program "Take Action" by the United Nations for all ordinary citizens [15], the practical steps about "What We Do" in the website of Greenpeace [3] and the steps that can be taken at Home, School, the Office, and "On the Road" provided by the United States Environmental Protection Agency [16] all offer up interesting ideas.

In our case, as scientists, our role is needed more than ever to provide scientific and technical support in identifying new development approaches and to enable technological leaps. Furthermore, helping decision makers, influencers, investors, students and all members of the public to formulate best climate change policy options with sustainable development. An interesting interview with a number of experts about the role of science in sustainable development, and in particular its impact on progress towards the The Millennium Development Goals can be read in [9].

The emergent academic discipline of Sustainability Science is a research field that seeks to understand the fundamental character of interactions between Nature and Society. According to the report Sustainability Science in a Global Landscape [11] conducted by Elsevier in collaboration with SciDev.Net, 334,019 articles on Sustainability Science were produced in the period 2009–2013 with an annual growth rate of 7.6%. Because of the high degree of complexity of the problems and challenges it faces, this field requires the integration of many sources of knowledge, new methodological approaches, tools and techniques that enable decision-makers and stakeholders to better understand the dynamic interactions between these issues so they can evaluate and make decisions based on a wide range of uncertainty and little information [5].

It is here where Soft Computing methodologies [17] can play an important role in addressing many of these challenges. The inherent tolerance of uncertainty and imprecision, and the robustness of the techniques that make up this paradigm can

Table 1 Proposals and soft computing techniques used

Chapter	Proposal	Soft computing techniques
1	Introduction	
2	Vessels fuel consumption: a data analytics perspective to sustainability	Machine learning techniques
3	FuzzyCovering: a spatial decision support system for solving fuzzy covering location problems	Fuzzy linear programming, metaheuristics
4	A fuzzy location problem based upon georeferenced data	Fuzzy multi-objetive programming, metaheuristics
5	A review of the application to emergent subfields in viticulture	Discriminant analysis, neural networks, support vector machines
6	Consumer segmentation through multi-instance clustering time-series energy data from smart meters	Multi-instance clustering techniques
7	A multicriteria group decision model for ranking technology packages in agriculture	Fuzzy outranking relation, multicriteria decision analysis (ELECTRE)
8	Fuzzy degree of geographic appropriateness for social impact investing	TOPSIS, fuzzy appropriateness measure, IOWA operators
9	A new approach for information dissemination in VANETs based on covering location and metaheuristics	Metaheuristics
10	Product matching to determine the energy efficiency of used cars available at internet marketplaces	Fuzzy classification techniques
11	Fault diagnosis in a steam generator applying fuzzy clustering techniques	Fuzzy clustering techniques
12	An updated review on watershed algorithm	Metaheuristics
13	An application sample of machine learning tools, such as SVM and ANN, for data editing and imputation	Machine learning techniques
14	Multimodal transport network problem: classical and innovative approaches	Fuzzy non-linear programming, metaheuristics, neural networks
15	A linguistic 2-tuple based environmental impact assessment for maritime port projects: application to moa port	Computing with Words, 2-tuple linguistic model

help solve or reduce the impact of these problems. Committed to this idea, the aim of this book is to show recent and novel studies and applications of Soft Computing in this new and complex field providing a catalogue of the most successful Soft Computing methodologies applied to Sustainability Science.

Thus, fourteen chapters solving problems that range from vessel fuel consumption, location of the best sites to place new malls considering sustainability criteria, etc. to the environmental impact assessment for maritime port projects are presented in this book. As one can see in Table 1 there is a set of Soft Computing techniques that can be used to solve appropriately this kind of problems.

The chapters are listed in alphabetical order by the first author's surname. Each chapter is briefly introduced below:

Andrea Coraddu, Luca Oneto, Francesco Baldi and Davide Anguita present in their paper **Vessels Fuel Consumption: a Data Analytics Perspective to Sustainability** a proposal to reduce vessel fuel consumption by means of optimising vessel operational conditions. They used Machine Learning techniques based on kernel methods and ensemble techniques predicting the influence of independent variables measured from the on-board monitoring system and the fuel consumption of a specific case study vessel. Specifically, they proposed an innovative Gray Box Model able to exploit the high prediction accuracy of black-box (non-parametric) models while reducing the amount of data required for training the model by adding a white-box model component. Thus, two different types of gray, "hybrid" modeling approaches combining elements of white and black box models were presented: the naive, and the advanced approach. In addition, the Regularised Least Squares, Lasso Regression and Random Forest methods were proposed for the construction of black box models. The results of this work showed the superiority of statistical methods over mechanistic models in their ability to accurately predict the performance of the vessel, and highlighted that application of gray box models to the problem of trim optimisation allowed for identifying the possibility of decreasing fuel consumption by up to 2.3% without the need for installing further equipment on board.

The location problem has been historically related to the economic dimension of sustainability, either explicitly or implicitly. Considering the usefulness and relevance in determining a suitable location for the facilities of a specific practical problem, Virgilio Cruz-Guzmán, David A. Pelta and José Luis Verdegay present **FuzzyCovering: a Spatial Decision Support System for solving fuzzy covering location problems**. The two covering location problems more commonly used, the Set Covering Location Problem (SCLP) and the Maximal Covering Location Problem (MCLP) are integrated in this tool, evaluating and modeling the uncertainty in the formulation of these problems. Hence, the models are more adjusted to the characteristics of real problems. Assuming that the decision maker allows some violations in the accomplishment of the distance constraint, a fuzzy approach is designed. The resulting fuzzy problem is converted into a crisp set of problems that can be solved by metaheuristics provided in the tool, from whose results a fuzzy solution is built for the starting problem. The modular architecture of FuzzyCovering allows it to extend its functionality by including of other location models and solution algorithms.

The location of the best sites to place new malls taking into account sustainability criteria is solved by Expósito-Márquez, Expósito-Izquierdo, Melián-Batista and Moreno-Vega using a fuzzy multi-objective optimization problem in their proposal **A Fuzzy Location Problem based upon Georeferenced Data**. The problem is modeled considering a real-environment by using open georeferenced data extracted from the OpenStreetMap project and handled by a Geographic Information System (GIS). Each point provided by the GIS receives a certain degree of membership for each criterion, which indicates the degree to which they satisfy the relevant criterion. Thus, each point on the map under analysis is considered a feasible location for a new requested mall. The authors propose fuzzy membership functions for representing the level of possible membership to the desired locations defined by the problem criteria. Specifically, Gaussian functions are applied to the points contained in the data layers associated with motorways, railways, and cities. The resulting locations are appropriately combined to determine the most interesting areas to locate the requested malls by using the weighted linear combination with weights assigned to the individual functions. Such tools are useful to decision-makers for storing, displaying, and analysing the existing information.

The integration of spectroscopic techniques with Soft Computing and multivariate analysis algorithms has emerged as a promising method for non-destructive analysis in different areas. Armando Fernandes, Véronique Gomes and Pedro Melo-Pinto present **A review of the application to emergent subfields in viticulture of local reflectance and interactance spectroscopy combined with soft computing and multivariate analysis**. This work is focused on the measurement of three enological parameters (sugar content, pH and anthocyanin content) in samples composed of a small number of whole berries and the identification of different varieties or clones of grapevine. This review analyses soft computing algorithms such as discriminant analysis, neural networks isolated or associated in committees and support vector machines. Also, multivariate analysis algorithms such as partial least squares and multiple linear regression were included. Some comparisons between the performance of discriminant analysis, neural networks, support vector machines and partial least squares for the same problems are described in order to obtain a complete overview of the works published in these subfields. Finally, among other suggestions, the authors point out as a priority the expansion of the models validity by including, for each model, data from more varieties, vintages and locations, and the use of more test sets containing a large enough number of samples for these sets to be representative of all possible samples.

A general approach to applying multi-instance clustering algorithms applied to real-life data of Belgian energy consumption is proposed by Alejandro Gómez-Boix, Leticia Arco and Ann Now in the chapter **Consumer segmentation through multi-instance clustering time-series energy data from smart meters**. The general schema of the proposed methodology has seven stages and applies BAMIC (BAg-level Multi-Instance Clustering) as multi-instance clustering algorithm with three distance functions: minimal Hausdorff distance, maximal Hausdorff distance and average Hausdorff distance. Internal indices are used for evaluating the clustering process. The consumer clustering problem is proposed as a multi-instance clustering problem, in which the authors model a consumer as a bag and each bag consists of some instances, where each instance will represent a day or a month of energy consumption. Using the electricity consumption of several homes from different substations in Belgium over a period of time as data, the approach was enable to identify clusters of houses according to their behavior and the most typical consumers considering their daily and monthly consumption behavior. The results obtained are generally applicable, and could be useful in a general business analysis context.

The need for qualifying technology packages to achieve the goals of agricultural companies in terms of economic, productivity and the environmental criteria is a difficult problem for farmers in particular when there is more than one decision-maker evaluating the alternatives. In this line, Juan Carlos Leyva López, Pavel Anselmo Álvarez Carrillo and Omar Ahumada Valenzuela propose **A multicriteria group decision model for ranking technology packages in agriculture** by means of a fuzzy outranking approach developed to work on cases with a great divergence among the decision-makers. It is based on individual solutions generated by decision-makers with ELECTRE III (valued outranking relation and ranking of alternatives) and a global solution generated with ELECTRE GD (collective ranking) representing every decision-maker. A feedback process based on an indirect procedure of preference disaggregation analysis is used to support the proposal of a temporary collective solution in which individual inter-criteria parameters are inferred concerning individual and global preferences.

Vicente Liern and Blanca Pérez-Gladish in their chapter **Fuzzy degree of geographic appropriateness for Social Impact Investing** define the degree of appropriateness of a country in terms of impact investment with respect to two dimensions: the ease of doing business and the level of human development. Bearing in mind the imprecise, ambiguous and uncertain nature of data related to social impact investment, an overall fuzzy measure of the appropriateness of a geographic area in terms of impact investing is designed, taking into account the variability and different nature of the scores in each topic for the dimensions considered. TOPSIS, Technique for Order Preference by Similarity to Ideal Solution, is used as an aggregation method for both appropriateness degrees, and therefore a ranking of the countries, which shows their suitability in terms of several indicators related to these two dimensions, ease of doing business and human development, is obtained.

Vehicular Ad-Hoc Networks (VANETs) have become an active area of research and development, among other things, due to the huge number of innovative applications that they can enable. For example, these have a high impact on reducing Greenhouse Gas emissions produced by vehicles related to traffic management and driver assistance. Antonio D. Masegosa, Idoia de la Iglesia, Unai Hernández-Jayo, Luis Enrique Diez, Alfonso Bahillo and Enrique Onieva in the chapter **A new approach for information dissemination in VANETs based on covering location and metaheuristics** present a new approach for information dissemination in VANETs where the process for selecting the Virtual Infrastructures (VI) is formulated as a Covering Location Problem. This optimization problem aims at maximizing the covering while minimizing the number of facilities, represented in this case by the area covered and the number of vehicles used as VI, respectively. A generational Genetic Algorithm with elitism is used to solve the problem and it was tested in a real scenario consisting of 45 vehicles moving in a rectangular area of 600 m $\times$ 700 m in the downtown area of the city of Málaga, Spain.

Product Matching aims at disambiguating descriptions of products belonging to several different data sources in order to be able to recognize identical products and to merge the content from those identical items. Related to this, Mario Rivas-Snchez, Maria P. Guerrero-Lebrero, Elisa Guerrero, Guillermo Bárcena-González, Jaime Martel and Pedro L. Galindo in their chapter **Product matching to determine the energy efficiency of used cars available at Internet marketplaces** evaluate some similarity measures for string matching and describe the complete procedure for obtaining a product linkage between the offers in the retail market of used cars and the Ministry of Industry, Energy and Tourism Agency dataset in order to determine the real efficiency index of these cars. They use a process comprising for four stages: pre-processing, indexing, matching and supervision. A fuzzy k-nearest neighbor algorithm is used as a classifier. The results obtained could be used as a starting point for further analysis of energy efficiency in the Spanish used cars market

The design of a fault diagnosis system using fuzzy clustering techniques for a BKZ-340-140 29M steam generator in a thermoelectric power station is presented by Adrián Rodríguez Ramos, Rayner Domínguez García, José Luis Verdegay Galdeano and Orestes Llanes Santiago in the chapter **Fault diagnosis in a steam generator applying fuzzy clustering techniques**. The application aims to study the advantages of these techniques in the development of a fault diagnosis method with the following characteristic to be robust to external disturbances and sensitive to small magnitude faults. The wavelet transform (WT) is used for isolating noise present in measurements. The fault diagnosis system was designed for the water-steam circuit of the steam generator for its great incidence in the correct operation of the generation blocks. The results obtained indicate the feasibility of the proposal.

The watershed transformation, at the basis of the morphological approach to segmentation, is a fast image segmentation method, resembling a topographic region growing process by its construction mode. Most of the research in watershed algorithms is specifically devoted to image segmentation, but there are some applications to real topographical watersheds, sustainability and flood risk evaluation. Rocio Romero-Zaliz and Juan Francisco Reinoso-Gordo review the most important works done on watershed algorithms in the chapter **An updated review on watershed algorithm**. In this work they analyze separately the two main strategies for determining watersheds: by inmersion and by rainfall concluding that it is still an open problem and that a lot of improvement must be made in this field, especially in areas related to the problem of the over-segmentation and parallel approaches.

One of the challenges facing the study of sustainability as science is the absence of data or availability of inconsistent data or unreliable data. Esther-Lydia Silva-Ramírez, Manuel López-Coello and Rafael Pino-Mejías in their chapter titled **An application sample of machine learning tools, such as SVM and ANN, for data editing and imputation** present an automatic procedure based on machine learning models for managing the data imputation to estimate missing values and the data editing and imputation process to identify and correct values erroneously recorded. The method performance was empirically assessed considering multilayer perceptron and support vector machines for data editing, and for data imputation a multilayer perceptron and a multiple imputation technique combining multilayer perceptron with k-nearest neighbours was used. The results obtained demonstrate that the models proposed improve the automation level and data quality offering a satisfactory performance in comparison with traditional tools.

In order to achieve the integrated and sustainable transport of passengers and freight an efficient multi-modal transport system that optimize the advantages of each mode of transport is necessary. Juliana Verga, Ricardo C. Silva and Akebo Yamakami present in the chapter **Multimodal Transport Network Problem: classical and innovative approaches** an extensive review of the multimodal transport network problem. Classical methods and soft computing methodologies, which combine approximate reasoning, such as fuzzy logic and functional, such as metaheuristics and neural networks, are described in detail. Finally, a novel approach to solving this problem in a fuzzy environment is modeled using graph theory and considering the intrinsic uncertainty of the real world. Thus, this problem is formulated as a fuzzy non-linear programming problem in which the cost of each edge depend on its flow.

The Environmental Impact Assessment (EIA) process has been developed for evaluating the impact of port operations on the environment, including its natural, social and economic aspects. Yeleny Zulueta, Rosa M. Rodríguez and Luis Martínez propose a 2-tuple linguistic model in their chapter **A Linguistic 2-tuple based Environmental Impact Assessment for Maritime Port Projects: Application to Moa Port**. This model is proposed to assess the overall environmental impact of Moa Port in Cuba by using a double matrix that represents impacts, which are characterized by multiple-criteria. The environmental parameters are defined through linguistic variables assessed by 2-tuple linguistic values and the different criteria can be assessed by using different linguistic scales according to experts' knowledge and nature of criteria. The use of linguistic 2-tuple extension, so-called linguistic hierarchies, enables the definition of an evaluation framework that models linguistic variables with different granularity according to the real situation.

In short, the current book can be seen as confirmation of the great potential of Soft Computing techniques to the formalization and robust resolution and low-cost of problems that arise in such a complex line of research as Sustainability Science.

We hope that students, researchers, engineers, and practitioners in the interface between Computer Science and Sustainable Development/Climate Change have in this book a reference catalogue for recent successful applications of Soft Computing methodologies in these fields, on the one hand, and a starting point to develop new research lines in these areas, on the other.

References

1. Alonso, G., Clark, I.: Cuba confronts climate change. MEDICC Rev. **17**(2), 10–3 (2015)
2. Füssel, H.M.: Vulnerability to Climate Change and Poverty, pp. 9–17. Springer, Dordrecht (2012). doi:10.1007/978-94-007-4540-7_2
3. Greenpeace. What We Do. http://www.greenpeace.org/international/en/campaigns/climate-change/Solutions/What-you-can-do/ (2016). Accessed 20 Jan 2017

4. IPCC, Climate Change 2014: Synthesis report. contribution of working groups I, II and III to the Fifth Assessment Report of the Intergovernmental Panel on Climate Change. In: Pachauri, R., Meyer, L. (eds.) Core Writing Team, p. 151. Geneva, Switzerland (2015)
5. Kates, R.W., Clark, W.C., Corell, R., Hall, J.M., Jaeger, C.C., Lowe, I., McCarthy, J.J., Schellnhuber, H.J., Bolin, B., Dickson, N.M., Faucheux, S., Gallopin, G.C., Grübler, A., Huntley, B., Jäger, J., Jodha, N.S., Kasperson, R.E., Mabogunje, A., Matson, P., Mooney, H., Moore, B., O'Riordan, T., Svedin, U.: Sustainability science. Science **292**(5517), 641–642 (2001). doi:10.1126/science.1059386, http://science.sciencemag.org/content/292/5517/641
6. Marrakech Action Proclamation. UN Climate Change Conference, 2016. http://newsroom.unfccc.int/unfccc-newsroom/marrakech-action-proclamation-expresses-irreversible-momentum-on-climate/ (2016). Accessed 20 Jan 2017
7. NCEI, National Centers for Environmental Information. Global Analysis - October 2016. https://www.ncdc.noaa.gov/sotc/global/201610 (2016). Accessed 20 Jan 2017
8. Otto, I.M., Reckien, D., Reyer, C.P.O., Marcus, R., Le Masson, V., Jones, L., Norton, A., Serdeczny, O.: Social vulnerability to climate change: a review of concepts and evidence. Reg. Environ. Change 1–12 (2017). doi:10.1007/s10113-017-1105-9
9. Science: What Has It Done For The Millennium Development Goals? http://www.scidev.net/global/health/feature/science-what-has-it-done-for-the-millennium-development-goals--1.html (2010). Accessed 20 Jan 2017
10. Stern, N.: Stern review on the economics of climate change (pre-publication edition). Archived from the original on 31 January 2010. Accessed 31 Jan 2010
11. Sustainability Science in a Global Landscape. A report conducted by Elsevier in collaboration with SciDev.Net. https://www.elsevier.com/__data/assets/pdf_file/0018/119061/SustainabilityScienceReport-Web.pdf (2015). Accessed 20 Jan 2017
12. Turn Down the Heat: Confronting the New Climate Normal. Washington, DC: World Bank Group. World Bank. License: CC BY-NC-ND 3.0 IGO. https://openknowledge.worldbank.org/handle/10986/20595 (2014). Accessed 20 Jan 2017
13. United Nations. Transforming our World: The 2030 Agenda for Sustainable Development. https://sustainabledevelopment.un.org/content/documents/21252030 (2015). Accessed 20 Jan 2017
14. United Nations. Report of the Secretary-General: Progress towards the Sustainable Development Goals. http://www.un.org/ga/search/view_doc.asp?symbol=E/2016/75&Lang=E (2016). Accessed 20 Jan 2017
15. United Nations. Take Action Programme. http://www.un.org/climatechange/take-action/ (2011). Accessed 20 Jan 2017
16. United States Environmental Protection Agency. What You Can Do about Climate Change. https://www.epa.gov/climatechange/what-you-can-do-about-climate-change (2016). Accessed 20 Jan 2017
17. Verdegay, J.L., Yager, R.R., Bonissone, P.P.: On heuristics as a fundamental constituent of soft computing. Fuzzy Sets Syst. **159**(7), 846–855 (2008). doi:10.1016/j.fss.2007.08.014

Vessels Fuel Consumption: A Data Analytics Perspective to Sustainability

Andrea Coraddu, Luca Oneto, Francesco Baldi and Davide Anguita

Abstract The shipping industry is today increasingly concerned with challenges related with sustainability. CO_2 emissions from shipping, although they today contribute to less than 3% of the total anthropogenic emissions, are expected to rise in the future as a consequence of increased cargo volumes. On the other hand, for the 2 °C climate goal to be achieved, emissions from shipping will be required to be reduced by as much as 80% by 2050. The power required to propel the ship through the water depends, among other parameters, on the trim of the vessel, i.e. on the difference between the ship's draft in the fore and the aft of the ship. The optimisation of the trim can, therefore, lead to a reduction of the ship's fuel consumption. Today, however, the trim is generally set to a fixed value depending on whether the ship is sailed in loaded or ballast conditions, based on results performed on model tests in basins. Nevertheless, the on-board monitoring systems, which produce a huge amount of historical data about the life of the vessels, lead to the application of state of the art data analytics techniques. The latter can be used to reduce the vessel consumption by means of optimising the vessel operational conditions. In this book chapter, we present the potential of data-driven based techniques for accurately predicting the influence of independent variables measured from the on board monitoring system and the fuel consumption of a specific case study vessel. In particular, we show that gray-box models (GBM) are able to combine the high prediction accuracy of black-box models (BBM) while reducing the amount of data required for training the

A. Coraddu (✉)
R&D Department, DAMEN Shipyard Singapore, Singapore, Singapore
e-mail: andrea.coraddu@damen.com

L. Oneto · D. Anguita
DIBRIS—University of Genoa, Genoa, Italy
e-mail: luca.oneto@unige.it

F. Baldi
Department of Shipping and Marine Technology, Chalmers University of Technology, Gothenburg, Sweden
e-mail: francesco.baldi@chalmers.se

D. Anguita
e-mail: davide.anguita@unige.it

C. Cruz Corona (ed.), *Soft Computing for Sustainability Science*,
Studies in Fuzziness and Soft Computing 358, DOI 10.1007/978-3-319-62359-7_2

model by adding a white-box model (WBM) component. The resulting GBM model is then used for optimising the trim of the vessel, suggesting that between 0.5 and 2.3% fuel savings can be obtained by appropriately trimming the ship, depending on the extent of the range for varying the trim.

1 Introduction

Shipping is a relatively efficient mean of transport when compared to other transport modes [75]. Despite its efficiency, however, shipping contributes significantly to air pollution [13], mainly in the form of sulphur oxides, nitrogen oxides, particulate matter, and carbon dioxide. For the latter, the contribution from shipping to global emissions is required to decrease significantly in the coming years [6, 20]. Since greenhouse gas emissions from the combustion of oil-based fuels are directly proportional to fuel consumption, improving ship energy efficiency is one of the possible solutions to this issue. Measures for the improvement of ship energy efficiency are normally divided into design and operational measures. While the former have been associated to larger saving potential, the latter can still provide a significant reduction in fuel consumption, while requiring a much more limited capital investment [6]. However, the large amount of variables influencing ship energy efficiency makes it hard to assess ship performance in relation to a standard baseline. Operational measures include, among others, improvement in voyage execution, reduction of auxiliary power consumption, weather routing, optimised hull and propeller polishing schedule, slow steaming, and trim optimisation [2, 42, 64].

Among the above mentioned fuel saving measures, trim optimisation has been extensively discussed in the past. It is well known, from hydrodynamics principles, that the trim of the vessel can significantly influence its fuel consumption [60]. In most cases principles of trim optimisation are applied roughly; the crew is provided with an indicative value for the trim to use when sailing laden and when sailing ballast, based on model tests. However, many more factors can influence the optimal value of the trim, such as draught, weather conditions, speed [38]. Taking these aspects into account when selecting the appropriate trim can therefore lead to significant, cost-free savings in terms of fuel required for ship propulsion. Previous work in scientific literature related to trim optimisation have focused on two main alternative strategies: White-Box numerical Models (WBMs), and black-box numerical models (BBMs). WBMs describe the behaviour of the ship resistance, propeller characteristics and engine performances based on governing physical laws and taking into account their mutual interactions [49]. The higher the detail in the modelling of the physical equations which describe the different phenomena, the higher the expected accuracy of the results and the computational time required for the simulation. WBMs are generally rather tolerant to extrapolation and do not require extensive amount of operational measurements; on the other hand, when employing models that are computationally fast enough to be used for online optimisation, the expected accuracy in the prediction of operational variables is relatively low. In addition, the construction of the model is a process that requires competence in the field, and availability of technical details which are often not easy to get access to. Examples of the use of WBMs for the optimisation of ship trim are [46], who employed advanced Computational Fluid Dynamics (CFD) methods, and [54] who employed simpler empirical models (Holtrop-Mannen) for the estimation of possible gains from trim optimisation. Differently from WBMs, BBMs (also known as data driven models [74]), make use of statistical inference procedures based on historical data collection. These methods do not require any a-priory knowledge of the physical system and allow exploiting even measurements whose role might be important for the calculation of the predicted variables but might not be captured by simple physical models. On the other hand, the model resulting from a black-box approach is not supported by any physical interpretation [65], and a significant amount of data (both in terms of number of different measured variables and of length of the time series) are required for building reliable models [12]. As an example, in [58] an application of BBMs is proposed (in particular of artificial neural network) to the prediction of the fuel consumption of a ferry and applied to the problem of trim optimisation. Gray-box models (GBMs) have been proposed as a way to combine the advantage of WBMs and BBMs [11]. According to the GBMs principles, an existing WBM is improved using data-driven techniques, either in order to calculate uncertain parameters or by adding a black-box component to the model output [48]. GBMs allow exploiting both the mechanistic knowledge of the underlying physical principles and available measurements. The proposed models are more accurate than WBMs with similar computational time requirements, and require a smaller amount of historical data when compared to a pure BBMs.

The aim of this book chapter is to propose the application of a gray-box modelling approach to the prediction of ship fuel consumption which can be used as a tool for online trim optimisation. In this framework the authors exploit Machine Learning techniques based on kernel methods and ensemble techniques [5, 65] so to improve an effective but simplified physical model [8] of the propulsion plant. The proposed model is tested on real data [3] collected from a vessel during two years of on board sensors data acquisitions (e.g. sheep speed, axis rotational speed, torque, wind intensity and direction, temperatures, pressure, etc.).

2 The Sustainability Challenge in Shipping

2.1 The Shipping Sector

International trade has been a major factor in the development of mankind all throughout the history. This can be seen in particular today, as shipping contributes to approximately 80–90% of global trade (in ton km, [52]) with an increase from 2.6 to 9.8 billion tons of cargo from 1970 to 2014. Today anything from coal, iron ore, oil and gas to grains, cars, and containerized cargo is transported by sea, thus making shipping the heart of global economy [73].

If compared to other transportation modes, shipping is relatively efficient if measured in terms of fuel consumed per unit of cargo transported and of distance covered [75]. Nevertheless, shipping is today under strong pressure for reducing its fuel consumption, both from an environmental and economical perspective.

2.2 *Shipping and Carbon Dioxide Emissions*

The main connection between energy efficiency and sustainability in shipping relates to the emissions of greenhouse gas (GHG), that are considered today to be the main contributor to global warming. Despite carbon dioxide (CO_2) emissions from the shipping sector were estimated to amount to less than 3% of the total in 2012, they are expected to grow in the future by between 50 and 250% in relation to the expected increase of transport volumes [67].

This pressure related to making shipping more sustainable will also have more and more impact on the shipping industry from an economical perspective. Not only environmental regulations are becoming stricter in many areas of the world (compliance often requires higher fuel expenses). In the particular case of CO_2 emissions, market based measures are being discussed, particularly but not only in the European Union, as a mean for incentivising the transition to low-carbon shipping.

In relation to the strive for sustainability, shipping is a very peculiar business, where conditions are not optimal for incentivising energy efficiency. Split incentives are often a hinder to implementing energy efficiency measures, as neither the owner of the ship, nor its operator pay for the fuel [39]. In addition, differently from e.g. planes and cars, ships are built on individual or small-series basis; this makes it particularly expensive to invest into research and development on an individual ship basis [18, 76]. Ships are very long-lasting products, whose operational life can range from 15 to more than 30 years [69].

In these conditions, although technical improvements to ship energy systems (both by retrofitting and in the design phase) are seen as the solutions with the largest potential for reducing ship fuel consumption, operational measures are of particular interest. In fact they do not require any initial investment and, therefore, are particularly easy to implement [6].

2.3 *Operational Efficiency in Shipping*

Operational measures is a category that includes different measures for energy efficiency on board that do not require the installation of new equipment. On most vessels, the energy demand for propulsion represents the largest share of the total energy demand [3]. For this reason, most of the measures that aim at reducing ship fuel consumption relate to the reduction of the fuel demand from the main engines. This, in turn, can be achieved by either reducing the thrust required for moving the ship's hull through the water, or by improving the efficiency of the most relevant conversion components, namely the propeller and the engine. An appropriate optimisation requires, however, an in-depth understanding of the influence of the speed of the vessel on its fuel consumption in different environmental and operational conditions.

As the power demand for propulsion roughly depends on the ship's speed to the third power (up to the fourth power for faster ships), reducing the speed of the vessel is often regarded as a possible solution for improving energy efficiency. Although the practice of slow-steaming has its inconveniences (e.g. demand for more ships to be built, longer time at sea, higher inventory costs), it has been shown that fuel can be saved by optimising the speed at each instant of the voyage, without changing the total voyage time. This practice is normally referred to as weather routing. Its correct application requires, however, not only the availability of reliable short-middle term predictions of the weather conditions, but also of an accurate understanding of the influence of given weather conditions on the ship's power demand for propulsion.

For given conditions of ship speed and weather, there are other operational parameters that influence the power demand for propulsion. In particular, the trim (defined as the difference between the draft at the ship's fore and aft, thereby measuring how much the position of the ship differentiates from that of being parallel to the sea surface) can be optimised in order to adapt to conditions of minimal demand for propulsive power. This is normally done on board starting from rules of thumb based on tests performed on ship physical models at reduced scale, where the ship's average draft and speed are the parameters that most influence the choice of the optimal trim. However, in real operating conditions, not only the influence of these variables can be different from what predicted by model tests, but also other conditions (e.g. the weather) can play a role in the determination of the optimal draft.

The added resistance coming from the growth of different types of organisms on the surface of the hull also plays a major role on the total power demand for ship propulsion, which can increase by up to 100% [63] as a consequence of the increased hull roughness. As a solution to this issue, most ships use a thin layer of poisonous paint (normally referred to as antifouling paint) which slowly releases substances which are poisonous for the organisms that grow on the surface of the hull. In addition, the hull is cleaned and, if necessary, re-painted at specific intervals. The choice of the hull cleaning intervals is, today, mostly based on rules of thumb (e.g. once a year), generally as a consequence of the difficulty of predicting the relative contribution of fouling on the total ship resistance. This practice could therefore be substantially improved if the contribution of added resistance due to fouling to the total ship resistance could be evaluated more accurately.

Finally, a significant share of ships are today equipped with a controllable pitch propeller (CPP), i.e. a propeller where the inclination of the blades in relation to the propeller axis can be changed according to the specific requirements as an operational variables. When this type of propeller is installed, the choice of the pitch is normally pre-set as a function of the propeller speed for optimising the efficiency of the propeller. However, not only the optimal propeller efficiency is also influenced by other factors (e.g. ship draft and weather conditions), but also the engine efficiency is influenced by the choice of its operating conditions in terms of speed and torque requirements. This shows potential for additional fuel savings if the propeller pitch is continuously optimised for optimal efficiency of the entire propulsion train.

It appears clearly from the previous section that an appropriate ability of predicting the influence of all the different environmental and operational variables on the performance of the ship is of utmost importance for achieving the most out of different operational measures for improving ship energy efficiency.

3 Problem Description

3.1 Ship Description

In this book chapter, the authors propose the utilisation of a predictive model of the fuel consumption for the online optimisation of the trim of a vessel. The proposed method has been tested on a Handymax chemical/product tanker in order to show its potential. A conceptual representation of the ship propulsion plant is shown in Fig. 1, while relevant ship features are presented in Table 1. The ship systems consists of two main engines (MaK 8M32C four-stroke Diesel engines) rated 3840 kW each and designed for operation at 600 rpm. The two engines are connected to one common gearbox; the gearbox has two outputs: a controllable pitch propeller designed for operations at 105 rpm for ship propulsion; and a shaft generator (rated 3200 kW)

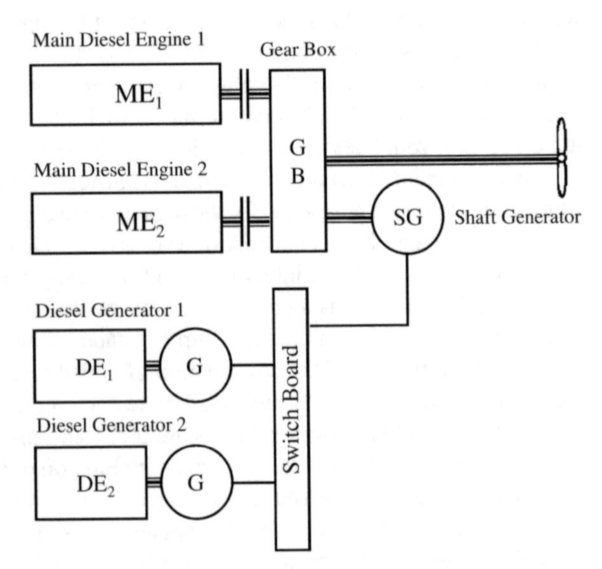

Fig. 1 Conceptual representation of the ship propulsion system

Table 1 Main features of the case study ship

Ship feature	Value	Unit
Deadweight	47000	(t)
Installed power (Main Engines)	3840 (x2)	[kW]
Installed power (Auxiliary Engines)	682 (x2)	[kW]
Shaft generator design power	3200	[kW]
Exhaust boilers design steam gen.	1400	[kg/h]
Auxiliary boilers design steam gen.	28000	[kg/h]

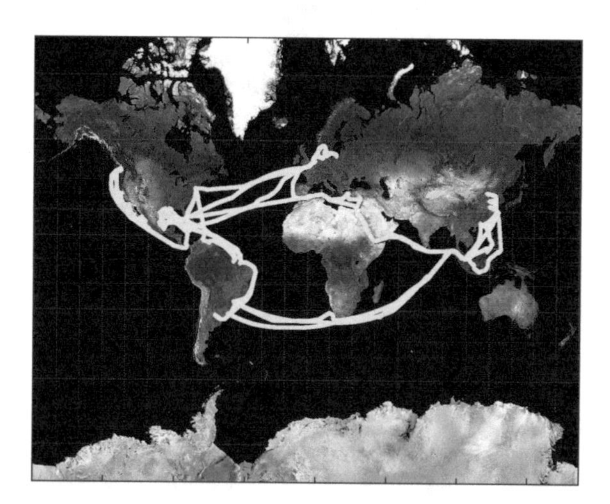

Fig. 2 Description of the ships routes

used for fulfilling on board auxiliary power demand. Auxiliary power can also be generated by two auxiliary engines rated 682 kW each. Auxiliary heat demand is fulfilled by a combination of exhaust gas boilers and auxiliary oil-fired boilers.

The ship is mainly used in the spot market (i.e. based on short-term planning of ship logistics, as opposed to long-term agreements with cargo owners on fixed schedules and routes) and therefore operates according to a variable schedule, both in terms of time spent at sea and of ports visited. The variety of different routes is shown in Fig. 2. Figures 3 and 4 represent the observed ship operations for the selected time period. It can be seen that although the ship spends a significant part of time in port, most of ship operations are related to open sea transport, either in laden or ballast mode (see Fig. 3). The focus of this work lies in the optimisation of ship trim; consequently, only data points related to sailing operations are considered in this study. Operations of manoeuvring, cargo loading, cargo unloading, and port stays were therefore excluded from the original dataset. These transport phases happen at a broad range of speeds, as shown in Fig. 4, which provides additional evidence of the need for an efficient tool for the optimisation of ship operations in different operational conditions. Because of their specific trading pattern, tankers are normally used in two very distinct operational modes: laden (i.e. with full cargo holds, delivering liquid bulk cargo to the destination port), and ballast (with empty

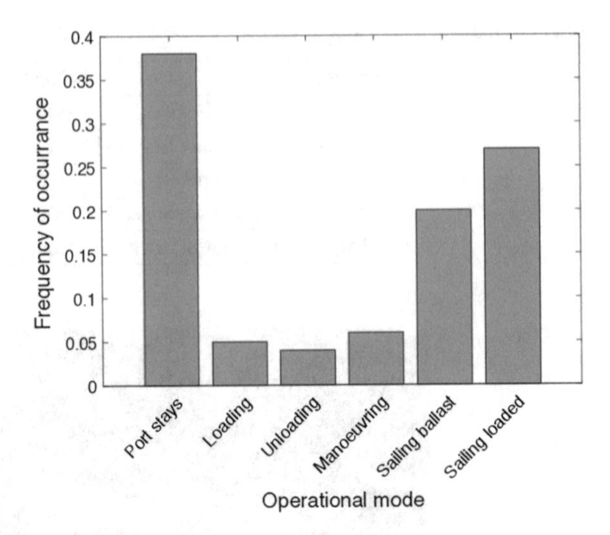

Fig. 3 Time spent in each operational mode for the selected vessel in the chosen period

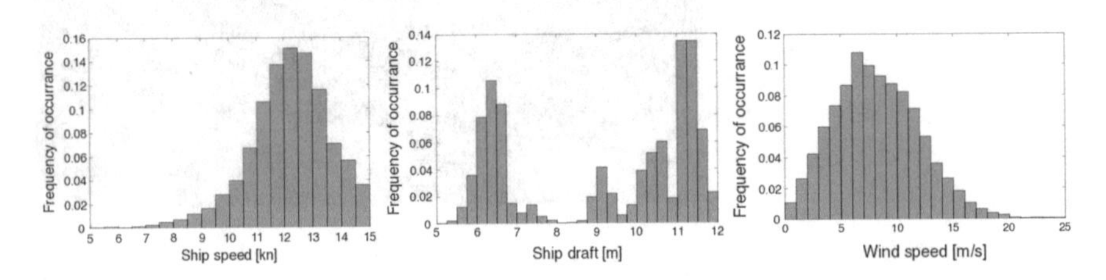

Fig. 4 Speed, draft and wind distributions during sailing time for the selected vessel in the chosen period

cargo holds, sailing to the port where the next cargo is available for loading). In reality, even when loaded, tankers vessels do not always sail with completely full holds due to differences in order sizes. The ship's draught can consequently vary, depending on the operation, from 11 m when the ship is fully loaded to 6 m when cargo holds are completely empty. The distribution of ship draught over the proposed dataset is presented in Fig. 4. In addition to ship speed and draught, weather conditions are also known to have an influence on the optimal trim to be used when sailing, and can vary during ship operations. Figure 4 represents wind speed which, in turn, is strongly correlated to ship added resistance.

3.2 *Data Logging System*

The ship under study is provided with a data logging system installed by an energy management provider which is used by the company both for on board monitoring and for land-based performance control. Table 2 summarises the available measurements from the continuous monitoring system.

The original data frequency measured by the monitoring system is of 1 point every 15 s. In order to provide easier data handling, the raw data are sent to the provider server, where they are processed into 15 min averages. The data processing is performed by the provider company and could not be influenced or modified by the authors.

Measured values come from on board sensors, whose accuracy and reliability cannot be ensured in the process. In particular, issues related to the measurement of speed through water (LOG speed) are well known. Such measurements are often partly unreliable since the flow through the measurement device can be easily disturbed by its interaction with the hull or by other environmental conditions. On the other hand measurements of speed over ground (GPS speed), although more reliable, do not include the influence of currents, which can be as strong as $2 \div 3$ knots depending on time and location and therefore influence ship power demand for propulsion. Fuel consumption is measured using a mass flow meter, which is known to be more accurate of the more common volume flow meters as it eliminates uncertainty on fuel density. It should be noted, however, that measurements of fuel specific energy content (LHV) were not available; variation of heavy fuel oil LHV is known to be in the order of ± 2 MJ/kg, which corresponds to a variation of $\pm 5\%$. Propeller speed, torque measurement and fuel mass flow accuracy were provided by the shipyard at respectively ± 0.1, ± 1 and $\pm 3\%$.

4 From Inference to Data Analytics

Inference is the act or process of deriving logical conclusions from premises known or assumed to be true [51]. There are two main families of inference processes: deterministic, and statistical inference. The former studies the laws of valid inference, while the latter allows to draw conclusions in the presence of uncertainty, and therefore represents a generalisation of the former. Several different types of inference are commonly used when dealing with the conceptual representation of reality as shown in Fig. 5:

- Modelling/approximation refers to the process of building a model of a real system based on the knowledge of the underlying laws of physics that are known to govern the behavior of the system. Depending on the expected use and needs of the model, as well a on the available information, different levels of approximation can be used. Modelling/approximation of a real system only based on mechanistic knowledge can be categorised as deterministic inference [37];

Table 2 Measured values available from the continuous monitoring system

Variable name	Unit
Time stamp	
Latitude	(°)
Longitude	(°)
Fuel consumption (Main engines)	(kg/15 mins)
Auxiliary engines power output	(kW)
Shaft generator power	(kW)
Propeller shaft power	(kW)
Propeller speed	(kW)
Ship draft (fore)	(m)
Ship draft (aft)	(m)
Draft port	(m)
Draft starboard	(m)
Relative wind speed	(m/s)
Relative wind direction	(°)
GPS heading	(°)
Speed over ground (GPS)	(kn)
Speed through water (LOG)	(kn)
Sea depth	(m)
Sea water temperature	(°C)
CPP setpoint	(°)
CPP feedback	(°)
Fuel density	(kg/m^3)
Fuel temperature	(°)
Ambient pressure	(mbar)
Humidity	(%)
Dew point temperature	(°C)
Shaft torque	(kNm)
Rudder angle	(°)
Acceleration X direction	(-)
Acceleration Y direction	(-)
Acceleration Z direction	(-)
GyroX	(-)
GyroY	(-)
GyroZ	(-)
Roll	(-)
Pitch	(-)
Yaw	(-)

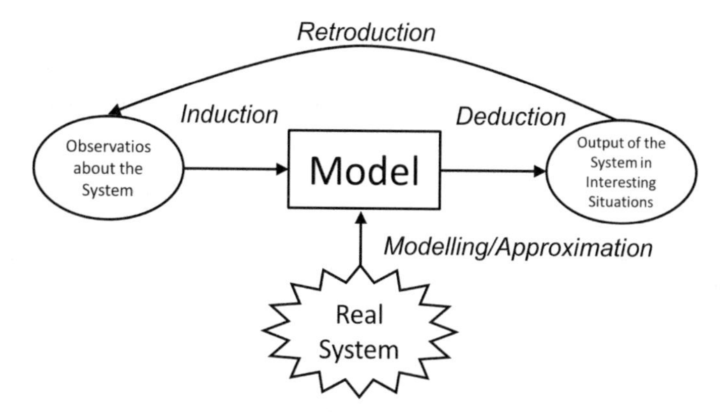

Fig. 5 Type of inference exploited in this book chapter

- When the model is built by statistically elaborating observations of system inputs and outputs, the process belongs to the category of statistical induction. As the model is inferred based on measurements affected by different types of noise, this process is intrinsically under the effect of uncertainty and therefore belongs to the category of statistical inference [74];
- The process of using an existing model to make predictions about the output of the system given a certain input is called deductive inference. This process can be both deterministic or probabilistic depending on how the model is formulated [14];
- The process of actively modifying model inputs in order to obtain a desired output is normally referred to as retroduction (or abduction) [41].

The subject of this book chapter can hence be seen as the application of a general category of problems to a specific case. The physical laws governing ship propulsion are known and widely used in the dedicated literature with the purpose of modelling the ship behaviour [11]. Moreover a series of historical data about the ship's propulsion system are available, and based on this it is possible to build a statistical model of the process [21, 32, 45, 68] which again can be exploited to predict the behaviour of the system. In particular data analytics tools allow performing different levels of statistical modelling [21]:

- descriptive analytics tools allow understanding what happened to the system (e.g. what was the temperature of the cylinders of the engine in the last days). Descriptive analytics answers to the question 'What happened?'
- diagnostic analytics tools allow understanding why something happened to the system (e.g. the fuel consumption it too high and this is due to a the decay of the hull). Diagnostic analytics answers to the question 'Why did it happen?'
- predictive analytics tools allow making predictions about the system (e.g. when a new propeller is installed to reduce fuel consumption). Predictive analytics answers to the question 'What will happen?'

- prescriptive analytics tools allow understanding why the system behave in a particular way and how to force the system to be in a particular state (e.g. what is the best possible way to steer the ship in order to save fuel). Prescriptive analytics answers to the question 'How can we make it happen?'

Descriptive analytics is something very simple to implement, for example in Sect. 3 authors showed some compressed information coming from the historical data collection which can be interpreted as a descriptive analytic process. These tools are the least interesting ones since there is no additional knowledge extracted from the data [21]. Diagnostic analytics is a step forward where the authors try to understand what happened in the past, searching correlation in the data in order to get additional information from the data itself. Examples of these approach in the context of naval transportation system can be found in [12, 40, 57, 78]. Finally, predictive and prescriptive analytics are the most complex approaches where a model of the system is built and studied in order to understand the accuracy and the properties of the model and make the system behave in a particular way. This is the most important analysis in practical applications since, even if diagnostic analytics allows improving the understanding of past and present conditions of the system, it is more important to predict the future and take action in order to prevent the occurrence of some event [12, 24] (substitute a component before it fails) or to make some event happen [58] (reduce the fuel consumption of a ship).

For these reasons in the next sections a more rigorous framework is depicted together with the description of the approaches adopted for building predictive models. An assessment of their accuracy and properties is performed and a complete description about how to use these models to force the system in producing an output is provided.

4.1 Supervised Learning

In the context of supervised machine learning, we are interested in a particular subproblem which is the regression one. Regression helps to understand how the value of a dependent variable changes when any one of the independent variables is varied. Using the conventional regression framework [65, 74] a set of data $\mathscr{D}_n = \{(\boldsymbol{x}_1, y_1), \cdots, (\boldsymbol{x}_n, y_n)\}$, with $\boldsymbol{x}_i \in \mathscr{X} \subseteq \mathbb{R}^d$ and $y_i \in \mathscr{Y} \subseteq \mathbb{R}$, are available from the automation system. Each tuple $(\boldsymbol{x}_i, y_i)$ is called sample and each element of the vector $\boldsymbol{x} \in \mathscr{X}$ is called feature.

When inferring a model starting from a real system, the goal is to provide an approximation $\mathfrak{M} : \mathscr{X} \to \mathscr{Y}$ of the unknown true model $\mathfrak{S} : \mathscr{X} \to \mathscr{Y}$. $\mathfrak{S}$ and $\mathfrak{M}$ are graphically represented in Fig. 6. It should be noted that the unknown model $\mathfrak{S}$ can be also seen, from a probabilistic point of view, as a conditional probability $\mathbb{P}(y|\boldsymbol{x})$ or, in other words, as the probability of the output y given the fact that we observed $\boldsymbol{x}$ as an input to $\mathfrak{S}$.

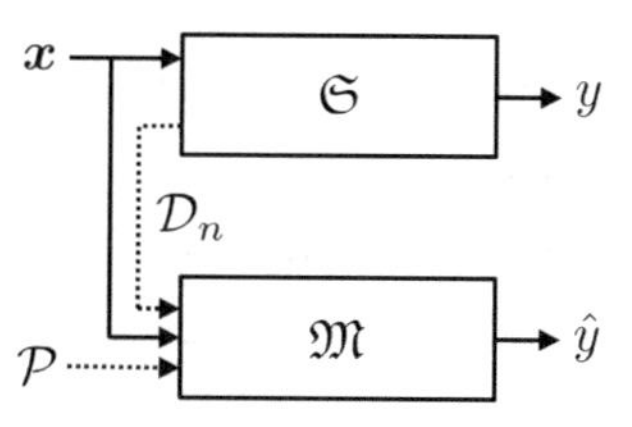

Fig. 6 The regression problem

As previously described, in this book chapter three alternative modelling strategies are compared: white-, black-, and gray-box models:

- White Box Model (WBM): in this case the model $\mathfrak{M}_{\text{WBM}}$ is built based on a priori, mechanistic knowledge of $\mathfrak{S}$ (numerical description of the body hull, propulsion plant configuration, design information of the ship). The implementation of a WBM in this specific case is described in Sect. 5.1.
- Black Box Model (BBM): in this case the model $\mathfrak{M}_{\text{BBM}}$ is built based on a series of historical observation of $\mathfrak{S}$ (or in other words $\mathscr{D}_n$). In this book chapter, this is done by exploiting state of the art Machine Learning techniques as described in Sect. 5.2.
- Gray Box Model (GBM): in this case the WBM and BBM are combined in order to build a model $\mathfrak{M}_{\text{GBM}}$ that takes into account both a priori information and historical data $\mathscr{D}_n$ so to improve the performances of both the WBM and BBM models. The implementation of the GBM principle to the specific case of this work is described in Sect. 5.3.

4.2 Estimation of Model Accuracy

The accuracy of the model $\mathfrak{M}$ as a representation of the unknown system $\mathfrak{S}$ can be evaluated using different measures of accuracy [26]. In particular, given a series data $\mathscr{T}_m = \{(\boldsymbol{x}_1, y_1), \cdots, (\boldsymbol{x}_m, y_m)\}$,[1] the model will predict a series of outputs $\{\widehat{y}_1, \cdots, \widehat{y}_m\}$ given the inputs $\{\boldsymbol{x}_1, \cdots, \boldsymbol{x}_m\}$. Based on these outputs it is possible to compute these performance indicators:

- mean absolute error (MAE) $\text{MAE} = \frac{1}{m} \sum_{i=1}^{m} |y_i - \widehat{y}_i|$
- mean absolute percentage error (MAPE) $\text{MAPE} = 100 \frac{1}{m} \sum_{i=1}^{m} \left| \frac{y_i - \widehat{y}_i}{y_i} \right|$
- mean square error (MSE) $\text{MSE} = \frac{1}{m} \sum_{i=1}^{m} (y_i - \widehat{y}_i)^2$
- normalised mean square error (NMSE) $\text{NMSE} = \frac{1}{m\Delta} \sum_{i=1}^{m} (y_i - \widehat{y}_i)^2$, $\Delta = \frac{1}{m} \sum_{i=1}^{m} (y_i - \bar{y})^2$, and $\bar{y} = \frac{1}{m} \sum_{i=1}^{m} y_i$

[1] The set $\mathscr{T}_m$ must be a different set respect to $\mathscr{D}_n$ which has been used to built the model $\mathfrak{M}$ in the case of BBMs and GBMs [1].

- relative error percentage (REP) $\mathrm{REP} = 100\sqrt{\frac{\sum_{i=1}^{m}(y_i - \widehat{y}_i)^2}{\sum_{i=1}^{m} y_i^2}}$
- Pearson product-moment correlation coefficient (PPMCC) which allows to compute the correlation between the output of the system and the output of the model $\mathrm{PPMCC} = \frac{\sum_{i=1}^{m}(y_i - \bar{y})(\widehat{y}_i - \bar{\widehat{y}})}{\sqrt{\sum_{i=1}^{m}(y_i - \bar{y})^2}\sqrt{\sum_{i=1}^{m}(\widehat{y}_i - \bar{\widehat{y}})^2}}$, and $\bar{\widehat{y}} = \frac{1}{m}\sum_{i=1}^{m}\widehat{y}_i$

Note that all these measures of accuracy are useful for giving an exhaustive description of the quality of the forecast [26].

4.3 *Prescriptive Analytics*

Once the model $\mathfrak{M}$ of the system $\mathfrak{S}$ is available, it is possible to control its inputs in order to produce a desired output. In this particular application, the goal is to find the minimum for the fuel consumption by acting on the ship's trim while keeping all other model inputs unchanged.

This approach, however, requires additional care and understanding of the underlying physics of the system:

- With reference to the previous work from the authors [11], not all variables available as measurements can be used as predictors. In this case, in particular, the power and torque at the propeller shaft had to be excluded from the input list (see Table 3). Changing the trim would consequently change ship resistance and, therefore, the power required for its propulsion. Therefore modifying the trim while keeping the propeller power constant would represent a conceptual error.
- Not all possible trim values are physically allowed, and therefore boundary values, based on a priori knowledge of the system, should be provided.
- Although GBMs are more reliable in the extrapolation phase, their accuracy is expected to be reduced if they are extrapolated too far for outside the boundaries of the original range $\mathscr{D}_n$. Extrapolation is therefore allowed (the use of GBMs proposed in this book chapter is also based on their improved performance for extrapolation compared to BBMs) but this operation should be performed with care.

Based on these considerations, in this book chapter a method for trim optimisation is proposed. WBM, BBM and GBM are presented and compared based on the accuracy metrics proposed in Sect. 4.2. Based on this comparison, one model is selected for further analysis, checked for physical plausibility (Sect. 6) and used for application to the problem of trim optimisation (Sect. 7).

Table 3 Variable of Table 2 exploited to built the $\mathfrak{M}$

Id	Name	Type
1	Latitude	Input
2	Longitude	Input
3	Volume	Input
4	State	Input
5	Auxiliary consumed	Input
6	Auxiliary electrical power output	Input
8	Shaft rpm	Input
9	Ship draft (fore)	Input
10	Ship draft (aft)	Input
11	Relative wind speed	Input
12	Relative wind direction	Input
13	GPS heading	Input
14	GPS speed	Input
15	Log speed	Input
16	Shaft generator power	Input
17	Sea depth	Input
18	Draft Port	Input
19	Draft Starboard	Input
20	Sea water temperature	Input
21	CPP setpoint	Input
22	CPP feedback	Input
23	Fuel density	Input
24	Fuel temperature	Input
25	Ambient pressure	Input
26	Humidity	Input
27	Dew point temperature	Input
29	Rudder angle	Input
30	Acceleration X direction	Input
32	Acceleration Y direction	Input
32	Acceleration Z direction	Input
33	GyroX	Input
34	GyroY	Input
35	GyroZ	Input
36	Roll	Input
37	Pitch	Input
38	Yaw	Input
39	True direction	Input
40	True speed	Input

(continued)

Table 3 (continued)

Id	Name	Type
41	Beaufort	Input
	Shaft power	Output
	Shaft torque	Output
	Main engine consumption	Output
	Shaft power predicted by the WBM	Input GBMs
	Shaft torque predicted by the WBM	Input GBMs
	Main engine consumption predicted by the WBM	Input GBMs

5 White, Black and Gray Box Models

5.1 *White Box Models*

A numerical model, the so called White Box Model (WBM), based on the knowledge of the physical processes was developed by the authors. The WBM model is able to evaluate the ship consumption, for different ship speed V and displacement Δ in calm water scenario.

The model is based on the knowledge of the ship's hull geometry, mass distribution, propeller characteristics and main Diesel engine consumption map. The selected control variables (i.e. the system input which is under the user's control) taken into account are: the main engine revolution N and the pitch ratio P/D. The control of these variables allow the ship to sail at the desired speed. The total ship's fuel consumption is used as model output.

The core of the procedure is the engine-propeller matching code utilised to evaluate the total ship fuel consumption and already tested as an effective tool in a previous work [9].

The prediction of ship resistance in calm water can be performed according to different approaches, normally divided in parametric approaches [4, 29, 34, 35], and approaches based on computational fluid dynamics (CFD), such as the Reynolds averaged Navier-Stokes (RANS) or boundary element methods (BEM) [33]. In this study only parametric methods were considered because of their lower computational requirements. In particular, the Guldhammer Harvald method [29] was employed for the prediction of calm water resistance and, in particular, of the coefficient of total hull resistance in calm water C_T in Eq. (1). The inputs related to ship geometry used in the Guldhammer Harvald method are summarised in Table 4.

$$R_{\text{tot}} = \frac{1}{2} C_T \rho S V^2 \tag{1}$$

where ρ is the sea water density.

Table 4 Main input quantities for ship resistance prediction

Input	Symbol	Unit
Length on waterline	L_{WL}	(m)
Breath on waterline	B_{WL}	(m)
Draught	T	(m)
Volume	V	(m^3)
Wetted surface	S	(m^2)
Longitudinal position of center of buoyancy	L_{CB}	(m)
Longitudinal position of center of buoyancy	L_{CB}	(m)
Bow shape coefficient		
Section shape coefficient		

For each displacement the equilibrium draft on even keel has been calculated, together with the necessary input variables [10] required by he Guldhammer Harvald method [29] to perform resistance prediction in calm waters. The propulsion coefficients have been corrected in magnitude as reported in [50].

Propeller thrust and torque were computed offline for different pitch settings by means of a viscous method and based on the knowledge of the geometrical features of the propeller. The calculated values were implemented in the matching code through the non dimensional thrust K_T and torque K_Q coefficients.

As reported in Fig. 1 a shaft generator is used for fulfilling on board auxiliary power demand. In order to optimise this feature the ship propulsion system has been set-up for working at fixed rpm using the pitch as control variable. Once the displacement, shaft rate of revolutions and vessel speed are fixed, the advance coefficient J is defined together with the non dimensional thrust coefficient according to the following equations:

$$J = \frac{V(1-w)}{nD}, \quad K_T = \frac{T}{\rho n^2 D^4} \tag{2}$$

where w is the wake factor, n is the propeller rate of revolution, D is the propeller diameter and T is the required thrust of the propeller. The engine-propeller matching code used in this work allows calculating the pitch ratio that provides the required thrust at the fixed shaft speed. Finally the delivered power P_d can be evaluated by means of the following quantities:

$$K_Q = \frac{Q}{\rho n^2 D^5}, \quad \eta_0 = \frac{J k_T}{2\pi K_Q} \tag{3}$$

A validation of the WBM model was performed based on the available measurements of delivered power at different displacement derived from model tests in calm

Table 5 White box model validation

$\Delta = 25000$ t			
Speed	P_{dh} (KW)	P_{dn} (KW)	Error (%)
10	1452.0	1548.9	6.7
12	2589.0	2653.8	2.5
14	4478.0	4465.9	0.3
16	7761.9	7430.3	4.3
$\Delta = 30000$ t			
10	1528.0	1637.6	7.2
12	2763.0	2805.3	1.5
14	4768.0	4818.1	1.1
16	18139.7	8139.1	0.0
$\Delta = 40000$ t			
Speed	P_{dh} (KW)	P_{dn} (KW)	Error (%)
10	1768.0	1804.9	2.1
12	3100.0	3091.1	0.3
14	5196.0	5403.4	4.0
16	8760.2	9359.5	6.8
$\Delta = 50800$ t			
Speed	P_{dh} (KW)	P_{dn} (KW)	Error (%)
10	1994.0	1815.5	9.0
12	3432.0	3106.3	9.5
14	5662.0	5458.5	3.6
16	9494.8	9546.7	0.5
$\Delta = 57100$ t			
Speed	P_{dh} (KW)	P_{dn} (KW)	Error (%)
10	2162.0	1908.6	11.7
12	3690.0	3265.1	11.5
14	6089.0	5786.9	5.0
16	10323.4	10283.7	0.4

water. The measured (P_{dh}) and predicted (P_{dn}) delivered power, together with the absolute percentage error of the model, are reported in Table 5. The results obtained with the WBM model are in good agreement with measured values: thus, the model tool is able to derive a general representation of the relationship between vessel speed, displacement and delivered power in calm water scenarios.

For a generic couple of ship displacement Δ_i and speed V_i values, the WBM model evaluates the propeller rate of revolution n, which ensures the propulsion equilibrium between delivered and required thrust, and finally the associated fuel consumption. Starting from propeller torque, the engine brake power P_b is computed by the global efficiency of the drivetrain and it is then possible to evaluate the corresponding specific fuel consumption.

5.2 Black Box Models

Machine Learning (ML) approaches play a central role in extracting information from raw data collected from ship data logging systems. The learning process for ML approaches usually consists of two phases: (i) during the training phase, a set of data is used to induce a model that best fits them, according to some criteria; (ii) the trained model is used for prediction and control of the real system (feed-forward phase).

As the authors are targeting a regression problem [74], the purpose is to find the best approximating function $h(\boldsymbol{x})$, where $h : \mathbb{R}^d \rightarrow \mathbb{R}$. During the training phase, the quality of the regressor $h(\boldsymbol{x})$ is measured according to a loss function $\ell(h(\boldsymbol{x}), y)$ [47], which calculates the discrepancy between the true and the estimated output (y and $\widehat{y}$). The empirical error then computes the average discrepancy, reported by a model over $\mathscr{D}_n$:

$$\widehat{L}_n(h) = \frac{1}{n} \sum_{i=1}^{n} \ell(h(\boldsymbol{x}_i), y_i). \tag{4}$$

A simple criterium for selecting the final model during the training phase consists in choosing the approximating function that minimises the empirical error $\widehat{L}_n(h)$: this approach is known as Empirical Risk Minimisation (ERM) [74]. However, ERM is usually avoided[2] in ML as it leads to severely overfitting the model on the training dataset [74]. A more effective approach consists in the minimisation of a cost function where the tradeoff between accuracy on the training data and a measure of the complexity of the selected approximating function is implemented [72]:

$$h^* : \quad \min_h \quad \widehat{L}_n(h) + \lambda \, \mathscr{C}(h). \tag{5}$$

where $\mathscr{C}(\cdot)$ is a complexity measure which depends on the selected ML approach and λ is a hyperparameter that must be set a priori and regulates the trade-off between the overfitting tendency, related to the minimisation of the empirical error, and the underfitting tendency, related to the minimisation of $\mathscr{C}(\cdot)$. The optimal value for λ is problem-dependent, and tuning this hyperparameter is a non-trivial task [1] and will be faced later in this section.

The approaches exploited in this book chapter are: the Regularised Least Squares (RLS) [31], the Lasso Regression (LAR) [71], and the Random Forrest (RF) [5].

[2]Note that some techniques use ERM and then, in order to improve the performance of the method, a post processing approach is adopted (i.e. pruning for Decision Tree [59]).

In RLS, approximation functions are defined as

$$h(\boldsymbol{x}) = \boldsymbol{w}^T \boldsymbol{\phi}(\boldsymbol{x}), \tag{6}$$

where a non-linear mapping $\boldsymbol{\phi} : \mathbb{R}^d \to \mathbb{R}^D$, $D \gg d$, is applied so that non-linearity is pursued while still coping with linear models.

For RLS, Problem (5) is configured as follows. The complexity of the approximation function is measured as

$$\mathscr{C}(h) = \|\boldsymbol{w}\|_2^2 \tag{7}$$

i.e. the Euclidean norm of the set of weights describing the regressor, which is a quite standard complexity measure in ML [72]. Regarding the loss function, the Mean Squared Error (MSE) loss is adopted:

$$\widehat{L}_n(h) = \frac{1}{n} \sum_{i=1}^{n} \ell(h(\boldsymbol{x}_i), y_i) = \frac{1}{n} \sum_{i=1}^{n} [h(\boldsymbol{x}_i) - y_i]^2 . \tag{8}$$

Consequently, Problem (5) can be reformulated as:

$$\boldsymbol{w}^* : \quad \min_{\boldsymbol{w}} \quad \frac{1}{n} \sum_{i=1}^{n} \left[\boldsymbol{w}^T \boldsymbol{\phi}(\boldsymbol{x}) - y_i\right]^2 + \lambda \|\boldsymbol{w}\|_2^2. \tag{9}$$

By exploiting the Representer Theorem [62], the solution h^* of the RLS Problem (9) can be expressed as a linear combination of the samples projected in the space defined by $\boldsymbol{\phi}$:

$$h^*(\boldsymbol{x}) = \sum_{i=1}^{n} \alpha_i \boldsymbol{\phi}(\boldsymbol{x}_i)^T \boldsymbol{\phi}(\boldsymbol{x}). \tag{10}$$

It is worth underlining that, according to the kernel trick [61], it is possible to reformulate $h^*(\boldsymbol{x})$ without an explicit knowledge of $\boldsymbol{\phi}$ by using a proper kernel function $K(\boldsymbol{x}_i, \boldsymbol{x}) = \boldsymbol{\phi}(\boldsymbol{x}_i)^T \boldsymbol{\phi}(\boldsymbol{x})$:

$$h^*(\boldsymbol{x}) = \sum_{i=1}^{n} \alpha_i K(\boldsymbol{x}_i, \boldsymbol{x}). \tag{11}$$

Of the several kernel functions which can be found in literature [15], the Gaussian kernel is often used as it enables learning every possible function [56]:

$$K(\boldsymbol{x}_i, \boldsymbol{x}_j) = e^{-\gamma \|\boldsymbol{x}_i - \boldsymbol{x}_j\|_2^2}, \tag{12}$$

where γ is an hyperparameter which regulates the non-linearity of the solution [56] and must be set a priori, analogously to λ. Small values of γ lead the optimisation to converge to simpler functions $h(x)$ (note that for $\gamma \to 0$ the optimisation converges to a linear regressor), while high values of γ allow higher complexity of $h(x)$.

Finally, the RLS Problem (9) can be reformulated by exploiting kernels:

$$\boldsymbol{\alpha}^* : \quad \min_{\boldsymbol{\alpha}} \frac{1}{n} \sum_{i=1}^{n} \left[\sum_{j=1}^{n} \alpha_j K(\boldsymbol{x}_j, \boldsymbol{x}_i) - y_i \right]^2 + \lambda \sum_{i=1}^{n} \sum_{j=1}^{n} \alpha_i \alpha_j K(\boldsymbol{x}_j, \boldsymbol{x}_i). \tag{13}$$

Given $\boldsymbol{y} = [y_1, \cdots, y_n]^T$, $\boldsymbol{\alpha} = [\alpha_1, \cdots, \alpha_n]^T$, the matrix K such that $K_{i,j} = K_{ji} = K(\boldsymbol{x}_j, \boldsymbol{x}_i)$, and the Identity matrix $I \in \mathbb{R}^{n \times n}$, a matrix-based formulation of Problem (13) can be obtained:

$$\boldsymbol{\alpha}^* : \quad \min_{\boldsymbol{\alpha}} \quad \frac{1}{n} \|K\boldsymbol{\alpha} - \boldsymbol{y}\|_2^2 + \lambda \boldsymbol{\alpha}^T K \boldsymbol{\alpha} \tag{14}$$

By setting the derivative with respect to $\boldsymbol{\alpha}$ equal to zero, $\boldsymbol{\alpha}$ can be found by solving the following linear system:

$$(K + n\lambda I)\,\boldsymbol{\alpha}^* = \boldsymbol{y}. \tag{15}$$

Effective solvers have been developed throughout the years, allowing to efficiently solve the problem of Eq. (15) even when very large sets of training data are available [80].

In LAR, instead, approximation functions are defined as

$$h(\boldsymbol{x}) = \boldsymbol{w}^T \boldsymbol{x} + b, \tag{16}$$

which are linear functions in the original space $\mathbb{R}^d$.

For LAR, Problem (5) is configured as follows. The complexity of the approximation function is measured as

$$\mathscr{C}(h) = \|\boldsymbol{w}\|_1 \tag{17}$$

i.e. the Manhattan norm of the set of weights describing the regressor [71].

Regarding the loss function, the Mean Squared Error (MSE) loss is again adopted. Consequently, Problem (5) can be reformulated as:

$$\boldsymbol{w}^* : \quad \min_{\boldsymbol{w}} \quad \frac{1}{n} \sum_{i=1}^{n} \left[\boldsymbol{w}^T \boldsymbol{\phi}(\boldsymbol{x}) - y_i \right]^2 + \lambda \|\boldsymbol{w}\|_1. \tag{18}$$

As depicted in Fig. 7 the Manhattan norm is quite different from the Euclidean one since it allows increasing the sparsity of the solution. In other words the solution will

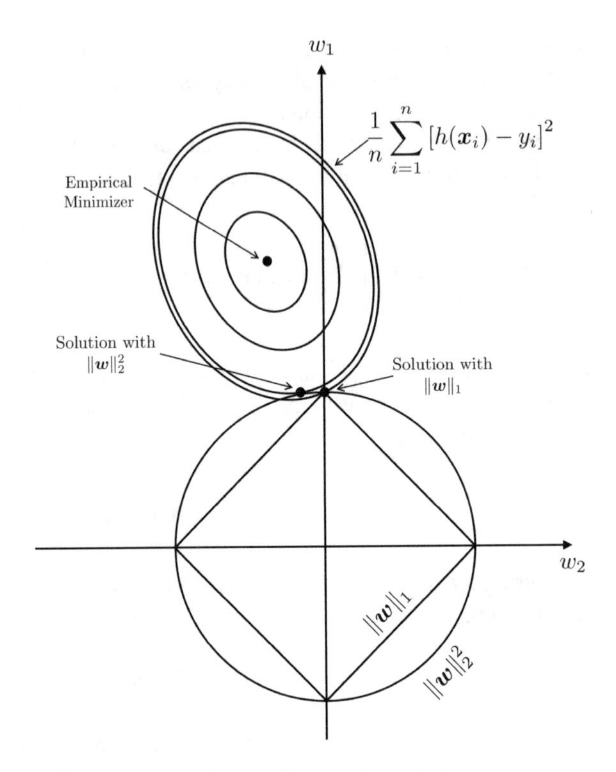

Fig. 7 Manhattan norm Versus euclidean norm

tend to fall on the edge of the square, forcing some weights of $\boldsymbol{w}$ to be zero. Hence, the Manhattan norm allows both regularising the function and discarding features that are not sufficiently relevant to the model. This property is particularly useful in the feature selection process [71].

Two main approaches can be used to compute the solutions of Problem (18): the LARS algorithms [19] and the pathwise coordinate descent [23]. In this book chapter, the LARS algorithm is exploited because of its straight-forward implementation [19].

The performance of RLS (or LAR) models depends on the quality of the hyperparameters tuning procedure. As highlighted while presenting this approach, the parameters $\boldsymbol{\alpha}^*$, $\widehat{\boldsymbol{\alpha}}^*$, and $\check{\boldsymbol{\alpha}}^*$ (or $\boldsymbol{w}$) result from an optimisation procedure which requires the a priori setting of the tuples of hyperparameters (λ, γ) (or λ). The phase in which the problem of selecting the best value of the hyperparameter is addressed is called model selection phase [1]. The most effective model selection approaches consist in performing an exhaustive hyperparameters grid search: the optimisation problem for RLS (or LAR) is solved several times for different values of γ and λ, and the best pair of hyperparameters is chosen according to some criteria.

For the optimal choice of the hyperparameters γ and λ, in this book chapter the authors exploit the Bootstrap technique (BOO) [1]. This technique represents an improvement of the well–known k–Fold Cross Validation (KCV) [44] where the original dataset is split into k independent subsets (namely, the folds), each one consisting of n/k samples: $(k-1)$ parts are used, in turn, as a training set, and the remaining fold is exploited as a validation set. The procedure is iterated k times.

The standard Bootstrap [1] method is a pure resampling technique: at each j-th step, a training set $\mathscr{D}_{\text{TR}}^j$, with the same cardinality of the original one, is built by sampling the patterns in $\mathscr{D}_n$ with replacement. The remaining data $\mathscr{D}_{\text{VL}}^j$, which consists, on average, of approximately 36.8% of the original dataset, are used as validation set. The procedure is then repeated several times $N_B \in [1, \binom{2n-1}{n}]$ in order to obtain statistically sound results [1].

According to the Bootstrap technique, at each j-th step the available dataset $\mathscr{D}_n$ is split in two sets:

- A training Set: $\mathscr{D}_{\text{TR}}^j$
- A validation Set: $\mathscr{D}_{\text{VL}}^j$

In order to select the best pair of hyperparameters (λ^*, γ^*) (or λ^*) among all the available ones $\mathscr{G} = \{(\lambda_1, \gamma_1), (\lambda_2, \gamma_2), \cdots\}$ (or $\mathscr{G} = \{\lambda_1, \lambda_2, \cdots\}$) for the algorithm for RLS (or LAR) the following optimisation procedure is required:

- for each $\mathscr{D}_{\text{TR}}^j$ and for each tuple (λ_i, γ_i) (or λ_i) with $i \in \{1, 2, \cdots\}$ the optimisation problem of Eq. 15 (or Eq. 18) is solved and the solution $h_j^i(\boldsymbol{x})$ is found
- using the validation set $\mathscr{D}_{\text{VL}}^j$ for searching the (λ^*, γ^*) (or λ^*) $\in \mathscr{G}$

$$\begin{array}{c}(\lambda^*, \gamma^*) \\ \text{or } \lambda^*\end{array} = \arg \min_{\substack{\{(\lambda_1, \gamma_1), \cdots, (\lambda_i, \gamma_i), \cdots\} \\ \text{or } \{\lambda_1, \cdots, \lambda_i, \cdots\}}} \frac{1}{N_B} \sum_{j=1}^{N_B} \frac{1}{|\mathscr{D}_{\text{VL}}^j|} \sum_{(\boldsymbol{x}, y) \in \mathscr{D}_{\text{VL}}^j} [h(\boldsymbol{x}) - y]^2 . \tag{19}$$

Once the best tuple is found, the final model is trained on the whole set $\mathscr{D}_n$ by running the learning procedure with the best values of the hyperparameters [1].

Another learning algorithm tested for building the BBM is the Random Forest (RF) [5]. Random Forests grows many regression trees. To classify a new object from an input vector each of the trees of the forest is applied to the vector. Each tree gives an output and the forest chooses the mode of the votes (over all the trees in the forest). Each single tree is grown by following this procedure: (I) n samples are sampled (with replacement) from the original $\mathscr{D}_n$, (II) $d' \ll d$ features are chosen randomly out of the d and the best split on these d' is used to split the node, (III) each tree is grown to the largest possible extent, without any pruning. In the original paper [5] it was shown that the forest error rate depends on two elements: the correlation between any couples of trees in the forest (increasing the correlation increases the forest error rate) and the strength of each individual tree in the forest (reducing the error rate of each tree decreases the forest error rate). Reducing d' reduces both the correlation and the strength. Increasing it increases both. Somewhere in between is an optimal range of d' - usually quite wide so this is not usually considered as an hyperparameter. Note that, since we used a bootstrap procedure by sampling n

samples with replacement from the original $\mathscr{D}_n$, we can use the error on the remaining part of the data (this is called out-of-bag error) to chose the best d'.

5.3 Gray Box Models

GBMs are a combination of a WBMs and BBMs. This requires to modify the BBMs as defined in the previous section in a way to include the mechanistic knowledge of the system. Two approaches are tested and compared in this book chapter:

- a Naive approach (N-GBM) where the output of the WBM is used as a new feature that the BBM can use for training the model.
- an Advanced approach (A-GBM) where the regularisation process is changed in order to include some a-priori information [1].

In the N-GBM case, the WBM can be seen as a function of the input $\boldsymbol{x}$. The WBM, that we call here $h_{\text{WBM}}(\boldsymbol{x})$, allows the creation of a new dataset:

$$\mathscr{D}_n^{\text{WBM},\mathscr{X}} = \left\{ \left(\begin{bmatrix} \boldsymbol{x}_1 \\ h_{\text{WBM}}(\boldsymbol{x}_1) \end{bmatrix}, y_1 \right), \cdots, \left(\begin{bmatrix} \boldsymbol{x}_n \\ h_{\text{WBM}}(\boldsymbol{x}_n) \end{bmatrix}, y_n \right) \right\}$$

Based on this new dataset a BBM can be generated $h_{\text{BBM}}\left(\left[\boldsymbol{x}^T | h_{\text{WBM}}(\boldsymbol{x})\right]^T\right)$.

According to this approach, every run of the GBM requires an initial run of the WBM in order to compute its output $h_{\text{WBM}}(\boldsymbol{x})$, which allows evaluating the model $h_{\text{BBM}}\left(\left[\boldsymbol{x}^T | h_{\text{WBM}}(\boldsymbol{x})\right]^T\right)$. This is the simplest approach for including new information into the learning process. Note that with this approach any of the previously cited BBMs (e.g. RLS, LAR or RM) can be used for building the corresponding N-GBM.

In the A-GBM case the WBM part of the model is assumed to be included in the $\boldsymbol{w}$ vector:

$$h_{\text{WBM}}(\boldsymbol{x}) = \boldsymbol{w}_{\text{WBM}}^T \boldsymbol{\phi}(\boldsymbol{x}), \tag{20}$$

According to [1], the regularisation process of Eq. (9) is modified to:

$$\boldsymbol{w}^*: \min_{\boldsymbol{w}} \frac{1}{n} \sum_{i=1}^{n} \left[\boldsymbol{w}^T \boldsymbol{\phi}(\boldsymbol{x}) - y_i\right]^2 + \lambda \|\boldsymbol{w} - \boldsymbol{w}_{\text{WBM}}\|_2^2. \tag{21}$$

It is possible to prove that by exploiting the kernel trick the solution to this problem can be rewritten as:

$$h^*(\boldsymbol{x}) = h_{\text{WBM}}(\boldsymbol{x}) + \sum_{i=1}^{n} \alpha_i^* K(\boldsymbol{x}_i, \boldsymbol{x}). \tag{22}$$

where

$$\boldsymbol{\alpha}^* : \min_{\boldsymbol{\alpha}} \frac{1}{n} \sum_{i=1}^{n} \left[\sum_{j=1}^{n} \alpha_j K(\boldsymbol{x}_j, \boldsymbol{x}_i) + h_{\text{WBM}}(\boldsymbol{x}_i) - y_i \right]^2 + \lambda \sum_{i=1}^{n} \sum_{j=1}^{n} \alpha_i \alpha_j K(\boldsymbol{x}_j, \boldsymbol{x}_i), \tag{23}$$

The solution to this problem can be computed by solving the following linear system:

$$(K + n\lambda I)\boldsymbol{\alpha}^* = \boldsymbol{y} - \boldsymbol{h}_{\text{WBM}}, \tag{24}$$

where $\boldsymbol{h}_{\text{WBM}} = [h_{\text{WBM}}(\boldsymbol{x}_1), \cdots, h_{\text{WBM}}(\boldsymbol{x}_n)]^T$. Note that the solution does not depend on the form of $h_{\text{WBM}}(\boldsymbol{x})$ so that any WBM can be used as $h_{\text{WBM}}(\boldsymbol{x})$.

Another possible way of achieving the same solution of the problem of Eq. (24) is to create a new dataset:

$$\mathscr{D}_n^{\text{WBM},\mathscr{Y}} = \{(\boldsymbol{x}_1, y_1 - h_{\text{WBM}}(\boldsymbol{x}_1)), \cdots, (\boldsymbol{x}_n, y_n - h_{\text{WBM}}(\boldsymbol{x}_n))\}$$

where the target is no longer the true label y but the true label minus the hint given by the a priori information included in $h_{\text{WBM}}(\boldsymbol{x})$. This means finding a BBM that minimises the error of the WBM prediction.

It should be noted that the A-GBM is more theoretically justified in the regularisation context while the N-GBM is more intuitive since all the available knowledge is given as input to the BBM learning process. From a probabilistic point of view, the A-GMB changes the $\mathbb{P}(y|\boldsymbol{x})$ while the N-GBM modifies the whole joint probability $\mathbb{P}(y, \boldsymbol{x})$, hence deeply influencing the nature of the problem.

The (λ, γ) for RLS, the λ for LAR and d' for RF of the N-GBM and A-GBM are tuned with the BOO as described for the BBM, since both N-GBM and A-GBM basically require to build a BBM over a modified training set.

5.4 *Model Validation*

The WBM was validated using the data described in Sect. 3 versus propeller shaft power, shaft torque, and total fuel consumption. The results of the validation are presented in Table 6. The results show that the WBM does not show sufficient accuracy when compared with operational measurements. The inability of the model to take into account the influence of the sea state (i.e. wind and waves) on the required propulsion power is considered to be the largest source of error for this model.

The BBMs built according to the RLS, LAR and RF methods were validated versus the same dataset as for the WBM validation procedure. However, in the case of the BBMs the $\mathscr{D}_n$ was divided in two sets $\mathscr{L}_{n_l}$ and $\mathscr{T}_{n_t}$ respectively for learning and test. The two sets were defined so that $\mathscr{D}_n = \mathscr{L}_{n_l} \cup \mathscr{T}_{n_t}$ and $\mathscr{L}_{n_l} \cap \mathscr{T}_{n_t} = \oslash$ in order to maintain the independence of the two sets.

Table 6 Indexes of performance of the WBM in predicting the shaft power, shaft torque, and fuel consumption

Shaft power					
MAE (KW)	MAPE (%)	MSE $\left(\text{KW}^2\right)$	NMSE	REP (%)	PPMCC
7.69e+02	17.85	1.00e+06	1.13	23.59	0.65
Shaft torque					
MAE (Nm)	MAPE (%)	MSE $\left(\text{N}^2\text{m}^2\right)$	NMSE	REP (%)	PPMCC
6.54e+01	18.13	6.92e+03	0.94	22.01	0.22
Fuel consumption					
MAE $\left(\frac{\text{g}}{\text{KWh}}\right)$	MAPE (%)	MSE $\left(\frac{\text{g}^2}{\text{KW}^2\text{h}^2}\right)$	NMSE	REP (%)	PPMCC
5.14e-02	20.95	3.94e-03	1.98	25.40	0.63

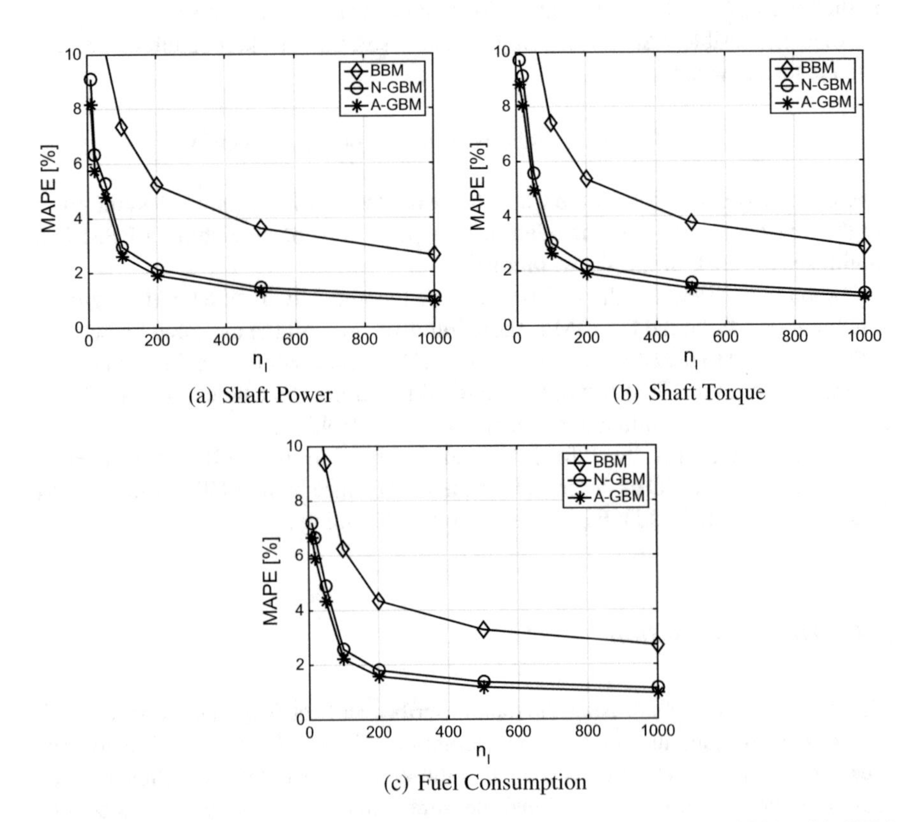

(a) Shaft Power

(b) Shaft Torque

(c) Fuel Consumption

Fig. 8 Shaft Power, Shaft Torque, and Fuel Consumption MAPE of the BBM, N-GBM and A-GBM for RLS and different n_l

The process of splitting the full dataset in a learning set and test set is repeated 30 times in order to obtain statistical relevant results. We always underline in bold the best results which are statistically significant [25]. The results are reported for different sizes of $\mathscr{L}_{n_l}$ with $n_l \in \{10, 20, 50, 100, 200, 500, 1000, 2000, 5000\}$. The optimisation procedure is repeated for different values of both hyperparameters (γ and λ), where their values are taken based on a 60 points equally spaced in logarithmic scale in the range $[10^{-6}, 10^{3}]$ and the best set of hyperparameters is selected according to the BOO (Sect. 5.2). The same has been done for d' in RF.

Also in the case of the GBM, analogously to the procedure adopted for the BBM, the original dataset $\mathscr{D}_n$ is divided in two sets $\mathscr{L}_{n_l}$ and $\mathscr{T}_{n_t}$ and λ, γ and d' are chosen according to the BOO procedure.

The entire set of results is not reported here because of space constraints but it can be retrieved in the technical report available at http://www.smartlab.ws/TR.pdf

From the results it is possible to note that the WBM, as expected, has the lowest performance in terms of prediction accuracy. On the other hand, the GBMs outperform the BBMs by a smaller percentage. The MAPE of the BBM, N-GBM and A-GBM for different values of n_l are reported in Figs. 8, 9, and 10.

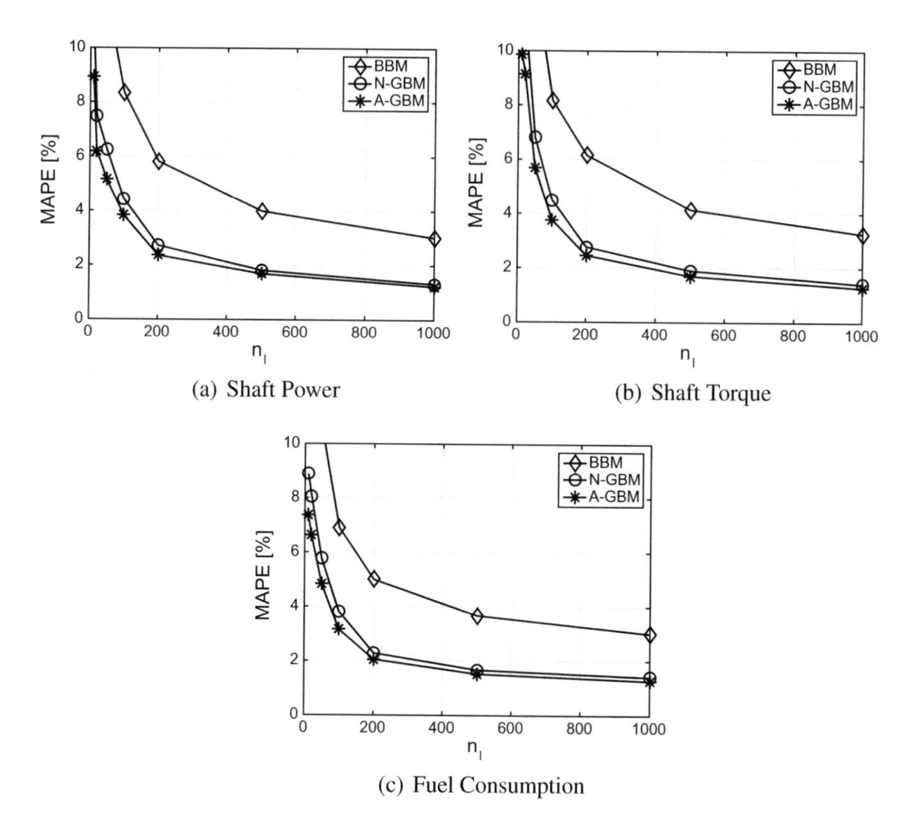

(a) Shaft Power

(b) Shaft Torque

(c) Fuel Consumption

Fig. 9 Shaft Power, Shaft Torque, and Fuel Consumption MAPE of the BBM, N-GBM and A-GBM for LAR and different n_l

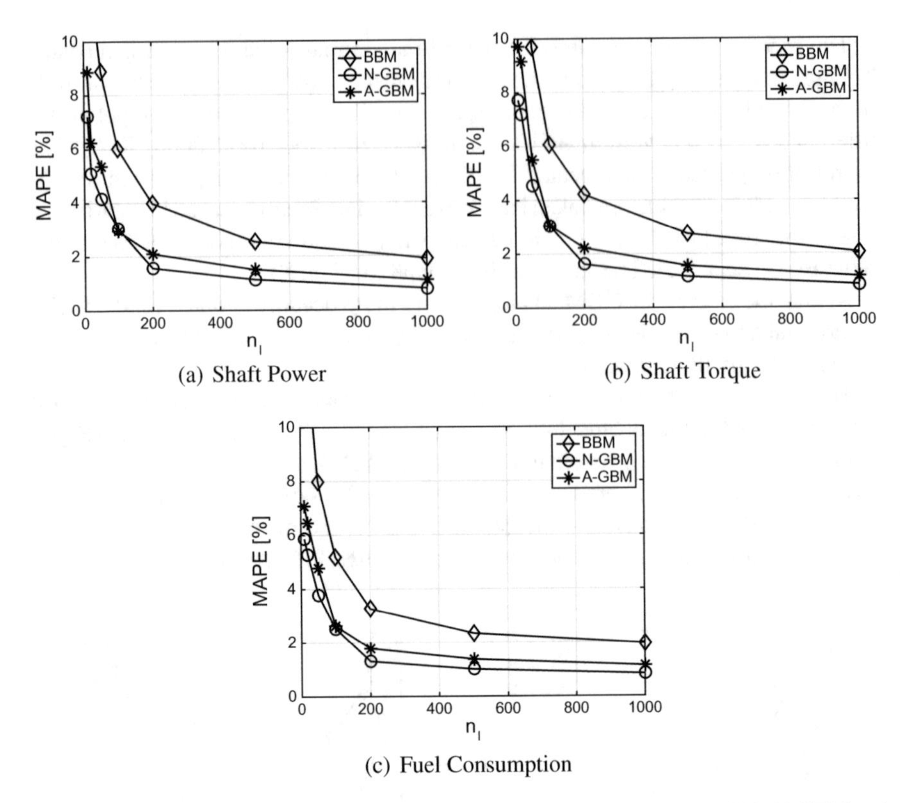

(a) Shaft Power

(b) Shaft Torque

(c) Fuel Consumption

Fig. 10 Shaft Power, Shaft Torque, and Fuel Consumption MAPE of the BBM, N-GBM and A-GBM for RF and different n_l

From the results of Figs. 8, 9, and 10 it is possible to note how the WBM, even if not so accurate, can help the GBM in obtaining higher accuracy, with respect to the BBM, by using almost half of the data given a required accuracy. This is a critical issue in real word applications where the collection of labeled data can be expensive or at least requires a long period of in-service operational time of the vessel [11].

6 Feature Selection

Once a model is built and has been confirmed to be a sufficiently accurate representation of the real system of interest, it can be interesting to investigate how the model $\mathfrak{M}$ is affected by the different features that have been used in the model identification phase.

In data analytics this procedure is called feature selection or feature ranking [7, 22, 30, 36, 79]. This process allows detecting if the importance of those features, that are known to be relevant from a theoretical perspective, is appropriately described by $\mathfrak{M}$.

The failure of the statistical model to properly account for the relevant features might indicate poor quality in the measurements. Feature selection therefore represents an important step of model verification, since the proposed model $\mathfrak{M}$ should generate results consistently with the available knowledge of the physical system under exam. This is particularly important in the case of BBM (and, to a more limited extent, for GBM), since they do not make use of any mechanistic knowledge of the system and might therefore lead to non-physical results (e.g. mass or energy unbalances). Feature selection also allows checking the statistical robustness of the employed methods.

In this book chapter, three different methods for feature ranking are applied:

- Brute Force Method (BFM), which searches for the optimal solution. This is the most accurate method but also the most computationally expensive (see Sect. 6.1) [22, 27].
- Regularisation Based Method (RBM) which works by building the BBM which automatically discarding the features that do not significantly contribute to the model output (for example by building an ad-hoc regularisers [16, 22, 53, 55, 66, 81, 82]). In this book chapter, the Lasso Regularisation technique was used (see Sect. 6.2).
- Random Forest based method (RFM) uses a combination of Decision Tree methods together with the permutation test [28] in order to perform the selection and the ranking of the features [22, 43, 70].

6.1 Brute Force Method

According to the Brute Force method (BFM) for feature selection, the k most important features of the model can be identified as follows:

- a first version of the model $\mathfrak{M}$ including all the available features is built. The full model is tested against a test set $\widehat{L}_{\text{Test}}$;
- for a given k, a set of new models is built for all possible configurations including feature k. For every possible configuration, which are $\binom{d}{k}$, a new model $\mathfrak{M}^j$ is built where $j \in \left\{1, \cdots, \binom{d}{k}\right\}$ together with its error on the test set $\widehat{L}^j_{\text{Test}}$;
- the smaller is the difference between $\widehat{L}^j_{\text{Test}}$ and $\widehat{L}_{\text{Test}}$, the greater is the importance of that set of features.

Given its high computational demands, this approach is not feasible for $d > 15 \div 20$. A solution for reducing the required computational time is to adopt a greedy procedure:

- a first version of the model $\mathfrak{M}$ including all the available features is built. The full model is tested against a test set $\widehat{L}_{\text{Test}}$;
- given a feature j_1, the model $\mathfrak{M}$ is built which only includes that feature. The error against the test set ($\widehat{L}^{j_1}_{\text{Test}}$) can now be calculated;
- the same procedure is performed for each feature $j_1 \in \{1, \cdots, d\}$;

- the smaller is the difference between $\widehat{L}_{\text{Test}}^{j_1}$ and $\widehat{L}_{\text{Test}}$ the grater is the importance of the features j_1

$$j_1^* = \arg \min_{j_1 \in \in \{1, \cdots, d\}} \widehat{L}_{\text{Test}} - \widehat{L}_{\text{Test}}^{j_1} \tag{25}$$

- this procedure is repeated by adding to j_1^* all the other features one at the time for finding the second most important feature $j_2^* \in \{1, \cdots, d\} \setminus j_1^*$. This operation is repeated until the required size (k) of the ranking is achieved.

Greedy methods are more time efficient compared to brute force methods, but do not ensure the full correctness of the result.

In this book chapter, several different models were proposed and are here tested for feature ranking. These models are: BBM, N-GBM and A-GBM with RLS, LAR and RF for a total of nine possibilities. It should be noted that for the N-GBM there is another feature which is the WBM (see Table 3).

6.2 Regularisation Based Method

The brute force method is a quite powerful approach but it requires a significant computational effort. The Lasso Regression can be used for ranking the importance of the features with lower computational demand.

However, the results of the Lasso Regression method are strongly influenced by the training dataset and by the choice of the hyperparameters used in the learning phase [53, 81, 82]. For this reason given the best value λ^* of the hyperparameter selected with the BOO procedure another bootstrap procedure is applied in order to improve the reliability of the feature selection method: n samples are extracted from $\mathscr{D}_n$, the model is built with LAR and λ^* and the features are selected. The bootstrap is repeated several times and features are ranked based on how many times each feature is selected as important by the LAR method [53].

In this work, the LAR method for feature selection was used in three different kind of models (BBM, N-GBM and A-GBM). Similarly to the case of Brute Force Methods, in the N-GBM case the WBM represents an additional feature (see Table 3).

6.3 Random Forest Based Method

In addition to its use for regression models, the Random Forest (RF) method can also be used to perform a very stable Feature Selection procedure. The procedure can be described as follows: in every tree grown in the forest the error on the out-of-bag must be kept. Then a random permutation of the values of variable j must be performed in the samples of the out-of-bag and the error on the out-of-bag must be kept again. Subtract the error on the untouched out-of-bag data with the error over the permuted out-of-bag samples. The average of this value over all the trees in the forest is the raw importance score for variable j. This approach is inspired by the permutation test [28] which is quite used in literature, is computationally inexpensive in the case of Random Forest, and has shown to be quite effective in real wold applications [17, 77]. The results for the RF feature selection method are reported for all the models (BBM, N-BBM and A-GBM).

6.4 Results

In Table 7 all the results of the feature selection method are reported. Authors decided to provide just the seven most informative features not to compromise the readability of the tables. From the tables it is possible to draw the following considerations:

- all methods identify the same variables as the most relevant for the model, thus confirming the validity of the modelling procedure. This also allows to trust the reliability of the information contained historical data.
- the BF methods are the most stable, closely followed by the RF methods.
- the WBM is always among the seven most important features for GBMs. This suggests that the N-GBM is able to take into account the information generated by the WBM and use it appropriately, confirming the results of the previous section which underlined the improved performance of GBMs compared to BBMs.

From a physical point of view the results of the feature selection identify the propeller pitch (both setpoint and feedback) and the ship speed (both GPS and LOG) as the most important variables for the prediction, which is what to be expected from this type of ship propulsion system. Propeller speed is not among the most important features, as it is normally kept constant during ship operations and therefore has very limited impact from a modelling perspective. The ship draft (fore and aft) are normally selected as important variables (5th-6th), which also reflects physical expectations from the system as the draft influences both ship resistance and, to a minor extent, propeller performance. As expected the shaft generator power, for this propulsion plants configuration, plays an important role for the prediction. In addition to this, some variables that could be expected to contribute significantly to the overall performance are missing. In particular, wind speed and direction are generally used for estimating the impact of the sea state, but are not included among the five most relevant features by any feature selection method. This suggests that either the sea state has a less significant impact on the ship's fuel consumption compared to what originally expected, or that wind speed and direction are not appropriate predictors for modelling this type of effects, contrarily to what often assumed in relevant literature. One possible additional explanation to the absence of wind speed and direction from the important variables is that the influence of the sea state is already accounted for by the propeller pitch ratio, which is expected to vary as a consequence of both ship speed and ship added resistance. As matter of fact in order to keep constant speed

Table 7 Feature ranking with the BF, LAR and RF with just the seven most informative features

	BFM									RBM			RFM		
	RLS			LAR			RF								
Ranking	BBM	N-GBM	A-GBM	BBM	N-GBM	A-GBM	BBM	N-GBM	A-GBM	BBM	N-GBM	A-GBM	BBM	N-GBM	A-GBM
	Shaft rpm														
1°	22	22	22	22	22	22	22	22	22	22	22	22	22	22	22
2°	15	15	15	15	15	15	15	15	15	15	15	15	15	15	15
3°	14	14	14	14	14	14	14	14	14	14	14	14	14	14	14
4°	16	16	16	16	16	16	16	16	16	21	21	9	16	16	16
5°	21	WBM	21	21	WBM	21	21	WBM	21	9	WBM	21	21	WBM	21
6°	9	9	9	9	9	9	9	9	9	16	9	10	9	9	9
7°	10	10	10	10	10	10	10	10	10	10	10	16	10	10	10
	Shaft Power														
1°	22	22	22	22	22	22	22	22	22	22	22	22	22	22	22
2°	14	14	14	14	14	14	14	14	14	14	14	14	14	14	14
3°	15	15	15	15	15	15	15	15	15	15	15	15	15	15	15
4°	16	16	16	16	16	16	16	16	16	19	9	16	16	16	16
5°	9	9	9	9	9	9	9	9	9	9	16	10	9	9	9
6°	19	WBM	19	19	WBM	19	19	WBM	19	16	10	19	19	WBM	19
7°	10	10	10	10	10	10	10	10	10	10	WBM	9	10	10	10
	Fuel Consumption														
1°	22	22	22	22	22	22	22	22	22	22	22	22	22	22	22
2°	14	14	14	14	14	14	14	14	14	14	14	14	14	14	14
3°	15	15	15	15	15	15	15	15	15	15	15	15	15	15	15

(continued)

Table 7 (continued)

	BFM									RBM			RFM		
	RLS			LAR			RF								
Ranking	BBM	N-GBM	A-GBM	BBM	N-GBM	A-GBM	BBM	N-GBM	A-GBM	BBM	N-GBM	A-GBM	BBM	N-GBM	A-GBM
4°	21	21	21	21	21	21	21	21	21	21	9	16	21	21	21
5°	9	9	9	9	9	9	9	9	9	9	16	9	9	9	9
6°	16	WBM	16	16	WBM	16	16	WBM	16	16	WBM	21	16	WBM	16
7°	10	10	10	10	10	10	10	10	10	27	27	27	10	10	10

profile, the on board automation system should be designed to change the pitch settings and the fuel consumption rate to take into account time domain variation of boundary conditions such as wind and sea state conditions. Under this assumption the relevant information about added resistance and wind intensity are already included in the propeller pitch ratio.

7 Using Machine Learning for Operational Energy Savings: Trim Optimisation

Of all the models proposed in the previous part of this book chapter, the N-GBM based on RF features the best accuracy properties and best physical plausibility and is therefore used for the trim optimisation problem. In order to meet the requirements expressed in Sect. 4.3 the following is considered:

- Variables that are influenced by the trim, such as propeller power and torque, were excluded from the model (see Table 3).
- For each pair of ship speed and displacement, the trim is only allowed to vary in the range observed from the available dataset, extended by $\delta\%$. This allows, for every pair, to limit the extrapolation and therefore to ensure additional reliability of the optimisation results.

In Table 8 the Fuel Consumption percentage reduction with the trim Optimisation technique is reported for different values of δ. As expected, the optimisation procedure always leads to a reduction in fuel consumption. The improvement that can be achieved via trim optimisation increases when δ is increased, although this tendency seems to stabilise for $\delta > 5\%$.

According to the results of this model, improvements exceeding 2% in fuel consumption can be achieved by applying the model for trim optimisation to the selected vessel. It should be noted that trim optimisation can be performed at near to zero cost on board, since it does not require the installation of any additional equipment. Future work in this area will include testing trim optimisation system here proposed on a real vessel, in order to check the validity of the model and the performance of the optimisation tool.

Table 8 Fuel consumption percentage reduction with the trim optimisation technique

$\delta(\%)$	% reduction
0	0.52 ± 0.12
1	1.45 ± 0.32
2	1.72 ± 0.51
5	2.22 ± 0.67
10	2.30 ± 0.64

8 Summary

This chapter focused on the utilisation of methods of Machine Learning for making ship operations more sustainable. Shipping is today facing large challenges in terms of its impact on the climate, and the reduction of CO_2 emissions that are expected to be achieved in the future will require a significant effort.

The achievement of such goals will require, among others, to improve today's ability to accurately model and predict the influence of environmental and operational variables on ship performance, and in particular on the fuel consumption of the ship. In this chapter, alongside with the white-box models commonly used today in this industry, black and gray box models were introduced as modelling approaches that can improve the accuracy of the prediction by making use of extensive measured data from ship operations. The regularised least squares, Lasso Regression and Random forest methods for the construction of black box models were proposed. In addition, two different types of gray, "hybrid" modelling approaches combining elements of white and black box models were also presented: the naive, and the advanced approach. Finally, feature selection methods were introduced, that can be used for testing the physical consistency of black and gray box models.

The book chapter was concluded with the application of the proposed methods to a case study, a chemical tanker, with the aim of testing their ability of predicting fuel consumption and of optimising the trim of the vessel. The results of this application case confirmed the superiority of statistical methods over mechanistic models in their ability of accurately predict the performance of the vessel, and highlighted that gray box models, although improving the performance of black box models only marginally, show an increased predictive ability with small sizes of the training dataset. The application of a naive-gray box model to the problem of trim optimisation allowed identifying the possibility of decreasing fuel consumption by up to 2.3% without the need of installing further equipment on board.

References

1. Anguita, D., Ghio, A., Oneto, L., Ridella, S.: In-sample and out-of-sample model selection and error estimation for support vector machines. IEEE Trans. Neural Netw. Learn. Syst. **23**(9), 1390–1406 (2012)
2. Armstrong, V.N.: Vessel optimisation for low carbon shipping. Ocean Eng. **73**, 195–207 (2013)
3. Baldi, F., Johnson, H., Gabrielii, C., Andersson, K.: Energy and exergy analysis of ship energy systems-the case study of a chemical tanker. In: 27th ECOS, International Conference on Efficiency, Cost, Optimization, Simulation and Environmental Impact of Energy Systems (2014)
4. Basin., D.W.T.M., Todd, F.H.: Series 60 Methodical Experiments with Models of Single-Screw Merchant Ships. Washington (1964)
5. Breiman, L.: Random forests. Mach. Learn. **45**(1), 5–32 (2001)
6. Buhaug, O., Corbett, J.J., Endersen, O., Eyring, V., Faber, J., Hanayama, S., Lee, D.S., Lee, D., Lindstad, H., Markowska, A.Z., Mjelde, A., Nilsen, J., Palsson, C., Winebrake, J.J., Wu, W., Yoshida, K.: Second IMO GHG study 2009. Technical reports, International Maritime Organization (IMO) (2009)

7. Chang, Y.W., Lin, C.J.: Feature ranking using linear svm. Causation Predict. Chall. Chall. Mach. Learn. **2**, 47 (2008)
8. Coraddu, A., Figari, M., Savio, S., Villa, D., Orlandi, A.: Integration of seakeeping and powering computational techniques with meteo-marine forecasting data for in-service ship energy assessment. In: Developments in Maritime Transportation and Exploitation of Sea Resources (2013)
9. Coraddu, A., Gaggero, S., Figari, M., Villa, D.: A new approach in engine-propeller matching. In: Sustainable Maritime Transportation and Exploitation of Sea Resources, vol. 1, pp. 631–637. CRC Press —Taylor and Francis Group (2011)
10. Coraddu, A., Gualeni, P., Villa, D.: Investigation about wave profile effects on ship sability. IMAM 2011 international maritime association of the mediterranean - sustainable maritime transportation and exploration of the sea. Resources **1**, 143–149 (2011)
11. Coraddu, A., Oneto, L., Baldi, F., Anguita, D.: A ship efficiency forecast based on sensors data collection: improving numerical models through data analytics. In: OCEANS (2015)
12. Coraddu, A., Oneto, L., Ghio, A., Savio, S., Anguita, D., Figari, M.: Machine learning approaches for improving condition-based maintenance of naval propulsion plants. Proceedings of the Institution of Mechanical Engineers, Part M: J. Eng. Marit. Environ. doi:10.1177/1475090214540874 (2014)
13. Corbett, J.J., Koehler, H.W.: Updated emissions from ocean shipping. J. Geophys. Res. Atmos. **108**(D20) (2003)
14. Cox, D.R.: Principles of Statistical Inference. Cambridge University Press (2006)
15. Cristianini, N., Shawe-Taylor, J.: An Introduction to Support Vector Machines and Other Kernel-Based Learning Methods. Cambridge Uiversity Pess (2000)
16. De Mol, C., De Vito, E., Rosasco, L.: Elastic-net regularization in learning theory. J. Complex. **25**(2), 201–230 (2009)
17. Deng, H., Runger, G., Tuv, E.: Bias of importance measures for multi-valued attributes and solutions. Artif.l Neural Netw. Mach. Learn. ICANN **2011**, 293–300 (2011)
18. Devanney, J.: The impact of the energy efficiency design index on very large crude carrier design and CO 2 emissions. Ships Offshore Struct. **6**(4), 355–368 (2011)
19. Efron, B., Hastie, T., Johnstone, I., Tibshirani, R.: Least angle regression. Ann. Stat. **32**(2), 407–499 (2004)
20. European Commission: Integrating maritime transport emissions in the EU's greenhouse gas reduction policies (2013)
21. Evans, J.R., Lindner, C.H.: Business analytics: the next frontier for decision sciences. Decis. Line **43**(2), 4–6 (2012)
22. Friedman, J., Hastie, T., Tibshirani, R.: The Elements of Statistical Learning. Springer series in statistics Springer, Berlin (2001)
23. Friedman, J., Hastie, T., Tibshirani, R.: Regularization paths for generalized linear models via coordinate descent. J. Stat. Softw. **33**(1), 1 (2010)
24. Fumeo, E., Oneto, L., Anguita, D.: Condition based maintenance in railway transportation systems based on big data streaming analysis. In: INNS Conference on Big Data (2015)
25. García, S., Fernández, A., Luengo, J., Herrera, F.: Advanced nonparametric tests for multiple comparisons in the design of experiments in computational intelligence and data mining: experimental analysis of power. Inf. Sci. **180**(10), 2044–2064 (2010)
26. Ghelardoni, L., Ghio, A., Anguita, D.: Energy load forecasting using empirical mode decomposition and support vector regression. IEEE Trans. Smart Grid **4**(1), 549–556 (2013)
27. Gieseke, F., Polsterer, K.L., Oancea, C.E., Igel, C.: Speedy greedy feature selection: better redshift estimation via massive parallelism. In: European Symposium on Artificial Neural Networks, Computational Intelligence and Machine Learning (2014)
28. Good, P.: Permutation Tests: A Practical Guide To Resampling Methods For Testing Hypotheses. Springer Science & Business Media (2013)
29. Guldhammer, H., Harvald, S.A.: Ship Resistance: Effect of Form and Principal Dimensions. Akademisk Forlag (1974)

30. Guyon, I., Elisseeff, A.: An introduction to variable and feature selection. J. Mach. Learn. Res. **3**, 1157–1182 (2003)
31. Györfi, L.: A Distribution-free Theory of Nonparametric Regression. Springer (2002)
32. Haas, P.J., Maglio, P.P., Selinger, P.G., Tan, W.C.: Data is dead... without what-if models. In: International Conference on Very Large Database (2011)
33. Hochkirch, K., Mallol, B.: On the importance of fullscale cfd simulations for ships. In: International Conference on Computer Applications and Information Technology in the Maritime Industries (2013)
34. Holtrop, J.: A statistical re-analysis of resistance and propulsion data. Int. Shipbuild. Prog. **31**(363), 272–276 (1984)
35. Holtrop, J., Mennen, G.G.: An approximate power prediction method. Int. Shipbuild. Prog. **29**, 166–171 (1982)
36. Hong, S.J.: Use of contextual information for feature ranking and discretization. IEEE Trans. Knowl. Data Eng. **9**(5), 718–730 (1997)
37. Howison, S.: Practical Applied Mathematics: Modelling, Analysis, Approximation. 38. Cambridge University Press (2005)
38. Iakovatos, M.N., Liarokapis, D.E., Tzabiras, G.D.: Experimental investigation of the trim influence on the resistance characteristics of five ship models. In: Developments in Maritime Transportation and Exploitation of Sea Resources—Proceedings of IMAM 2013, 15th International Congress of the International Maritime Association of the Mediterranean (2014)
39. Jafarzadeh, S., Utne, I.B.: A framework to bridge the energy efficiency gap in shipping. Energy **69**, 603–612 (2014)
40. Jardine, A.K., Lin, D., Banjevic, D.: A review on machinery diagnostics and prognostics implementing condition-based maintenance. Mech. Syst. Signal Process. **20**(7), 1483–1510 (2006)
41. Josephson, J.R., Josephson, S.G.: AbductIve Inference: Computation, Philosophy, Technology. Cambridge University Press (1996)
42. Khor, Y.S., Xiao, Q.: CFD simulations of the effects of fouling and antifouling. Ocean Eng. **38**(10), 1065–1079 (2011)
43. Kohavi, R., John, G.H.: Wrappers for feature subset selection. Artif. Intell. **97**(1), 273–324 (1997)
44. Kohavi, R., et al.: A study of cross-validation and bootstrap for accuracy estimation and model selection. In: International Joint Conference on Artificial Intelligence (1995)
45. LaValle, S., Lesser, E., Shockley, R., Hopkins, M.S., Kruschwitz, N.: Big data, analytics and the path from insights to value. In: MIT Sloan Management Review (2013)
46. Lee, J., Yoo, S., Choi, S., Kim, H., Hong, C., Seo, J.: Development and application of trim optimization and parametric study using an evaluation system (solution) based on the rans for improvement of eeoi. In: International Conference on Ocean, Offshore and Arctic Engineering (2014)
47. Lee, W.S., Bartlett, P.L., Williamson, R.C.: The importance of convexity in learning with squared loss. IEEE Trans. Inf. Theory **44**(5), 1974–1980 (1998)
48. Leifsson, L., Saevarsdottir, H., Sigurdsson, S., Vesteinsson, A.: Grey-box modeling of an ocean vessel for operational optimization. Simul. Model. Pract. Theory **16**, 923–932 (2008)
49. Lewis, E.V.: Principles of Naval Architecture. Society of Naval Architects (1988)
50. Lützen, M., Kristensen, H.: A model for prediction of propulsion power and emissions—tankers and bulk carriers. In: World Maritime Technology Conference (2012)
51. MacKay, D.J.C.: Information Theory, Inference and Learning Algorithms. Cambridge University Press (2003)
52. Maritime Knowledge Centre: International shipping facts and figures - Information resources on trade, safety, security, environment. Technical reports, IMO (2012)
53. Meinshausen, N., Bühlmann, P.: Stability selection. J. R. Stat. Soc. Ser. B (Stat. Methodol.) **72**(4), 417–473 (2010)
54. Moustafa, M.M., Yehia, W., Hussein, A.W.: Energy efficient operation of bulk carriers by trim optimization. In: International Conference on Ships and Shipping Research (2015)

55. Ng, A.Y.: Feature selection, l 1 vs. l 2 regularization, and rotational invariance. In: International Conference on Machine Learning (2004)
56. Oneto, L., Ghio, A., Ridella, S., Anguita, D.: Support vector machines and strictly positive definite kernel: the regularization hyperparameter is more important than the kernel hyperparameters. In: International Joint Conference on Neural Networks (2015)
57. Palmé, T., Breuhaus, P., Assadi, M., Klein, A., Kim, M.: New alstom monitoring tools leveraging artificial neural network technologies. In: Turbo Expo: Turbine Technical Conference and Exposition (2011)
58. Petersen, J.P., Winther, O., Jacobsen, D.J.: A machine-learning approach to predict main energy consumption under realistic operational conditions. Ship Tech. Res. **59**(1), 64–72 (2012)
59. Quinlan, J.R.: Simplifying decision trees. Int. J. Man-Mach. Stud. **27**(3), 221–234 (1987)
60. Reichel, M., Minchev, A., Larsen, N.: Trim optimisation—theory and practice. Int. J. Marine Navig. Saf. Sea Transp. **8**(3), 387–392 (2014)
61. Scholkopf, B.: The kernel trick for distances. In: Neural Information Processing Systems (2001)
62. Schölkopf, B., Herbrich, R., Smola, A.J.: A generalized representer theorem. In: Computational Learning Theory (2001)
63. Schultz, M.P., Bendick, J., Holm, E.R., Hertel, W.M.: Economic impact of biofouling on a naval surface ship. Biofouling **27**(1), 87–98 (2011)
64. Shao, W., Zhou, P., Thong, S.K.: Development of a novel forward dynamic programming method for weather routing. J. Mar. Sci. Tech. **17**(2), 239–251 (2011)
65. Shawe-Taylor, J., Cristianini, N.: Kernel Methods for Pattern Analysis. Cambridge University Press (2004)
66. Simon, N., Friedman, J., Hastie, T., Tibshirani, R.: A sparse-group lasso. J. Comput. Graph. Stat. **22**(2), 231–245 (2013)
67. Smith, T.W.P., Jalkanen, J.P., Anderson, B.A., Corbett, J.J., Faber, J., Hanayama, S., OKeeffe, E., Parker, S., Johansson, L., Aldous, L.: Third imo ghg study 2014. Technical report, International Maritime Organisation (2014)
68. Stewart, T.R., McMillan Jr, C.: Descriptive and prescriptive models for judgment and decision making: implications for knowledge engineering. In: Expert Judgment and Expert Systems (1987)
69. Stopford, M.: Maritime Economics. Routeledge, New York (2009)
70. Sugumaran, V., Muralidharan, V., Ramachandran, K.: Feature selection using decision tree and classification through proximal support vector machine for fault diagnostics of roller bearing. Mech. Syst. Signal Process. **21**(2), 930–942 (2007)
71. Tibshirani, R.: Regression shrinkage and selection via the lasso. J. Royal Stat. Soc. Ser. B (Methodol.) pp. 267–288 (1996)
72. Tikhonov, A., Arsenin, V.Y.: Methods for solving ill-posed problems. Nauka, Moscow (1979)
73. UNCTAD: Review of maritime transport. Technical report, United Conference on Trade and Development (2012)
74. Vapnik, V.N.: Statistical Learning Theory. Wiley–Interscience (1998)
75. Von Karman, T., Gabrielli, G.: What price speed? specific power required for propulsion of vehicles. Mech. Eng. **72**, 775–781 (1950)
76. Wang, H., Faber, J., Nelissen, D., Russell, B., St Amand, D.: Marginal Abatement Costs and Cost Effectiveness of Energy-Efficiency Measures. Technical report, Institute of Marine Engineering, Science and Technology (2010)
77. White, A.P., Liu, W.Z.: Technical note: bias in information-based measures in decision tree induction. Mach. Learn. **15**(3), 321–329 (1994)
78. Widodo, A., Yang, B.S.: Support vector machine in machine condition monitoring and fault diagnosis. Mech. Syst. Signal Process. **21**(6), 2560–2574 (2007)
79. Yoon, H., Yang, K., Shahabi, C.: Feature subset selection and feature ranking for multivariate time series. IEEE Trans. Knowl. Data Eng. **17**(9), 1186–1198 (2005)
80. Young, D.M.: Iterative Solution of Large Linear Systems. Dover Publications. Com (2003)
81. Zou, H., Hastie, T.: Regularization and variable selection via the elastic net. J. R. Stat. Soc.: Ser. B (Stat. Method.) **67**(2), 301–320 (2005)
82. Zou, H., Hastie, T., Tibshirani, R.: On the degrees of freedom of the lasso. Ann. Stat. **35**(5), 2173–2192 (2007)

FuzzyCovering: A Spatial Decision Support System for Solving Fuzzy Covering Location Problems

Virgilio C. Guzmán, David A. Pelta and José Luis Verdegay

Abstract This chapter presents a spatial decision support system called FuzzyCovering, which is designed to support the decision-making process related to facility location problems. Different components that facilitate modeling, the solution and results display, specifically about covering location problems are integrated in FuzzyCovering. FuzzyCovering allows the study of various scenarios of facilities location and provides a range of solutions that allow the users to make the best decisions. To treat the uncertainty inherent to some underlying parameters of the real location problems, FuzzyCovering integrates a fuzzy approach in which the problem constraints can be imprecisely defined. A detailed description of the architecture and functionality of the system is presented, and a simulated practical case of a maximal covering location problem with fuzzy constraints is shown to demonstrate the benefits of FuzzyCovering.

1 Introduction

Spatial Decision Support Systems (SDSS) have become useful tools in decision-making processes in which spatial data are involved. A SDSS is an information system that integrates technologies of Geographical Information Systems (GIS) and Decision Support Systems (DSS), whose main purpose is to support the decision maker in spatial dimension problems [21]. Densham [9] points out that a SDSS provides a framework in which a database management system with a set of analytical models, graphical visualizers and tabular report representation capacity are

V.C. Guzmán (✉)
Unidad Académica de Ciencias y Tecnologías de la Información, Universidad Autónoma de Guerrero, Mexico, Mexico
e-mail: vguzman@uagro.mx

D.A. Pelta · J.L. Verdegay
Universidad de Granada, Granada, Spain
e-mail: dpelta@decsai.ugr.es

J.L. Verdegay
e-mail: verdegay@decsai.ugr.es

C. Cruz Corona (ed.), *Soft Computing for Sustainability Science*,
Studies in Fuzziness and Soft Computing 358, DOI 10.1007/978-3-319-62359-7_3

mainly integrated. A SDSS is usually aimed at solving problems of a specific application domain. In a general sense, by means of a graphical interface, a SDSS allows the interaction of its integrated optimization models with the decision maker, and provides a set of alternatives (solutions) feasible to the problem under study.

This chapter presents a spatial decision support system, which we have called FuzzyCovering. FuzzyCovering is specifically aimed at supporting decision-makers to solve complex covering location problems. By using FuzzyCovering it is possible to study different scenarios with the aim of finding the best solutions for facilities location.

Different real world situations can be modeled by means of a covering location problem (CLP). For instance, some models have been proposed to determine the best locations for police stations [6], ambulance location [3], fire stations [22], facility location of medical services for large-scale emergencies generated by natural or man-made disasters [12]. In a general context, all these models are focused on finding the best locations for a certain number of facilities, taking into account that the distance or travel time between demand points which require services and the facilities offering such services does not exceed an established threshold.

The set covering location problem (SCLP), introduced by Toregas [19], and the maximal covering location problem (MCLP) proposed by Church [5] are the most commonly used models. The SCLP tries to find a minimum number of facilities that cover all demand points (full coverage); whereas MCLP tries to maximize the demand covered with a fixed number of facilities known a priori, without the need to cover all points (partial coverage).

Both SCLP and MCLP can be precisely defined in FuzzyCovering. However, due to the need to treat the uncertainty associated in some of the elements of these types of problems, FuzzyCovering also contains fuzzy models that allows to consider and address this uncertainty. On the other hand, since the distance or time of service is one of the most important elements in location problems, the models implemented in FuzzyCovering consider this item as an imprecise value (probably linguistically established). This distance is modeled by fuzzy constraints. The resulting fuzzy problem is converted into a crisp set of problems that can be solved by conventional optimization techniques, from whose solutions can build a solution to the starting problem. In this sense, FuzzyCovering provides metaheuristic methods to solve these models.

Like any SDSS, FuzzyCovering provides a interactive and easy-to-use graphical interface for the decision maker. Specifically, this interface allows the configuration of scenarios and solution visualization for a given location problem. Furthermore, due to the modular way in which FuzzyCovering was implemented, it is possible and easy to include extensions or other location models, as well as other solution methods.

Consequently, this chapter is organized as follows. Section 2 describes the covering location problems, specifically focusing on the MCLP. Furthermore, a fuzzy extension of this problem is described in detail. The architecture and functionality of

FuzzyCovering is described in Sect. 3. Section 4 shows the results obtained of a simulated practical case solved in FuzzyCovering. Finally, our conclusions are presented in Sect. 5.

2 The Covering Location Problem

The need to find the best locations for service facilities is essential to improve the quality of these services required in different contexts of our society. This situation has generated the emergence of numerous proposals aimed at resolving the various existing facility location problems. In the context of the emergency services, the need arises to optimally locate service units that respond efficiently to incidents generated among the population. In this direction, some models have been proposed to determine the best locations of ambulances [3], police stations [6], fire stations [22], location of hospitals, among others. Also, there have been other works oriented to model large-scale emergency services, which are caused by natural or man-made disasters (see, for example, [12, 15]). Other important application domains are the location of telecommunication antennas [14], the transport of passengers [11], postal services [23] and in distribution system design [16], etc.

Among the location problems there are the covering location problems (CLP). As its name suggests, the coverage is the most important aspect in this type of problem, which can be defined as the distance or critical time (travel time or response time) within which a demand point is considered covered. In this sense, it is clear to assume that the main purpose of a CLP is to try to locate one or more service facilities, taking into account that the distance or critical time from a facility to a demand point is less than a established threshold. Under this coverage criterion, we should consider that establishing a minimum distance value, could significantly improve the quality of service offered. However, this situation would obviously force to locate a large number of facilities, which could lead to an expensive and difficult solution to implement in a real environment. On the other hand, obviously, considering a large distance value, it reduces the number of facilities, however, in this case, the quality of service can be deteriorated. Given this reality, and considering the type of location problem that is being addressed, the distance value and the number of facilities to locate must be compensated.

In a discrete spatial representation of the CLP, it can be assumed that the facilities can be located on the same demand points, therefore, some or all of the demand points can be considered as potential locations for facilities.

The set covering location problem (SCLP), formulated by Toregas [19] was the first problem proposed within the category of covering location problems. In this problem the amount of demand generated at the demand points is not considered and its aim is to find the minimum number of facilities needed to ensure to cover all demand points within a pre-defined distance. In practice, this requirement of full coverage could lead to undesirable situations. On the one hand, the number of required facilities could be very high. On the other hand, since the problem does

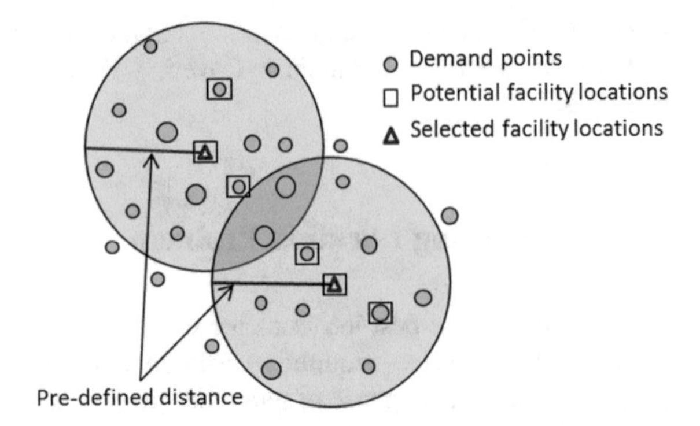

Fig. 1 A solution of the MCLP with two facilities, six potential facility locations, and twenty five demand points

not distinguish points of higher and lower demand, it could locate a facility to cover lower demand points.

In some cases, if the points of greater demand are considered the most important, it seems clear to prefer covering these points and leave uncovered those with lower demand. Under these considerations, Church and Revelle [5] posed the Maximal Covering Location Problem (MCLP). In this problem, a set of demand points and a set of facilities to be located are considered. Each demand point has an associated value, which represent its level of importance (demand generated at the point). The objective of this problem is to find the best locations for a fixed number of facilities known a priori that maximizes the total demand coverage. Since it is considered the demand generated at the points, the points with greater demand will be always covered, and some points can be uncovered. Figure 1 shows a solution for a MCLP with two facilities, which cover twenty five of a total of 30 demand points. The size of the point indicates its demand. Here, there are six potential facility locations, two demand points are covered by the two facilities, and there are points uncovered.

In this chapter we will focus specifically on the MCLP, so we present its formulation below.

Sets

i, I index and set of the demand points.

j, J index and set of potential locations for the facilities.

$N_i = \{j \in J | d_{ij} \leqslant S\}$ the set of potential facility locations that can cover the point i within the time or distance S, d_{ij} is the distance between the point i and the potential location for the facility j.

Input parameters

p the number of facilities to be located.

S the maximum allowed time or distance to respond to a request.

w_i value that represent the demand associated to point i.

Decision variables

x_j 1 if a facility is located at the point j, 0 otherwise.

y_i represent the coverage of point i, 1 if point is covered $(\exists_j | x_j = 1 \wedge j \in N_i)$, 0 otherwise.

Mathematical model

$$\max Z = \sum_{i \in I} w_i y_i \tag{1}$$

subject to

$$\sum_{j \in N_i} x_j \geqslant y_i \quad \forall i \in I \tag{2}$$

$$\sum_{j \in J} x_j = p \tag{3}$$

$$x_j = \{0, 1\} \quad \forall j \in J \tag{4}$$

$$y_i = \{0, 1\} \quad \forall i \in I \tag{5}$$

The objective function (1) maximizes the covered demand by the set of established facilities. Constraint (2) guarantees that one or more facilities will be located within the distance or travel time pre-defined S from the demand point i. Constraint (3) estates that the number of facilities to be located be p. Finally, the constraint (4) and (5) indicate binary restriction on the decision variables x_i and y_i.

2.1 Uncertainty in MCLP

In most of the works that have addressed a MCLP, it has been considered an approach in which all the parameters of the problem are known with certainty. However, we can find many real scenarios in which these problems present imprecision or uncertainty in some of these parameters. This situation requires consideration of such uncertainty in the modeling of the problem. C. Guzmán et al. [10] describe some of the parameters of the MCLP in which there may exist uncertainty. In this sense, an assumption that is often made in a MCLP is to define the distance as precise or exact value and assume that all demand points that are within such distance from the facilities are fully covered, while those points that are beyond that distance are uncovered. Under this assumption, we can say, for instance, that the demand points which are at a distance of 3.0 kms. from the nearest facility, are covered, while those demand points that are at a distance of 3.01 are not covered. In this regard, it is noteworthy that in real applications, this rigidity in the distance may lead to an unrealistic and inappropriate modeling of the problem. Therefore considering the uncertainty in the formulation of the problem is a crucial element in the MCLP. In this line, Berman et al. [2], Karasakal and Karasakal [13] extend the MCLP in which they assumed a gradually decreasing coverage depending of the distance.

In general, considering and modeling uncertainty in covering location problems allows to obtain models more adjusted to the characteristics of the real problems. In this context, we can point the relevance and acceptance of the fuzzy approach as an appropriate technique to solve these types of problems. Under this approach, fuzzy sets have been commonly used to represent the different elements of the problem with the presence of linguistic uncertainty, that is, non-probabilistic one. Batanovic et al. [1] present a MCLP applied to networks in which the demand values are described by linguistic terms and represented by triangular fuzzy numbers. Takaĉi et al. [18] proposed a fuzzy maximal covering location model in which the travel time from a facility to demand points is modeled by fuzzy sets. Davari et al. [7] formulate a large-scale MCLP with fuzzy coverage radius whose value is represented by a triangular fuzzy number. Shavandi and Mahlooji [17] present a fuzzy approach based on the Queuing Theory in which the demand, server capacity, average number of customers at the queue, and service times of server are considered as triangular fuzzy numbers. For further details, the interested readers may refer to C. Guzmán et al. [10] where a review of recent literature on the fuzzy approach applied to the MCLP is presented.

2.2 A Fuzzy Extension of the MCLP

Therefore, as it seems clear that the distance is a matter of degree, in this chapter, a fuzzy extension of MCLP is presented. In this extension, we relax the distance value S, thus allowing to obtain solutions beyond this value, specifically to a value $S + \tau$, where τ represents a value of tolerance permitted by the decision maker. Since the distance is the most important element in the MCLP, any changes made to it, directly affects the solution of the problem. In this regard, by relaxing the value of distance, solutions with different coverage values can be obtained. Figure 2 graphically represents this idea. The dashed line represents the border generated by the pre-defined distance S, while the solid line shows border generated by $S + \tau$.

In the range $[S, S + \tau]$, there is a set of solutions with different distance values and, therefore, with different coverages. Considering all demand points of same importance, we can see in Fig. 2 that with a distance S only 6 demand points can be covered, and with $S + \tau$, up to 13 demand points can be covered, thereby, with distances in this range, we can find other coverage values. With this idea, it is provides a wide range of solutions, which significantly facilitate the decision-making process.

In order to model this situation, we have used a fuzzy approach, where it is assumed that the decision maker allows some violations in the accomplishment of the distance constraint. In this sense, we consider the following fuzzy constraint used to construct the N_i set:

$$N_i : \{j | d_{ij} \leqslant_f S\} \tag{6}$$

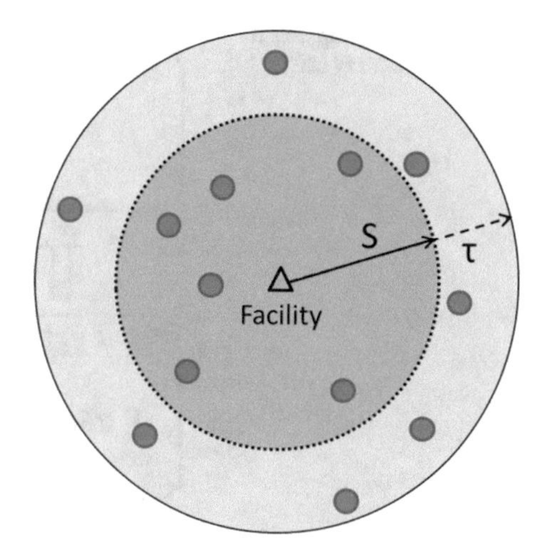

Fig. 2 Schematic representation of a MCLP with fuzzy border

where $\leqslant_f$ stands for the constraint (6) could be slightly violated. It should be noted that although d_{ij} and S do not appear directly in the formulation of the problem, they are included in the definition of the set N_i. Thus, the membership function that represents the satisfaction degree of the constraint (6), is the following piece-wise linear function:

$$\mu(d_{ij}) = \begin{cases} 1 & \text{if } d_{ij} \leqslant S, \\ 1 - \frac{d_{ij}-S}{\tau} & \text{if } S < d_{ij} \leqslant S + \tau, \\ 0 & \text{if } d_{ij} > S + \tau \end{cases} \quad (7)$$

where $\tau \in \mathbb{R}$ is the maximum tolerance allowed by the decision maker.

There are several approaches proposed to transform fuzzy models into classical ones, i.e. crisp models, see, for example, [4, 8, 24]. In this chapter, the parametric approach posed by Verdegay [20] is used to solve our fuzzy extension of the MCLP. Basically, this approach employs two phases. The first phase transforms the fuzzy problem into several crisp problems by using α-cuts, where the parameter α-cut represents the satisfaction degree of the decision maker. As a consequence, for each α-cut considered, a classical MCLP (α-MCLP) is obtained. In the second phase, all these problems are solved by using conventional well known techniques. The results obtained for the different α values generate a set of solutions, which are integrated by the Representation Theorem for fuzzy sets. Thus, we can say that the solution provided by the parametric approach is a solution of our fuzzy MCLP extension. Figure 3 shows the steps involved in this approach.

Following this approach, the Eq. (6) is transformed into the following constraint:

$$N_i : \{j | d_{ij} \leqslant S + \tau(1 - \alpha)\}$$

where $\alpha \in [0, 1]$ is the degree of relaxation in the distance and τ is the tolerance level allowed by the decision maker.

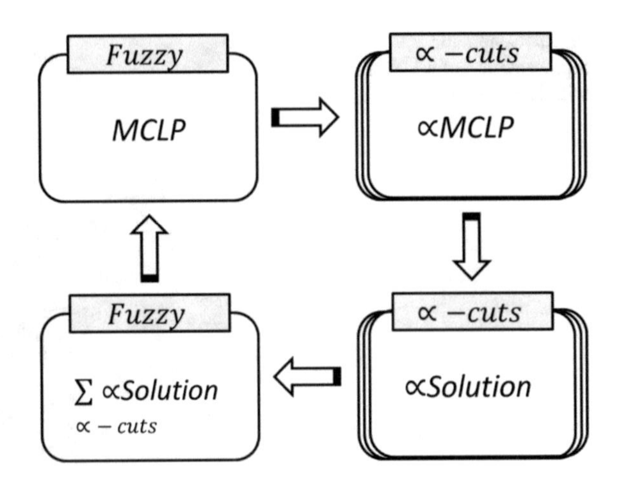

Fig. 3 Parametric approach applied to the fuzzy MCLP

3 FuzzyCovering: A Spatial Decision Support System

The SDSS proposed herein, which we have named FuzzyCovering is directed towards covering location problems. Specifically, FuzzyCovering integrates the two covering location problems more commonly used, SCLP and MCLP. These problems can be precisely defined in FuzzyCovering. However, because of the need to treat uncertainty associated with these types of problems, FuzzyCovering contains fuzzy models (for both MCLP and SCLP), in which the constraints of the problem can be imprecisely defined. This model with fuzzy constraints described above, treats the distance or response time as imprecise values, which allows solving location problems closest to reality. On the other hand, FuzzyCovering includes some metaheuristics aimed to solve its incorporated location models.

In addition, like any SDSS, FuzzyCovering provides a interactive and easy-to-use graphical interface, which allows, among other things, the configuration of scenarios and cartographic display of solutions for a specific facility location problem.

In this section, the architecture and functionality of FuzzyCovering is described with more detail. In addition, it also describes the workflow to be followed in Fuzzy-Covering for setting up and solving a problem.

3.1 Architecture and Description

FuzzyCovering architecture (see Fig. 4) is based on the architecture scheme proposed by Densham [9] for a SDSS. However, considering the rapid advance of technology and, on the other hand, the formulation of new extensions or location models, we have considered a scalable architecture of FuzzyCovering. In this sense, we have made a modular design basically consisting of a three-layer system. This architecture

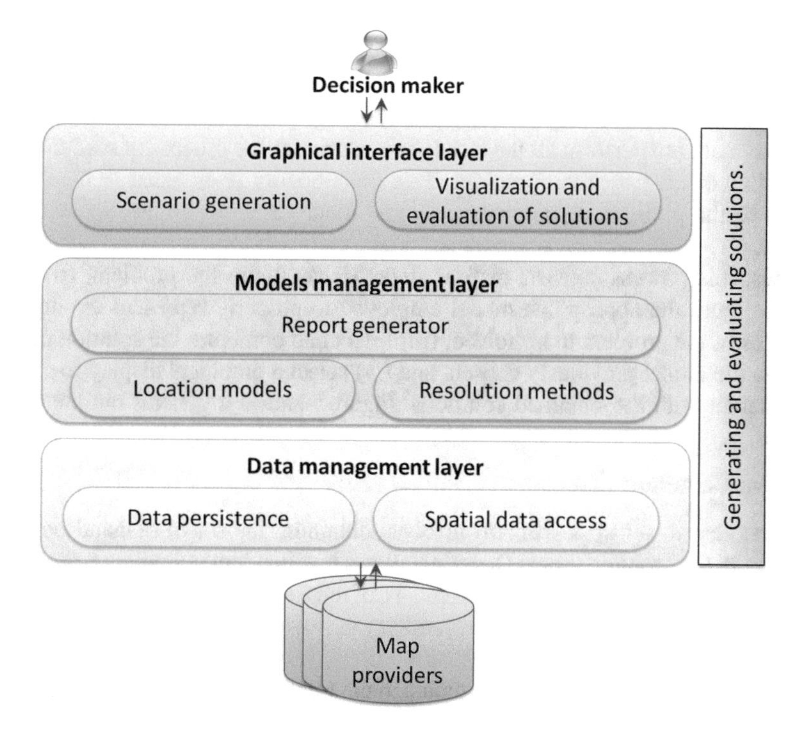

Fig. 4 FuzzyCovering architecture

facilitates the integration of other location models and resolution methods. Each of these layers are detailed below.

- **Graphical interface layer** provides the graphical components needed to analyze various scenarios for a given location problem. Besides these components, this layer contains components designed for a simultaneous display of both the results obtained by mathematical models, which can be displayed in graphs or tables as well as a cartographic representation of the solution.
- **Models management layer** manages the facility location models as well as resolution methods implemented in FuzzyCovering. Also this layer contains the components that integrate the results obtained by the resolution methods, and other components that perform the report generation process.
- **Data management layer** provides all functionality necessary for spatial data manipulation. FuzzyCovering uses unfolding for a simplified access to geographic map providers such as OpenStreetMap, Google Maps, Bing, among others. Unfolding is a library that allows to create interactive maps and customized geovisualizaciones for a specific application domain. It is noteworthy that the modular design of FuzzyCovering allows to relatively simply change this library for any other.

3.2 *FuzzyCovering Environment*

As mentioned above, FuzzyCovering provides a graphical interface that allows the decision maker to perform all the tasks associated with the process of modeling and solving a covering location problem. This functionality is included in the screens related to the problem definition, model and metaheuristic settings, and solutions display.

Thus, this process consists of four steps: (i) configure the problem, (ii) select and configure the appropriate model employed to properly represent the different variables of the problem to be solved, (iii) select and configure the solution method to solve the model previously chosen, and (iv) select a graphical display to analyze and interact with the obtained solutions. Figure 5 shows a general outline of this process.

Problem Definition

The first step to set up a scenario involves obtaining the set of demand points to consider in the problem. FuzzyCovering allows reading demand points from a CSV or XML file. Each demand point must consist of three attributes: an identifier, location (x, y), and a numeric value that represents the demand. The identifier uniquely determines the demand point. The location (x, y) may be a geographical location if we are working with maps or a coordinate in the Euclidean plane if we are analyzing other types of scenarios. The demand value represents the importance of the demand point. It is noteworthy that this last parameter is only needed for MCLP because

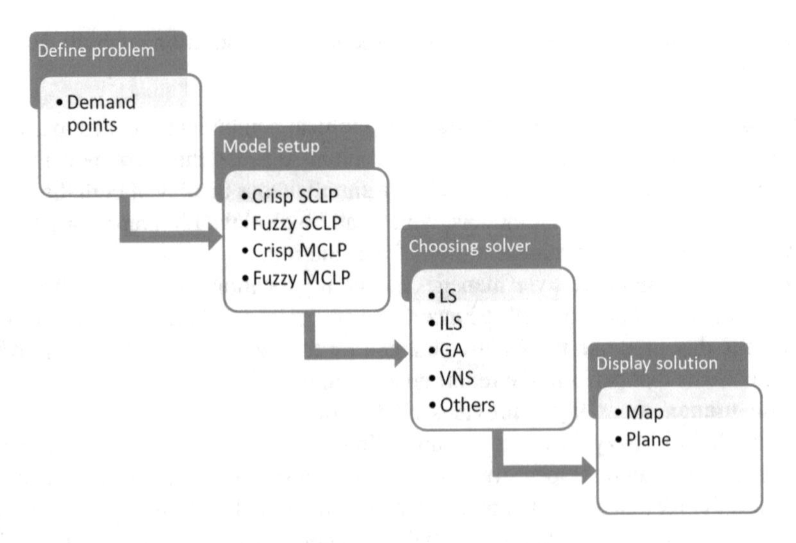

Fig. 5 Workflow to solve a location problem in FuzzyCovering

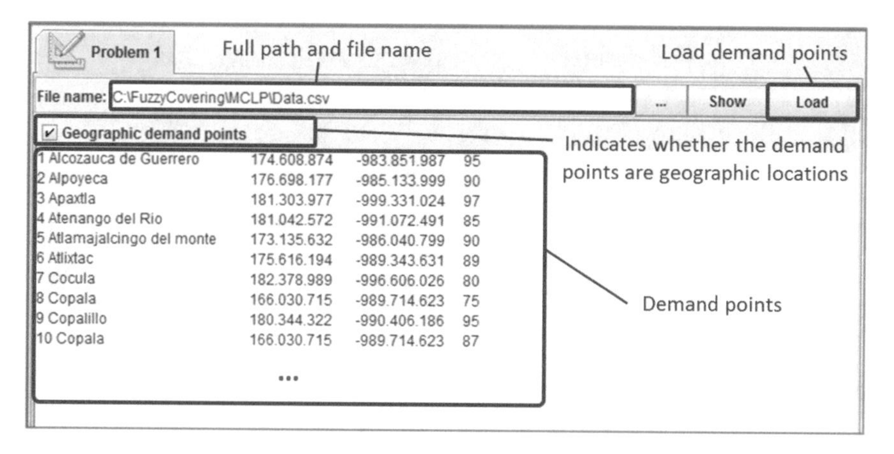

Fig. 6 Screenshot for obtaining demand points contained in a text file (CSV)

demand is considered in the nature of the problem, whereas for the SCLP is not necessary. When a file is selected, FuzzyCovering displays the contents of this file, and creates a set of objects representing the demand points of the problem. These objects are persistently stored in a XML-format file, which can be used in other instances of FuzzyCovering in order to define new scenarios based on the same data.

Figure 6 shows the screenshot that allows obtaining the demand points of the problem from a (CSV) text file. In this screen, the user must activate the option *Geographical demand points* if the demand points are geographic coordinates. With this action, FuzzyCovering will treat the value of x as latitude, and value of y as longitude.

Model Configuration

To generate a model, it is necessary to have previously configured a problem, on which the model will be applied. Figure 7 illustrates the screenshot used for entering parameters of a MCLP with fuzzy constraints. This screen consists of two parts. On top, the basic parameters of the model are introduced. These parameters are the problem to be solved; the pre-defined distance value between facilities and demand points; the distance measure, which can be geographic type, if we are working with geographical locations, or Euclidean type if we are working with other distance values; and the number of facilities to be located. The parameters required for the fuzzy constraint are introduced into the bottom of the screen. Such parameters are the percentage of tolerance permitted to violate the fuzzy constraint, and the corresponding values for each $\alpha \in [0.0, 0.1, \ldots, 1.0]$.

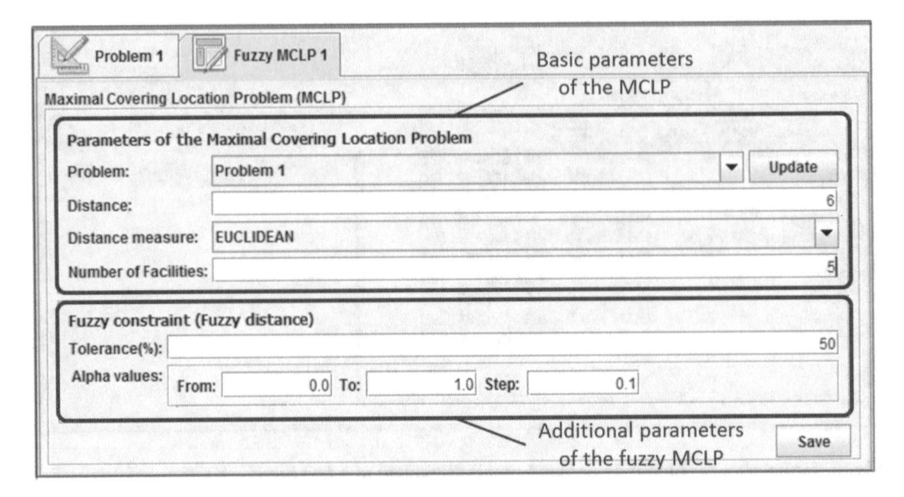

Fig. 7 Screenshot for entering the parameters of the fuzzy MCLP

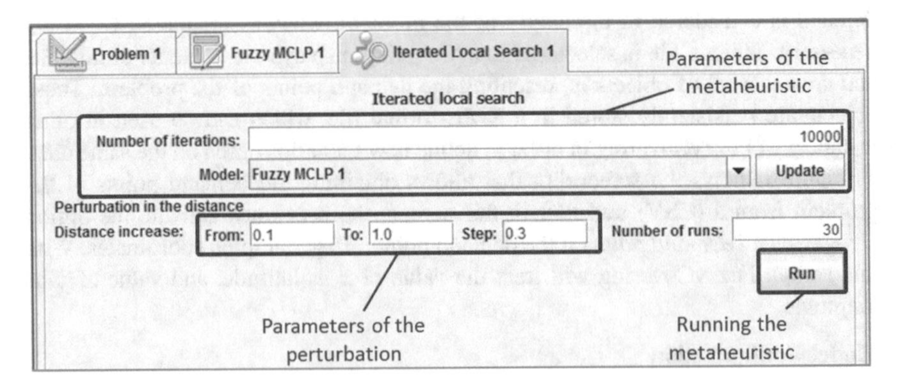

Fig. 8 Screenshot for configuring and executing a metaheuristic

Configuration and Execution of the Resolution Method

As mentioned previously, FuzzyCovering has metaheuristics methods aimed to solve the location models contained in its *Models management layer*. Specifically, in this version there are three metaheuristics implemented: a Local Search (LS), an Iterated Local Search (ILS) and Genetic Algorithm (GA). All models can be solved with any of these metaheuristics. In general, a metaheuristic receives the model to be resolved as input parameters, and the stopping criterion corresponding to the chosen metaheuristic. For example, the number of iterations for methods based on local search or the number of generations reached by genetic algorithm, in addition to other specific parameters of the chosen metaheuristic.

Figure 8 shows the screenshot corresponding to the configuration and execution of iterated local search. Besides the parameters mentioned above, for the particular

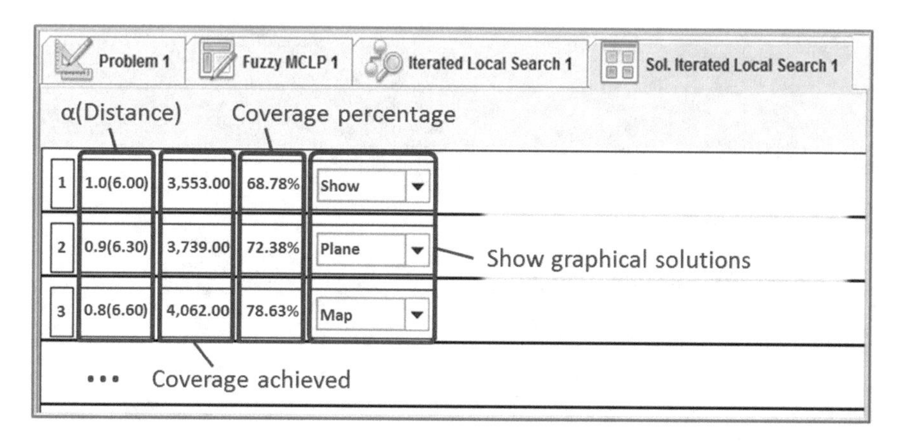

Fig. 9 Screenshot showing the different solutions found for a specific scenario of a covering location problem

case of ILS, one must set up the parameters required for the perturbation procedure used by this method to escape from local optima. The number of runs is also entered on this screen. Once the configuration of the metaheuristic is done, it can be run through the *Run* option. This action generates a solution for each set value of α.

Solution Visualizer

Running a metaheuristic in FuzzyCovering automatically generates different solutions to the problem under study, as many as α-cuts have been defined. These solutions can be observed in the screenshot shown in Fig. 9. In this case, the solutions for a MCLP with fuzzy constraints are presented. In this screen, the solutions are organized in tabular form, where each row corresponds to a crisp solution obtained for a problem with a specific α value. For each solution, five different values are shown: a sequential value that uniquely identifies a solution; an α value associated to each problem, and its corresponding distance value (S for the first solution, $S + \tau$ for the last one, and a value in this range for other solutions); the value and percentage of coverage achieved.

In addition, Fuzzy Covering allows the decision maker to visualize each of these solutions graphically, on a geographic map if demand points are geographical locations or on spider map, otherwise. Figure 10 shows a solution with 5 facilities located on a Geographical map. On the map, the facilities are represented by square, the demand points are indicated by black dots, and arcs that connect the facilities with demand points symbolize the coverage obtained.

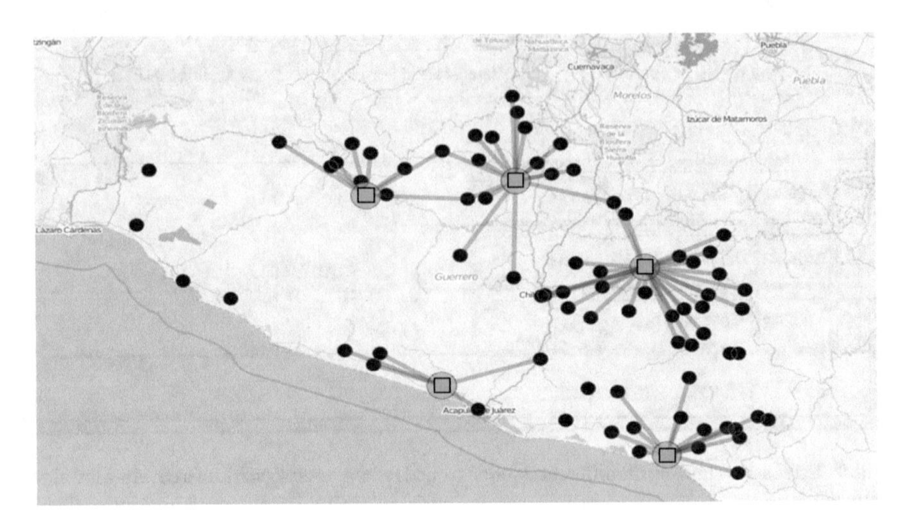

Fig. 10 Solution shown on a geographical map generated by FuzzyCovering

4 A Simulated Practical Case

In order to show the advantages offered by FuzzyCovering to solve a covering location problem, in this section, a simulated practical case is presented. Specifically, in this study, a MCLP with fuzzy constraints applied to random data is modeled and solved.

In this scenario, we assume an area of dimension 30×30 units in which 200 demand points were located. The coordinates for each point were randomly selected. Firstly, a x-coordinate value was chosen by following a uniform distribution [0, 30] and then a y-coordinate value was also chosen on a uniform distribution of [0, 30]. The demand assigned to each point is a value that also follows a uniform distribution in the range $[1, 50]$. Thus, points whose demand is 1, are of lesser importance, while points with demand $= 50$ are of greater importance. The number of facilities to locate is 5 and the distance pre-defined standard is 6 units. We also consider that all demand points are potential locations to locate facilities, and assume that the installation cost is the same for all the facilities, therefore, it is not considered in the model. In addition, all facilities to be located are homogeneous, namely, it is assumed that they have the same available service capacity. The Euclidean distance was used to calculate the distance between facilities and demand points.

In the fuzzy sense, we consider that the tolerance level in the distance constraint is up to 50% more about the pre-defined distance standard, that is, the decision maker allows violations of the distance up to a value 9 units, with respect to the value 6. This scenario was solved for each value of $\alpha \in [0.0, 0.1, \ldots, 1.0]$ where for each α, 30 runs of the ILS (Method integrated in FuzzyCovering) were independently carried out with 10,000 evaluations of the objective function. All these parameters were entered in FuzzyCovering through the screens described above. Then demand

Table 1 Coverage percentages obtained (maximum, minimum, average, and standard deviation) for different α-cut

α-cut	Coverage (percentage)			
	Max.	Min.	Avg.	Std.
1.0	**68.78**	**59.78**	**65.17**	2.20
0.9	72.38	66.26	70.05	1.71
0.8	78.63	71.04	74.59	1.85
0.7	79.21	68.39	74.97	2.44
0.6	82.19	75.40	80.18	1.96
0.5	89.88	75.11	83.82	3.35
0.4	90.40	80.97	86.47	2.31
0.3	94.35	84.53	89.39	2.85
0.2	95.78	79.38	90.93	4.38
0.1	99.01	90.65	95.97	2.26
0.0	**99.15**	**91.13**	**96.02**	2.21

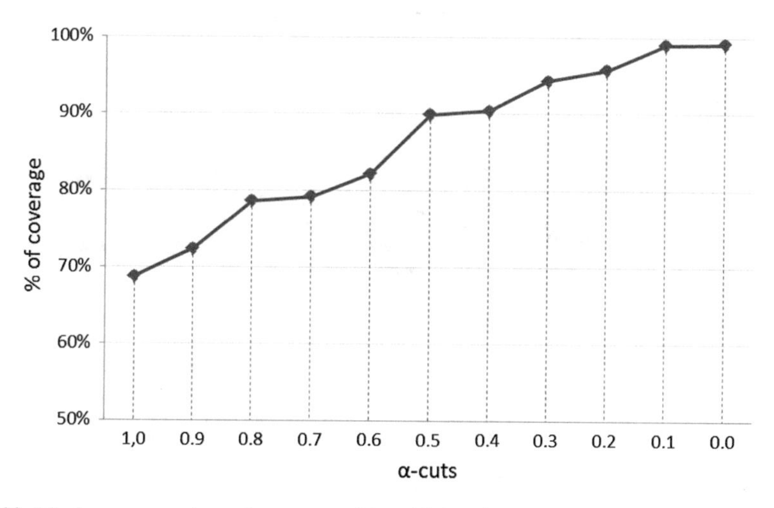

Fig. 11 Maximum percentage of coverage achieved for each α-cut

points used in this scenario were obtained from a (CSV) text file, which was also generated by FuzzyCovering.

Table 1 shows the maximum, minimum, average and standard deviation coverage percentages obtained by FuzzyCovering for each value of α-cut. The values shown in bold are the percentages of coverage obtained for both the starting problem ($\alpha = 1.0$) and the problem with maximum tolerance value allowed by the decision maker ($\alpha = 0.0$). In general, we see an increase of coverage when the problem becomes more relaxed, which is obvious, since the distance value increases. These results provide a range of solutions with different percentages of coverage and with various

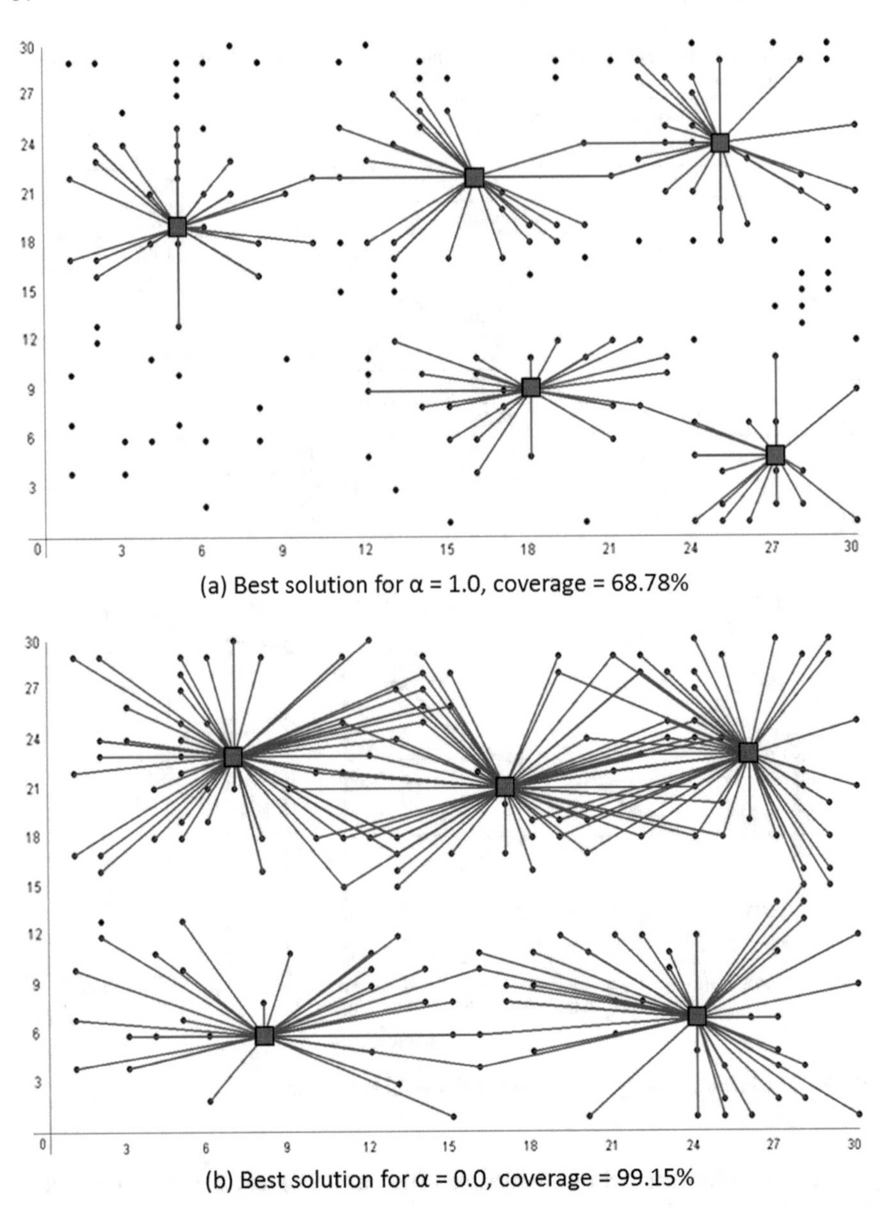

Fig. 12 Best solutions found by FuzzyCovering for a fuzzy MCLP with 5 facilities

degrees of tolerance in the distance, which help the decision maker to make the best decisions about the value of coverage to be achieved.

FuzzyCovering provides a CSV or XML file containing all these results. In addition, FuzzyCovering provides solution graphs that facilitate the decision maker to analyze the percentages, especially the maximum coverage for different values of α

(see Fig. 11). In this type of graph, it can be observed the increased coverage in terms of α-cut.

As mentioned above, FuzzyCovering allows to graphically show the solutions found. In this analysis, since random data were used, the solutions were shown in a spider map (see Fig. 12). The solutions visualized are for $\alpha = 1.0$ (Fig. 12a) and for $\alpha = 0.0$ (Fig. 12b). The percentage of maximum coverage reached for each value of α is shown. The demand points are represented with black dots, the squares indicate the facilities locations, and the line segments that connect the demand points with the facilities represent the coverage. When the constraint is relaxed ($\alpha = 0.0$ vs $\alpha = 1.0$), the coverage percentage is increased. This increase is not only due to the use of a greater distance coverage, but also to the relocation of the facilities. It should also be noted that when $\alpha = 0.0$, there is a greater amount of demand points covered by more than one facility. This side effect is appropriate in real applications where it is important to locate backup facilities that respond to customers when the main facilities are not available. Using these kind of plots, FuzzyCovering greatly facilitates the analysis for these type of situations.

5 Conclusions

FuzzyCovering is a spatial decision support system designed to assist decision-making process related to facility location problems. Specifically, FuzzyCovering is focused on modeling and solving covering location problems in an interactive way. Because of the need to address the uncertainty associated in some of the parameters of these problems, FuzzyCovering integrates fuzzy models in which the constraints of the problem can be defined imprecisely.

Considering the usefulness and relevance of facility location problems in the current social and technological context, and the inherent complexity to determine a suitable location for the facilities of a specific practical problem, FuzzyCovering facilitates the analysis of various scenarios of the problem under study, and provides a wide range of solutions that help the decision maker to make the best decisions.

The modular architecture of FuzzyCovering allows to extend its functionality by including of other location models and solution algorithms. In this sense, we can say that FuzzyCovering is a flexible and adaptable tool to the needs of the decision maker.

Acknowledgements V.C. Guzmán is supported by a scolarship from PROMEP, México, PROMEP /103.5/12/6059. D. Pelta and J.L. Verdegay acknowledge support through projects TIN2014-55024-P from the Spanish Ministry of Economy and Competitiveness, and P11-TIC-8001 from the Andalusian Government (both including FEDER funds).

References

1. Batanović, V., Petrović, D., Petrović, R.: Fuzzy logic based algorithms for maximum covering location problems. Inf. Sci. **179**(1–2), 120–129 (2009)
2. Berman, O., Krass, D., Drezner, Z.: The gradual covering decay location problem on a network. Eur. J. Oper. Res. **151**(3), 474–480 (2003)
3. Brotcorne, L., Laporte, G., Semet, F.: Ambulance location and relocation models. Eur. J. Oper. Res. **147**(3), 451–463 (2003)
4. Cadenas, J., Verdegay, J.: Towards a new strategy for solving fuzzy optimization problems. Fuzzy Optim. Decision Mak. **8**(3), 231–244 (2009)
5. Church, R., Revelle, C.: The maximal covering location problem. Pap. Reg. Sci. Assoc. **32**(1), 101–118 (1974)
6. Curtin, K.M., Hayslett-McCall, K., Qiu, F.: Determining optimal police patrol areas with maximal covering and backup covering location models. Netw. Spat. Econ. **10**(1), 125–145 (2007)
7. Davari, S., Zarandi, M.H.F., Turksen, I.B.: A greedy variable neighborhood search heuristic for the maximal covering location problem with fuzzy coverage radii. Knowl.-Based Syst. **41**, 68–76 (2013)
8. Delgado, M., Verdegay, J., Vila, M.: A general model for fuzzy linear programming. Fuzzy Sets Syst. **29**, 21–29 (1989)
9. Densham, P.J.: Spatial decision support systems. Geographical Information Systems: Principles and Applications, vol. 1, pp. 403–412. Longman, Publishing Group, London (1991)
10. Guzman, V.C., Verdegay, J.L., Pelta, D.A.: Fuzzy models and resolution methods for covering location problems: an annotated bibliography. Int. J. Uncertain. Fuzziness Knowl.-Based Syst. (2016) (In press)
11. Jaillet, P.: Airline network design and hub location problems. Locat. Sci. **4**(3), 195–212 (1996)
12. Jia, H., Ordóñez, F., Dessouky, M.: A modeling framework for facility location of medical services for large-scale emergencies. IIE Trans. **39**(1), 41–55 (2007)
13. Karasakal, O., Karasakal, E.K.: A maximal covering location model in the presence of partial coverage. Comput. Oper. Res. **31**(9), 1515–1526 (2004)
14. Lee, G., Murray, A.T.: Maximal covering with network survivability requirements in wireless mesh networks. Comput. Environ. Urban Syst. **34**(1), 49–57 (2010)
15. Murali, P., Ordóñez, F., Dessouky, M.M.: Facility location under demand uncertainty: response to a large-scale bio-terror attack. Socio-Econ. Plan. Sci. **46**(1), 78–87 (2012)
16. Selim, H., Ozkarahan, I.: A supply chain distribution network design model: an interactive fuzzy goal programming-based solution approach. Int. J. Adv. Manuf. Technol. **36**(3–4), 401–418 (2008)
17. Shavandi, H., Mahlooji, H.: A fuzzy queuing location model with a genetic algorithm for congested systems. Appl. Math. Comput. **181**(1), 440–456 (2006)
18. Takaĉi, A., Marić, M., Drakulić, D.: The role of fuzzy sets in improving maximal covering location problem (MCLP). In: IEEE 10th Jubilee International Symposium on Intelligent Systems and Informatics, pp. 103–106 (2012)
19. Toregas, C., Swain, R., Revelle, C., Bergman, L.: The location of emergency service facilities. Oper. Res. **6**, 1363–1373 (1971)
20. Verdegay, J.L.: Fuzzy mathematical programming. In: Gupta, M.M., Sanchez, E. (eds.) Fuzzy Information and Decision Processes, pp. 231–237 (1982)
21. Walsh, M.R.: Toward spatial decision support systems in water resources. J. Water Resour. Plan. Manag. **119**(2), 158–169 (1993)
22. Yang, L., Jones, B.F., Yang, S.H.: A fuzzy multi-objective programming for optimization of fire station locations through genetic algorithms. Eur. J. Oper. Res. **181**(2), 903–915 (2007)
23. Zarandi, M.H.F., Davari, S., Hamidifar, M., Turksen, B.: Locating post offices using fuzzy goal programming and geographical information system (GIS). In: 17th Americas Conference on Information Systems 2011, AMCIS 2011, pp. 74–81 (2011)
24. Zimmermann, H.: Fuzzy Sets, Decision Making and Expert Systems. Kluwer Academic Publishers, Boston (1987)

A Fuzzy Location Problem Based Upon Georeferenced Data

Airam Expósito-Márquez, Christopher Expósito-Izquierdo, Belén Melián-Batista and J. Marcos Moreno-Vega

Abstract Locating theory seeks to exploit geographic information in order to identify the best suited areas to place new facilities. This chapter tackles an optimization problem aimed at finding suitable locations for a new infrastructure in Spain by considering sustainability criteria and from a fuzzy perspective. In order to solve this problem, open georeferenced data provided through a Geographic Information System are used. Several fuzzy membership functions are proposed to represent the level of possible membership to the desired locations defined by the problem criteria. The resulting locations are appropriately combined under different choices of a given decision maker. In addition, the approach can be easily adapted to tackle similar location problems.

Keywords Fuzzy optimization · Location theory · Geographic information system

1 Introduction

Location theory studies how to discover the best sites out of a finite set of eligible candidates to establish a given number of new facilities (e.g., fire stations, schools, delivery centres, hospitals, factories, etc.). Its wide range of practical applications and its interdisciplinary have encouraged the interest of the scientific community in this field over the last decades. This fact is today evidenced by the large body of literature. Providing an exhaustive analysis of location theory is out of the scope

A. Expósito-Márquez (✉) · C. Expósito-Izquierdo · B. Melián-Batista ·
J.M. Moreno-Vega
University of La Laguna, San Cristóbal de La Laguna, Spain
e-mail: aexposim@ull.edu.es

C. Expósito-Izquierdo
e-mail: cexposit@ull.edu.es

B. Melián-Batista
e-mail: mbmelian@ull.edu.es

J.M. Moreno-Vega
e-mail: jmmoreno@ull.edu.es

C. Cruz Corona (ed.), *Soft Computing for Sustainability Science*,
Studies in Fuzziness and Soft Computing 358, DOI 10.1007/978-3-319-62359-7_4

of this chapter, however, the interested reader is referred to the books [7, 8] and the works [12, 21] to obtain exhaustive reviews of location theory in heterogeneous application fields. As done in the present chapter, most of these applications have been addressed as multi-objective optimization problems.

In general terms, the suitability of a given site when locating a new facility is subject to one or several conflicting goals [11] derived from the individual interests of the involved stakeholders (e.g., business leaders, politicians, potential customers, financial investors, etc.). This way, the models are intrinsically stated as optimization problems in which a given objective function is minimized or maximized in the presence of constraints to satisfy. These constraints represent conditions for the underlying variables. This is the case of, for instance, forbidden areas to place some kinds of infrastructures due to their high ecological value, ensuring at least an agreed quality of service when meeting the demand of customers in telecommunication contexts, or having nearby supply networks to fulfil transport, electric, or information requirements. The current literature brings together a large number of practical applications in which different criteria and constraints are considered. By way of example, [18] addresses a location-routing problem that seeks to minimize the facility opening costs. Reference [1] overviews the main objectives considered in health-care contexts, such as average travel times to reach hospitals or patient satisfaction. Reference [10] combines selecting suitable locations for new firm facilities and setting of product prices to maximize the market share in competitive environments.

Society today requires considering environmental, social, and economic aspects in location decisions [20]. At the same time, government policies promote the appearance of new sustainability development initiatives. These facts have given rise to that sustainability criteria in location theory has consolidated as an active research direction over the last years [5]. These criteria are especially relevant when placing dangerous (e.g., refineries, nuclear power plants, etc.) or unpopular facilities (e.g., cemeteries, factories, etc.) near living and natural areas [24]. The corporate sector is also concerned by this new global scenario. In this regard, it has undergone a fundamental change away from neoclassical approaches based upon commercial criteria in keeping with applicable legal principles and towards emerging business approaches in which sustainability criteria are seen as highlighted opportunities to strengthen the corporate image, reduce costs, and increase benefits [6].

In spite of the fact that, as previously indicated, location decisions are addressed as optimization problems, the criteria and constraints are usually subject to some degree of uncertainty and vagueness. The reason is usually found in the way the decision makers express their preferences. Specifically, in most of the cases, the expert knowledge given by a decision maker in spatial problems is expressed through imprecise terms, such as *maybe*, *not too far*, or *similar quality* [3]. In real-life contexts of location decisions, the potential sites to place a given facility will not usually verify a given constraint strictly. For example, this is the case of selecting a suitable site to place a new electrical generator aimed at serving a given population. In some cases, some people could think that a few hundred of meters are enough to avoid possible noise, whereas other people could think that at least several kilometres are required. Fuzzy logic [23] emerges in these environments as a more useful alternative than classic Boolean logic due to its closeness to human reasoning and its aptitude to deal with non-linearities and uncertainties.

This chapter presents an illustrative case study to locate new infrastructures in Spain in a competitive environment by considering several fuzzy criteria. The impact of several fuzzy membership functions to satisfy imprecise expert semantic descriptions is assessed. As described in Sect. 3, the solution of the proposed location problem is based upon open georeferenced data provided by a Geographic Information System [14].

The remainder of the present chapter is organized as follows. Section 2 introduces an optimization problem based upon fuzzy criteria. Later, Sect. 3 describes a solving approach to find suitable sites that satisfy the statement of the proposed problem. Finally, Sect. 4 presents the main concluding remarks and indicates several lines for further research.

2 Problem Description

In the present chapter, an illustrative multi-objective optimization problem belonging to the field of location theory is proposed. The main goal is to assess the applicability of different fuzzy membership functions to satisfy several imprecise criteria in practical environments. As described in Sect. 3, this is carried out through the usage of open georeferenced data provided by a Geographic Information System (GIS).

Without loss of generality, the optimization problem under analysis seeks to determine the best suited sites to place a given number of new malls in the mainland Spain. It should be noted that alternative contexts can be also tackled in the same vein. In this case and as analysed in the introduction of this chapter, the locations of the malls are subject to several fuzzy sustainability criteria. Specifically, their locations must simultaneously satisfy social, economic, and transportation criteria. That is, (i) the locations of the new malls in a competitive environment have to be firstly as far as possible from the pre-existing malls. In order to increase the market share in an environment in which competition already exists, (ii) the locations of the new malls have to be also close to the main population centres to count on the largest set of potential customers. This fact is based upon considering that, from a game theoretic standpoint, customers prefer to buy in the closest facilities [13]. Lastly, the locations must be well communicated with the population with the aim of easing its access. In order to address this issue, the locations have to be close to (iii) the main motorways and (iv) railways.

It is worth mentioning that, as stated by the definition of the optimization problem here presented, each point on the map under analysis could be considered as a feasible location for a new requested mall. This means that the feasible solution space of the optimization problem is composed of all the points provided by the GIS. This set of points is hereafter denoted as $\mathscr{L}$. From a general point of view, each point $i \in \mathscr{L}$ is sufficiently defined in this context by its geographic coordinates on the map, denoted as (x_i, y_i). In practice, those points located on, for instance, the sea, regions with high ecological value, and so forth are intuitively discarded, and therefore should be removed from the feasible solution set through explicit constraints. Furthermore, the set of locations in which the new malls can be placed is denoted as $\mathscr{L}' \subseteq \mathscr{L}$. In this case, the set of pre-existing malls is denoted as $\mathscr{M}$. Their characteristics and locations are open georeferenced data. The number of malls to place, denoted as $k > 0$, is selected by the decision maker (i.e., $|\mathscr{L}'| = k$). It should be noted that it can be easily included into each corresponding mathematical model as a constraint. However, this issue is out of the scope of this study. In real cases, the number of potential locations to place the malls is much greater than k (i.e., $|\mathscr{L}| \gg k$).

According to the previous description, the proposed multi-objective optimization problem, $\mathscr{P}$, on a rectangular map with width $\mathscr{W}$ and height $\mathscr{H}$ can be defined as finding that subset of locations $\mathscr{L}'$ which minimizes at the same time the distance to the population centres, motorways, and railways, and maximizes the distance to the pre-existing malls. This can be formally stated as follows:

$$\mathscr{P} : \min_{i \in \mathscr{L}'} (-f_1(i), f_2(i), f_3(i), f_4(i)) \tag{1}$$

subject to

$$1 \leq x_i \leq \mathscr{W} \tag{2}$$

$$1 \leq y_i \leq \mathscr{H} \tag{3}$$

In this case, $f_1(\cdot)$, $f_2(\cdot)$, $f_3(\cdot)$, and $f_4(\cdot)$ are scalar functions defined as follows:

$$f_1(i) = \sum_{j \in \mathscr{M}} d_{ij} \tag{4}$$

$$f_2(i) = \sum_{j \in \mathscr{C}} d_{ij} \tag{5}$$

$$f_3(i) = \sum_{i \in \mathscr{R}} d_{ij} \tag{6}$$

$$f_4(i) = \sum_{i \in \mathscr{T}} d_{ij} \tag{7}$$

where $\mathscr{M} \subseteq \mathscr{L}, \mathscr{C} \subseteq \mathscr{L}, \mathscr{R} \subseteq \mathscr{L}$, and $\mathscr{T} \subseteq \mathscr{L}$ are the sets of points that compose the pre-existing malls, population centres, motorways, and railways, respectively. Additionally, d_{ij} represents the distance between the points $i \in \mathscr{L}$ and $j \in \mathscr{L}$ on the map. This way, Eq. (4) represents the sum of distances between the selected locations and the pre-existing malls. Similarly, Eq. (5) represents the sum of distances towards the population centres and Eq. (6) represents the sum of distances towards the motorways. Lastly, Eq. (7) represents the sum of distances towards the railways.

As discussed in the following, several fuzzy membership functions are applied to represent the possibility of a given site of being suite for placing a new mall on the basis of the aforementioned criteria.

Due to the fact that the criteria imposed by the problem are imprecise in practice, the set of points on the map is here handle as fuzzy. The fuzzy sets are used in multi-objective optimization problems with the aim of standardizing the existing criteria. This is done by assigning to each point provided by the GIS a degree of membership for each criterion. A fuzzy set is denoted as $(\tilde{L}, \mu_{\tilde{L}})$, where $\tilde{L}$ is the set of points provided by the GIS from a fuzzy perspective and μ_L is a function with the form $\mu_L : L \rightarrow [0..1]$. $\mu_{\tilde{L}}(i)$ quantifies the strength of membership of the point $i \in L$ in the fuzzy set $\tilde{L}$. Here, a value $\mu_L(\cdot) = 1$ represents a full membership in the set, $\mu_{\tilde{L}}(\cdot) = 0$ indicates not membership to the set, whereas intermediate values (i.e., $0 < \mu_{\tilde{L}}(\cdot) < 1$) represent partial membership to the set. Other related concepts arisen in the field are *linguistic variable* and linguistic label. The former represents a variable whose values are provided by means of sentences in a given language, whereas the latter represents a fuzzy value from the domain under analysis.

Given a continuous map provided by a GIS, the proposed scenario can be stated as a multi-objective optimization problem that seeks to determine locations satisfying the imposed criteria. For each criterion, the points are designated by a particular fuzzy membership function, which indicates the degree in which they satisfy the relevant criterion. Lastly, these degrees of membership are appropriately combined. The literature covers many combination alternatives in GIS-based geospatial analysis. Some representative examples can be found in the works [9, 22, 25]. However, the points under the different fuzzy membership functions are in this chapter combined by using the weighted linear combination [4] with weights α_1, α_2, α_3, and α_4 associated with the individual functions $f_1(\cdot)$, $f_2(\cdot)$, $f_3(\cdot)$, and $f_4(\cdot)$, respectively. The values of the weights are appropriately set by the decision maker on the basis of his preferences.

3 Practical Experience

The present section is devoted to describe the practical experience carried out to solve the multi-objective optimization problem introduced in Sect. 2. With this goal in mind, the technologies used in the experience are firstly described. Afterwards, a short introduction to fuzzy logic and the proposed solution approach are presented.

3.1 Geographic Information Systems

In general terms, Geographic Information Systems (GISs) are computer-based systems composed of a set of tools aimed at integrating, storing, handling, and analysing large quantities of spatially referenced data [17]. These systems also enable to visualize and extract patterns, relationships, and trends in the data [22]. Typically, information in a GIS is vertically organized and portrayed by means of visual symbols as data layers, which allows the user to overlay data according to his particular requirements. Some excellent examples of GIS are GRASS[1] [19], QGIS,[2] and gvSIG.[3] However, due to its versatility, powerful tools, and functional features, ArcGIS[4] is undoubtedly the most outstanding software on the market.

The relevance of GISs in optimization problems belonging to the field of location theory arises from the fact that this information technology aids to facilitate problem understanding and decision-making process. In this regard, the main applications of GISs involve selecting a suitable site to place a new facility [16] and designing a corridor across a given landscape [15].

3.2 OpenStreetMap

One of the main inconveniences faced by a decision maker when dealing with a location problem is the availability of georeferenced data. Fortunately, nowadays, many free data sources are ready to be exploited by a GIS. For instance, Natural Earth Data,[5] supported by the North American Cartographic Information Society, provides public-domain data sets related to physical and cultural aspects of the countries. The Socioeconomic Data and Applications Center[6] of the NASA contains a wide variety of socioeconomic data related to agriculture, conservation, poverty, etc. Lastly, the Environmental Data Explore[7] of the United Nations Environment Programme provides data sets with more than 500 different variables, such as forests, climate, disasters, etc.

The case study addressed in this chapter is based upon georeferenced data taken from the OpenStreetMap project.[8] Broadly speaking, OpenStreetMap (OSM) is a collaborative project aimed at creating a free-editable map of the world. Due to its wide variety of information, it constitutes a great source of data to support case studies associated with location issues in which georeferenced data are massively exploited. The real power of OSM is the possibility to effectively access the data behind its map rendering. OSM is a database project, whose main purpose is to create an exhaustive database of every street, city, road, building, etc. on the planet, and not being only a map display project. For this, OSM differs from other well-known tools, such as

[1]https://grass.osgeo.org/.

[2]http://www.qgis.org/en/site/.

[3]http://www.gvsig.com/en/home.

[4]https://www.arcgis.com/.

[5]http://www.naturalearthdata.com/.

[6]http://sedac.ciesin.columbia.edu/.

[7]http://geodata.grid.unep.ch/.

[8]https://www.openstreetmap.org/

Google Maps or Bing Maps, among others. Reference [2] provides a detailed review of OSM in Geographical Information Science.

Georeferenced data can be downloaded from the OSM project in a variety of ways. Typically, data are available in XML-formatted .osm files. The full datasets are available from the download area in the website of OSM. This option is possible whenever relatively small areas to download are selected. In order to download huge amounts of georeferenced data (e.g., an entire continent, countries, or metropolitan centres), complementary tools such as Osmosis, osmconvert, and osmfilter should be used. In addition to these tools, there are different data repositories that contain specific areas aimed at easing the downloading of data maps. These repositories are frequently updated by the community.

3.3 Fuzzy Logic

One of the key concepts arisen in logic is that of *crisp set*. In general terms, a crisp set is a conventional set in which the degree of membership of their elements is binary. This means that an element is either a member of the set or not, which indicates that the boundaries of crisp sets are sharp. But frequently, one have to deal with environments in which imprecise and/or imperfect information appears. In these situations, crisp sets are not adequate due to the fact that they are not able to represent elements that are partially included in the set.

The human being possesses great skills to communicate his experience using fuzzy linguistic rules. Linguistic terms as *near*, *far*, *high*, *low*, etc. These could be followed without problem by a human, which is able to interpret these instructions quickly. But hardly representable in a language that can be understood by a computer. Conventional logic is not suitable for processing such rules. Generally, expert knowledge has often vague and imprecise characteristics. In a non-deterministic world sometimes it is not possible to completely set all the environment variables or how to determine the variables is not known. Even knowing all the variables, it may be difficult to obtain specific data associated for study. In addition, this information may be incomplete, and even be wrong.

According to the previous discussion, let consider in the following the case of modelling the concept of *proximity*. One may intuitively start by organizing the different distances into classes. For example, a starting point could be considering the three following classes: close, midway, and far. The boundaries of these classes could be [0–20], [20–60], and $[60, \infty)$ m, respectively. If the distance between two points is of 19.95 m, it would be classified in the close class. Furthermore, if the distance between two points is of 20.05 m, it would be classified in the midway class. It should be here noted that, with a difference of only 0.1 m between two distances, they are placed in two separate classes. Consequently, given the inflexibility of the proposed classifications, the full relationships between the distances cannot be precisely captured. The left part of Fig. 1 illustrates this example.

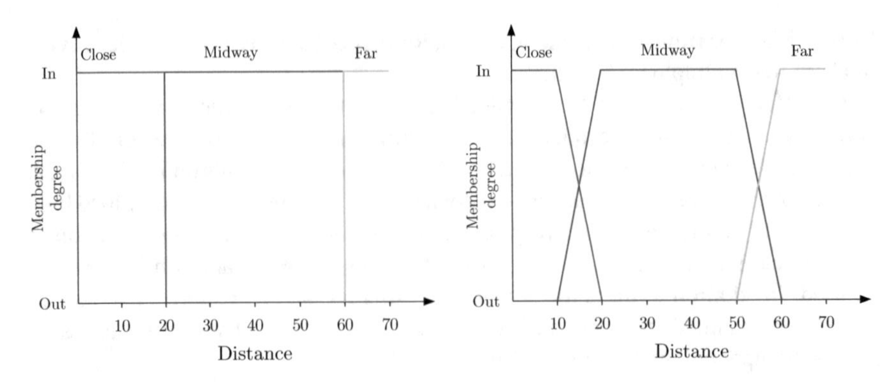

Fig. 1 Illustrative comparison between crisp sets and fuzzy sets

In order to represent more accurately the concept of proximity than in the case of crisp sets, more classes can be added. However, no matter how many classes are handled, there is still a generalization of the concept. It should be noted that there are some concepts that cannot be therefore easily classified into strictly defined classes. The imprecision in nature can be modelled through fuzzy logic models, where the classes are defined as fuzzy sets. For each fuzzy set, there is a membership function associated with its elements, indicating the extent to which each element is part of the fuzzy set. This way, a degree of membership is associated with the elements of the concepts under study.

The right part of Fig. 1 depicts the degree of membership in different fuzzy sets when measuring distance between points. As done in the previous example, in this case, there are three fuzzy sets that represent the following classes: close, midway, and far. As can be seen, the degree of membership of each distance can be partial. This is the case of 15 m, which is partially included into the sets representing close and midway distances.

3.4 Solution Approach

The optimization problem introduced in Sect. 2 is here addressed in the context of Spain, specifically in the Iberian Peninsula and the Balearic Islands. Data about Canary Islands have been not used in this experience due to their low quantity and poor quality. For this purpose, open georeferenced data available in the OSM project have been used. However, due to the large volume of data considered in the proposed approach, data could not be directly downloaded from the website of OSM. Instead, the repository called Geofabrik[9] has been used for this aim. This server contains daily extracts of the OSM project that can be downloaded directly by the user. The geographic information used in this practical experience is illustrated in Fig. 2.

[9]http://download.geofabrik.de/.

Fig. 2 Base map provided by the OpenStreetMap project with the georeferenced information about the Iberian Peninsula and the Balearic Islands

The open georeferenced data downloaded from the repository Geofabrik are divided into four basic data layers to be handled by a GIS. These include information regarding places, points, roads, and railways, respectively. The first layer contains different types of places such as towns, villages, suburbs, and other interest areas. The second layer has useful information about relevant points to be exploited by logistic applications. For example, swimming pools, schools, insurance agencies, agricultural centres, animals, hotels, banks, bars, restaurants, casinos, hospitals, pharmacies, parking areas, schools, universities, and many others points. The third layer contains data about different types of roads, such as cycleways, paths, foot-ways, lanes, motorways, pedestrian paths, residential roads, and many other ones. Lastly, the last layer contains information about the main railway lines. The information provided by the data layers can be seen in Fig. 3. The places and points are respectively represented as red and green dots in the subfigures displayed at the top. Furthermore, the roads and railways are represented as blue and orange lines in the subfigures displayed at the bottom.

The multi-objective optimization problem introduced in Sect. 2 describes the requirements to locate the requested malls. By way of reminder, these are locating the new malls as far as possible from other pre-existing ones while the locations of the malls have to be also as close as possible to the urban centres, main motorways, and railways (Eq. 1). In short, the locations of the new malls should be found away from competition and close to potential customers and infrastructures in order to facilitate their accessibility. However, the data provided by OSM involve a great amount of information that is irrelevant for the case study under analysis. Conse-

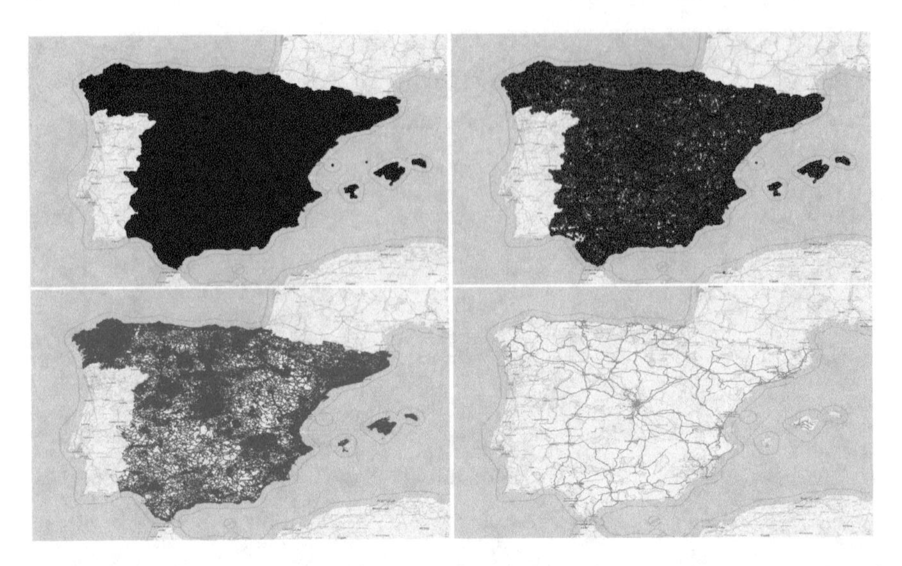

Fig. 3 Data layers downloaded from the Geofabrik repository related to places, points, roads, and railways, respectively

quently, a preliminary filtering process is firstly carried out to remove unwanted information from the original data layers. It is worth mentioning that it is essential to filter the data contained in the original data layers to get information regarding urban centres, main motorways, and pre-existing malls. Also, the information contained in the data layer about railways is in this case entirely used to address the optimization problem, in such a way that no filtering process is required.

With the goal of filtering the available information in the original data layers and selecting only relevant features from them, selection queries have been used. Particularly, selection queries by attributes, which allow to obtain subsets of data whose elements satisfy a given criterion. In the case of the point layer, the information has been filtered according to the type of point, by setting the attribute type to "mall". So, a new data layer has been generated with the information of the pre-existing malls in OSM exclusively. Also, to get a layer with information about urban centres, the same process on the places layer has been repeated but by setting the attribute type to "city", being city the largest urban centres in Spain recorded by OSM. Finally, in order to filter information about the main motorways, the selection query has been applied by setting the attribute type of the road layer to "motorway". The data layers containing the filtered information are represented in Fig. 4.

Once the available information in OSM about the pre-existing malls, cities, motorways, and railways has been gathered in different data layers, the four optimization criteria defined by the problem are tackled (Eq. 1). In this case, four scalar membership functions associated with the distance have been defined. These functions indicate the degree of suitability of each point on the map when fulfilling the imposed criteria on the basis of the distance to the pre-existing malls, cities, motorways, and

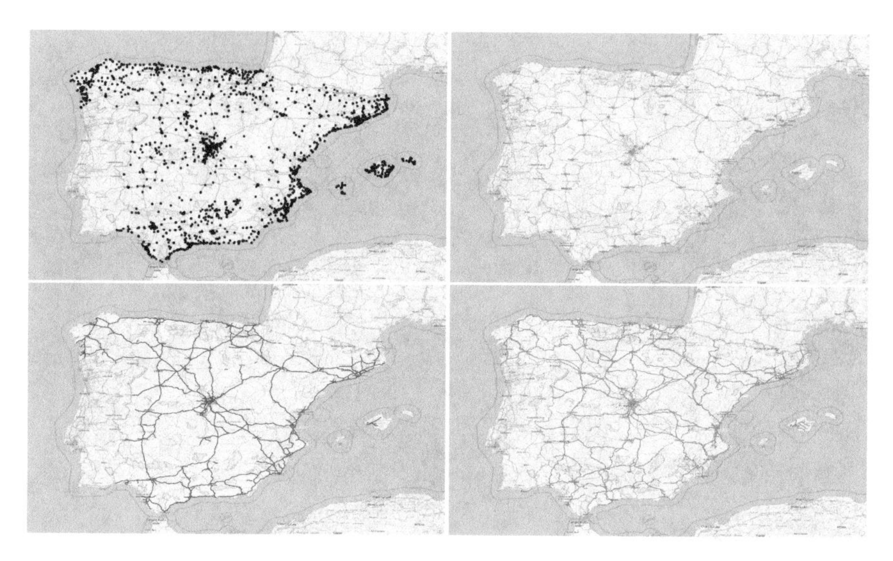

Fig. 4 Filtered data layers related to the pre-existing malls, cities, motorways, and railways, respectively

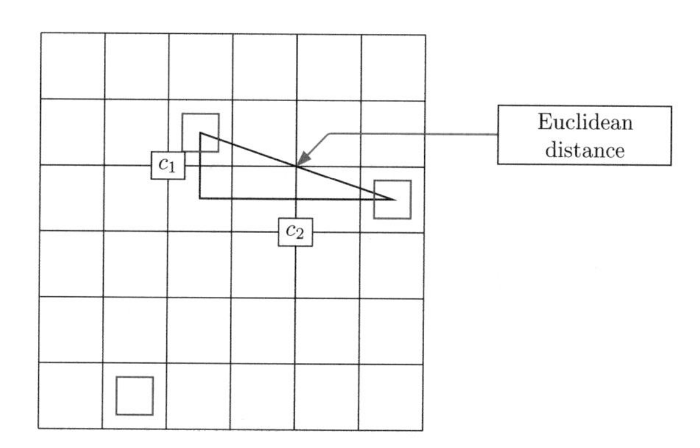

Fig. 5 Calculation of Euclidean distance in raster layers

railways, respectively. With this goal in mind, new data layers containing information about the Euclidean distance between each pair of points on the map have been firstly generated. To do this, the subsequent step has been to apply an algorithmic procedure to compute the Euclidean distance from each point on the map to the location of the interest elements, namely, pre-existing malls, cities, motorways, and railways. It should be here pointed out that the data layers in the GIS contain raster information. This means that they are made up of regular squared cells, where each one represents a single visual value. In this case, the algorithmic procedure used by the GIS calculates the Euclidean distance from the center of the cells to the center of its surrounding cells. This process is illustrated in Fig. 5. In this example, for each cell, the distance to each remaining cell is obtained by calculating the hypotenuse of the underlying triangle in which c_1 and c_2 are its cathetus.

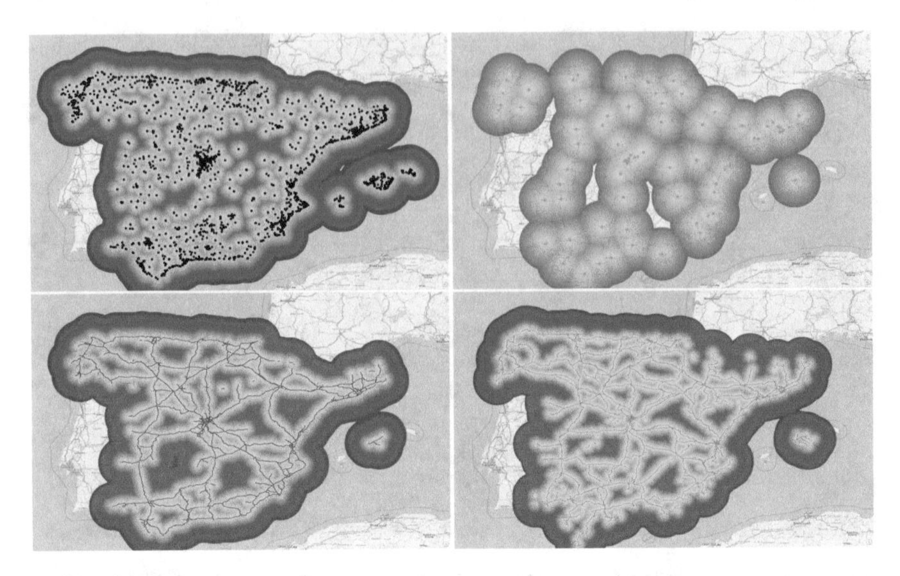

Fig. 6 Data layers containing information related to distances to pre-existing malls, cities, motorways, and railways

The output of the algorithmic procedure used by the GIS is presented by means of a new data layer for each criterion of the optimization problem. This layer contains the Euclidean distance from every cell towards its nearest element of interest (i.e., pre-existing malls, cities, motorways, and railways). This way, a suitability data layer is obtained for each type of element of interest in which a coloured ramp is used to represent the gradual distance between the relevant elements for the optimization problem and the remaining points. The resulting data layers reported by the algorithmic procedure when solving the proposed optimization problem at hand are displayed in Fig. 6. In this case, colours close to yellow are assigned to the lowest values of distance, whereas colours close to violet are assigned to the highest values of distance.

Given the highlighted role of the information display when distinguishing the most promising areas to locate a new mall on the map, a colour ramp composed of 32 colours is used in each data layer to represent the Euclidean distance from each point to its nearest elements of interest in the optimization problem. The colours in the ramp are distributed over the values contained in the cells of the data layers. The distribution of the colours encourages to prevail cells with high values in terms of distance. These cells are those of most interest to locate new malls. So, most colours are assigned to cells with high values. This encourages to determine which areas are considered as promising to locate a new mall in a graphical fashion. With the aim of illustrating this, the setting on the map display associated with the data layer of motorways is shown in Fig. 7. Specifically, the different shades of yellow delimiting the closest points to areas in which there is a motorway can be easily appreciated.

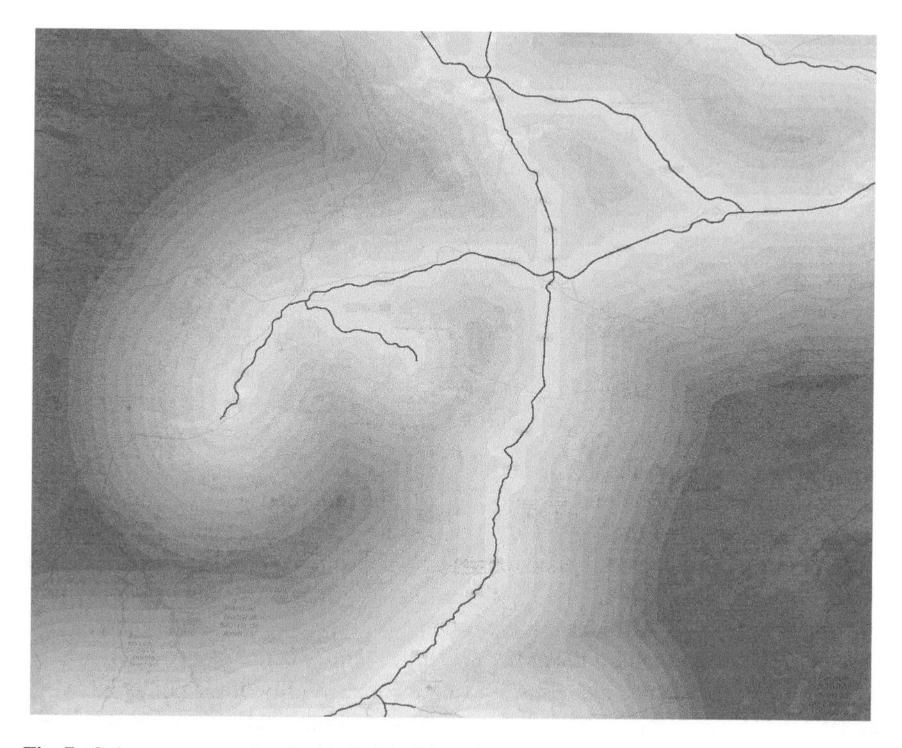

Fig. 7 Colour ramp associated with the Euclidean distances towards motorways

The subsequent step of the proposed solution approach involves to address the imprecise and/or imperfect nature of the information associated with the existing distances from the potential locations of the new malls and the relevant elements imposed by the definition of the optimization problem at hand. In this case, this has been carried out from a fuzzy perspective by using several fuzzy membership functions. A fuzzy membership function has been here defined for each data layer with the Euclidean distances reported by the GIS. This way, a certain individual degree of suitability can be easily defined for each point on the map under each optimization criterion. Lastly, the individual degrees of suitability should be thereafter combined to determine the most interesting areas to locate the requested malls.

Without loss of generality and in order to provide an illustrative example of applicability of fuzzy membership functions to data layers in GIS-based applications, different fuzzy membership functions have been considered to be applied to the available data layers. Specifically, Gaussian functions are applied to the points contained in the data layers associated with the motorways, railways, and cities. Broadly speaking, a Gaussian function is a characteristic symmetric bell-shaped curve that is controlled by three parameters: the height of the peak of its curve,

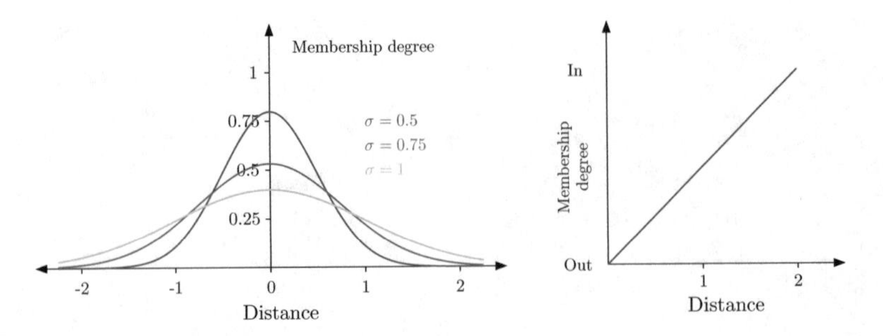

Fig. 8 Fuzzy membership functions used in the solution approach

which takes values from 0 to 1 with the aim of indicating the degree of membership of the elements, the position of the center of the peak, and the width of the bell, that is, the standard deviation of the population. In the proposed solution approach, the center of the peaks of the Gaussian functions are set to zero, which indicates that those points exactly located on the relevant elements of the problem has the highest degree of suitability and, as the Euclidean distance increases, this suitability progressively decreases. Also, the width of the bells are set to 0,1. Furthermore, in the case of pre-existing malls, a linear fuzzy membership function is applied. This way, as the Euclidean distance increases, the degree of suitability of the points increases linearly in the range [0..1]. This intuitively allows to represent that those points found at the furthest possible distance from the pre-existing malls are the most suitable for locating new malls under the competitiveness criterion. Figure 8 shows graphical representations of Gaussian and linear functions. In the case of the Gaussian function, the longer the distance, the lesser the membership. However, the longer the distance, the higher the membership when using the linear function.

The information related to the computed Euclidean distances that is contained in the data layers is used to generate new layers by applying the aforementioned fuzzy membership functions. The new data layers provide a graphical representation of the individual degrees of suitability associated with the existing points when satisfying the imposed optimization criteria. These data layers are illustrated in Fig. 9. As can be checked, shades of red represent the lowest degrees of suitability of the points under the different optimization criteria in the problem at hand. This way, those points close to the pre-existing malls and those located far from the cities, motorways, and railways are considered as inappropriate for placing the new malls. Simultaneously, shades of blue indicate the most suitable locations under the corresponding optimization criteria, whereas intermediate shades in the colour ramp represent medium degrees of suitability of the points.

As pointed out by the description of the optimization problem, the suitability of a given point on the map provided by the GIS depends directly on the existing Euclidean distance towards its nearest elements of interest. Unfortunately, the previ-

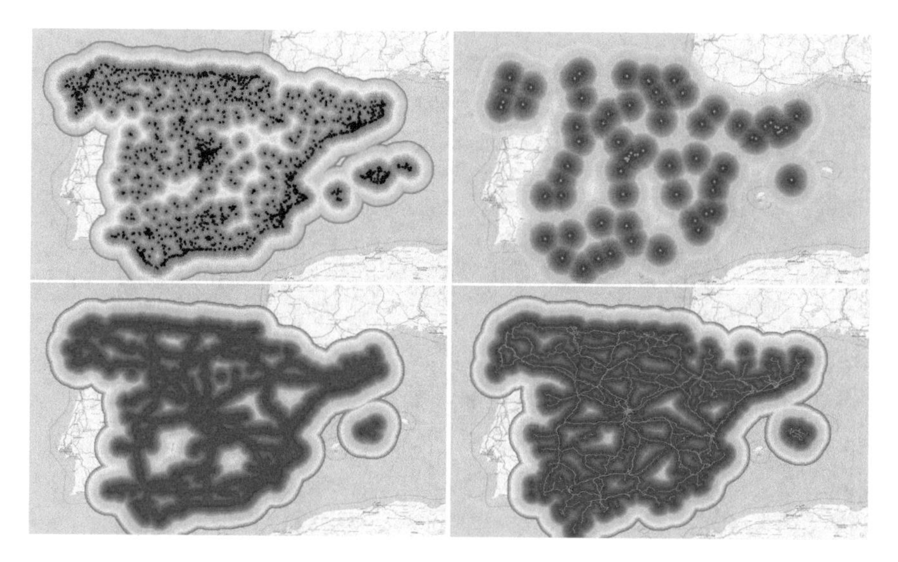

Fig. 9 Data layers containing information regarding the individual degrees of suitability of the points under the criteria related to pre-existing malls, cities, motorways, and railways

ous data layers only contain the individual suitability of the points under the specific optimization criteria. At this point, the locations of the requested malls could be appropriately found according to the competitiveness, accessibility, and transportation criteria. However, in order to identify the best suited sites to place the malls from the multi-criteria perspective imposed by the optimization problem at hand, a new data layer must be generated. This data layer should contain combined information of the points on the map under the individual criteria.

According to the previous discussion, a weighted linear combination of the source data layers has been here applied. The result is a new data layer in which the individual suitability of the points on the map provided by the GIS under the optimization criteria are adequately merged. Specifically, the weighted linear combination allows to combine data belonging to multiple sets based upon the set theory analysis. In this case, it determines the degree of membership of the existing points to multiple fuzzy sets, which are defined by the different criteria of the optimization problem. The resulting degree of membership associated with a point on the map is an increasing linear combination function whose value depends on the multiple optimization criteria considered in the problem. Those points with a zero degree of membership indicate they are completely inappropriate to place new malls, whereas those points with a degree of membership equals to one represent those points in which new malls must be safely placed. Lastly, it is worth mentioning that alternative combination procedures can be transparently used in the proposed solution approach.

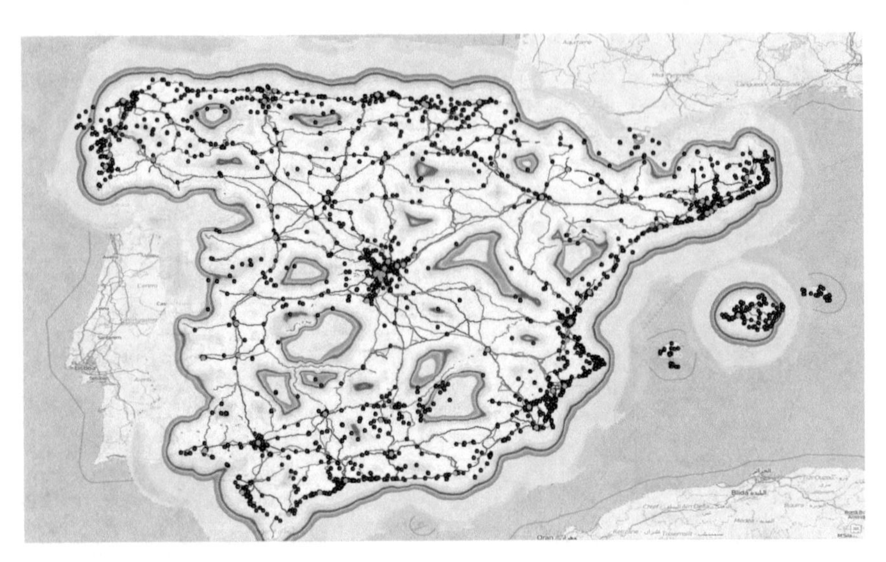

Fig. 10 Resulting data layer of the weighted linear combination of the individual layers with the fuzzy criteria

In the optimization problem addressed in this chapter, the source data layers are linearly combined in such a way that all the optimization criteria have the same impact on the resulting data layer. This means that $\alpha_1 = \alpha_2 = \alpha_3 = \alpha_4 = 1$ (see Sect. 2). A graphical representation of the resulting data layer is reported in Fig. 10. As can be checked, this resulting data layer allows to identify the best suited sites to place a new mall under the different optimization criteria imposed by the problem. With this goal in mind, a colour ramp ranging from white to blue is used. Particularly, the shades of white indicate the best suited sites to place new malls on the map. At the same time, the blue areas indicate the worst suited sites. Intermediate sites are highlighted by using complementary colours.

The great amount of visual information and the dimensions of the map enable only a cursory analysis in the decision-making process when identifying the best suited sites to place the malls requested by the statement of the optimization problem. With the aim of examining a reduced study region of concern on the map, a zoom to the extend of the data layer in the display is carried out. Figure 11 shows a reduced study region of the original map that can be used to find the locations of malls. It should be pointed out that alternative study regions can be similarly selected by the decision maker. On the study region of concern, the decision maker can identify sites on the map that are highlighted by using shades of white. These sites are highly influenced by the proximity of transportation infrastructures, such as motorways and railways.

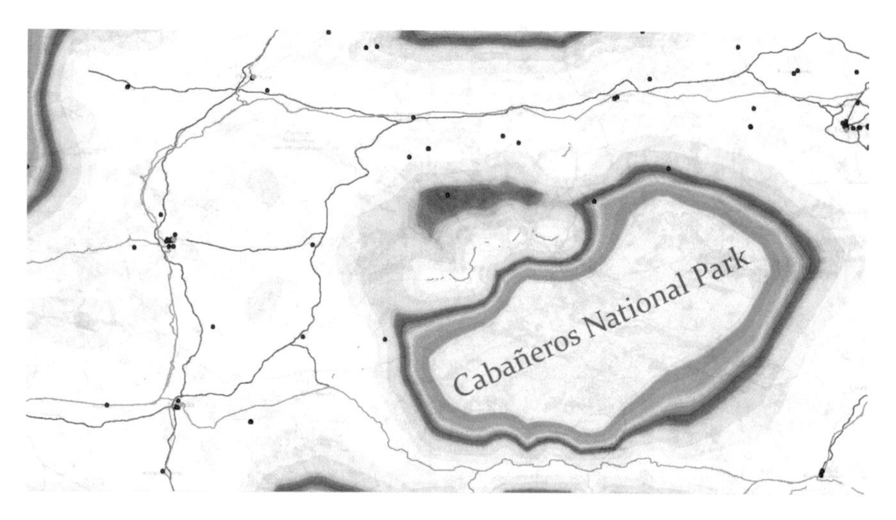

Fig. 11 Study region on the map containing information of the weighted linear combination of the individual layers with fuzzy criteria

Also, they are influenced by the main cities existing on the map. These are Toledo, Cáceres, Mérida, and Plasencia, situated on the corners of the map. Furthermore, those sites on the study region that are located either close to the pre-existing malls or far from the population centres and the main transportation infrastructures are inappropriate to place new malls. These sites are those with the lowest value reported by the weighted linear combination of fuzzy optimization criteria and are highlighted by using shades of colours such as red, green, or yellow. In the study case at hand, the large red-border area that comprises the Cabañeros National Park and its surroundings are clearly inappropriate to place new malls.

Another relevant aspect to be considered by the decision maker when placing new mall is the negative influence derived from the closeness to competition, the pre-existing malls. In this case, they are especially located at the north-west of the National Park previously mentioned. Specifically, there is a mall without nearby motorways or railways infrastructures. The area with low degree of membership to the combined is here highlighted by using shades of blue. Moreover, on the basis of the description of the proposed optimization problem, the decision maker must take into account only the degree of suitability of the points considered on the map provided by the GIS. For this purpose, it is easy to dismiss inappropriate areas to locate a mall at a first glance. Nevertheless, with the aim of identifying the most promising areas, proximity to urban centres and accessibility criteria are considered. As a consequence, according to the optimization criteria, the most suitable area to place a new mall is that highlighted in blue in Fig. 12.

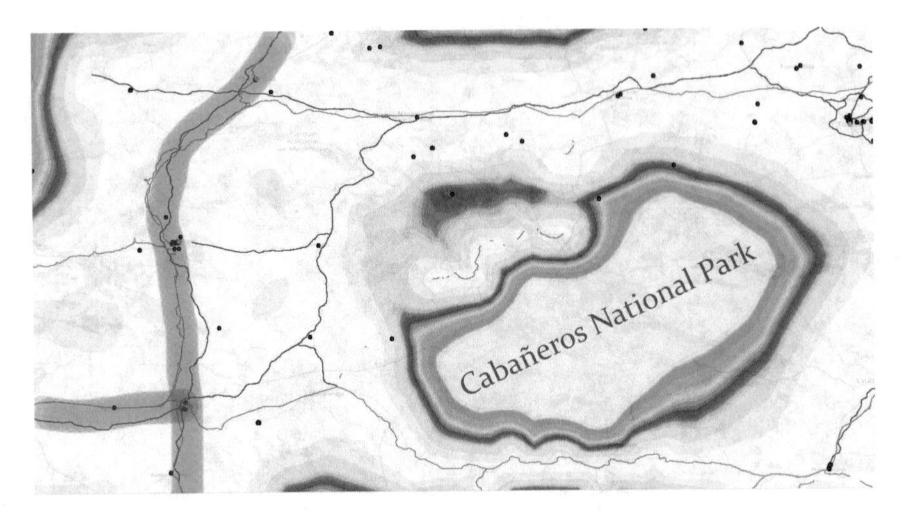

Fig. 12 Most suitable areas to place new malls in the selected study region

4 Conclusions and Further Research

A wide range of practical optimization problems is encompassed by the location theory. This research field especially covers those contexts in which the best suited sites to place new facilities must be appropriately selected on the basis of the particular requirements of the applications under analysis. Some representative examples arise in heterogeneous areas, such as locating new hospitals, fire stations, delivery centres, factories, and so forth. In this regard, sustainability criteria have taken an increasingly highlighted role over the last years. Particularly, environmental, social, and economic aspects in decision-making processes are more and more demanded by current society and supported by governmental measures.

In this chapter, an illustrative fuzzy multi-objective optimization problem that seeks to determine the best sites to place a given number of new malls in Spain is introduced. The criteria in the problem at hand involve to place the new malls as far as possible from the pre-existing ones and close to the main population centres, motorways, and railways. The problem is here studied in a practical environment by using open georeferenced data extracted from the OpenStreetMap project and handled by a Geographic Information System, in short GIS. GISs are powerful computer-based tools to make decisions whenever geospatial information is available. These tools allow the user to store, display, and analyse the existing information. Additionally, they allow to extract relevant patterns and trends.

In order to tackle the proposed optimization problem, several fuzzy membership functions are individually analysed. The rationale behind their use is to model the imprecise nature of the criteria in practical environments. This way, each point provided by the GIS receives a certain degree of membership for each criterion, which indicates the degree in which they satisfy the relevant criterion. Lastly, the different fuzzy membership functions are in this chapter combined by using a weighted linear combination that, as discussed, allows the decision maker to select the most suitable sites to place the new malls.

Finally, it is worth mentioning that several promising lines for further research are still open to be considered. In spite of the fact that a location problem has been addressed in this chapter, similar optimization problems belonging to other research fields can be solved from a similar fuzzy approach and by exploiting the functionalities of GISs and availability of open georeferenced data. Also, considering other fuzzy membership functions to model particular requirement of decision makers is not still sufficiently explored.

Acknowledgements This work has been partially funded by the Spanish Ministry of Economy and Competitiveness (projects TIN2012-32608 and TIN2015-70226-R). Airam Expósito-Márquez and Christopher Expósito-Izquierdo would like to thank the Canary Government for the financial support they receive through their post-graduate grants.

References

1. Arnolds, I., Nickel, S.: Layout planning problems in health care. In: Eiselt, H.A., Marianov, V. (ed.) Applications of Location Analysis. International Series in Operations Research and Management Science, vol. 232, pp. 109–152. Springer, Heidelberg (2015)
2. Arsanjani, J.J., Zipf, A., Mooney, P., Helbich, M. (eds.): OpenStreetMap in GIScience: Experiences, Research and Applications. Springer, Cham (2015)
3. Bashiri, M., Hosseininezhad, S.J.: A fuzzy group decision support system for multifacility location problems. Int. J. Advanced Manuf. Technol. **42**(5–6), 533–543 (2009)
4. Chang, N.-B., Parvathinathan, G., Breeden, J.B.: Combining GIS with fuzzy multicriteria decision-making for landfill siting in a fast-growing urban region. J. Environ. Manag. **87**(1), 139–153 (2008)
5. Chen, L., Olhager, J., Tang, O.: Manufacturing facility location and sustainability: a literature review and research agenda. Int. J. Prod. Econ. **149**, 154–163 (2014). The Economics of Industrial Production
6. Dombrowski, U., Riechel, C., Dring, H.: Sustainability in manufacturing facility location decisions: comparison of existing approaches. In: Grabot, B., Vallespir, B., Gomes, S., Bouras, A., Kiritsis, D. (eds.) Advances in Production Management Systems. Innovative and Knowledge-Based Production Management in a Global-Local World, volume 439 of IFIP Advances in Information and Communication Technology, pp. 246–253. Springer, Berlin (2014)
7. Drezner, Z., Hamacher, H.: Facility location: Applications and Theory. Springer, Berlin (2002)
8. Eiselt, H.A., Marianov, V. (eds.): Foundations of Location Analysis. International Series in Operations Research and Management Science, vol. 232. Springer, Berlin (2015)
9. Feizizadeh, B., Roodposhti, M.S., Jankowski, P., Blaschke, T.: A gis-based extended fuzzy multi-criteria evaluation for landslide susceptibility mapping. Comput. Geosci. **73**, 208–221 (2014)
10. Fernández, J., Salhi, S., Tóth, B.G.: Location equilibria for a continuous competitive facility location problem under delivered pricing. Comput. Oper. Res. **41**(3), 185–195 (2014)
11. Gang, J., Tu, Y., Lev, B., Xu, J., Shen, W., Yao, L.: A multi-objective bi-level location planning problem for stone industrial parks. Comput. Oper. Res. **56**, 8–21 (2015)
12. Hale, T.S., Moberg, C.R.: Location science research: a review. Ann. Oper. Res. **123**(1–4), 21–35 (2003)
13. Hotelling, H.: Stability in competition. Econ. J. **39**(153), 41–57 (1929)

14. Hwang, S., Thill, J.C.: Modeling localities with fuzzy sets and GIS. In: FrederickE, P., VincentB, R., MariaA, C. (eds.) Fuzzy Modeling with Spatial Information for Geographic Problems, pp. 71–104. Springer, Berlin (2005)
15. Jiang, L., Kang, J., Schroth, O.: Prediction of the visual impact of motorways using GIS. Environ. Impact Assess. Rev. **55**, 59–73 (2015)
16. Jiuping, X., Song, X., Yimin, W., Zeng, Z.: Gis-modelling based coal-fired power plant site identification and selection. Appl. Energy **159**, 520–539 (2015)
17. Kresse, W., Danko, D.M.: Springer Handbook of Geographic Information. Springer, Berlin (2012)
18. Labbé, M., Rodríguez-Martín, I., Salazar-Rodríguez, J.J.: A branch-and-cut algorithm for the plant-cycle location problem. J. Oper. Res. Soc. **55**(5), 513–520 (2004)
19. Neteler, M., Beaudette, D.E., Cavallini, P., Lami, L., Cepicky, J.: Grass GIS. In: Hall, G.B., MichaelG, L. (eds.) Open Source Approaches in Spatial Data Handling. Advances in Geographic Information Science, pp. 171–199. Springer, Berlin (2008)
20. Rao, C., Goh, M., Zhao, Y., Zheng, J.: Location selection of city logistics centers under sustainability. Transp. Res. Part D: Transp. Environ. **36**, 29–44 (2015)
21. ReVelle, C.S., Eiselt, H.A.: Location analysis: a synthesis and survey. Eur. J. Oper. Res. **165**(1), 1–19 (2005)
22. Yanar, T.A., Akyürek, Z.: The enhancement of the cell-based GIS analyses with fuzzy processing capabilities. Inf. Sci. **176**(8), 1067–1085 (2006)
23. Zadeh, L.A.: Fuzzy sets. Inf. Control. **8**(3), 338–353 (1965)
24. Zhao, J., Verter, V.: A bi-objective model for the used oil location-routing problem. Comput. Oper. Res. **62**, 157–168 (2015)
25. Zhu, A.-X., Qi, F., Moore, A., JamesE, B.: Prediction of soil properties using fuzzy membership values. Geoderma **158**(34), 199–206 (2010)

A Review of the Application to Emergent Subfields in Viticulture of Local Reflectance and Interactance Spectroscopy Combined with Soft Computing and Multivariate Analysis

Armando Fernandes, Véronique Gomes and Pedro Melo-Pinto

Abstract Spectroscopic techniques have shown great potential due to their quick response, cost-effective, non-destructive and non-invasive nature, and environmental friendliness. These characteristics make this technology very attractive for sustainable industry and research activities, being viticulture industry no exception. Spectroscopic techniques are an appealing alternative for ripeness assessment and harvest date determination as well as for plant variety and clone determination. Numerous recent works have clearly demonstrated that it is highly advantageous to process the high dimensionality spectroscopic data with soft computing or multivariate analysis techniques such as Partial Least Squares, Neural Networks or Support Vector Machines. In this review, focus will be given to two emergent subfields in viticulture where the combination of spectroscopy and soft computing is fundamental: (1) The difficult measurement of enological parameters, namely sugar content, pH and anthocyanin content, in samples containing a small number of grape berries, with the aim of assessing grapes' ripeness; (2) The multiclass problem of identifying plant varieties and clones. The results of the various works in these subfields will be presented. The present article starts with a brief description of the spectroscopy principles and continues by making an overview of the scientific literature considering the number of berries per sample and the total number of samples in the various works. The use of different varieties, vintages and harvest locations in the same model will also be addressed. Special attention is given to the validation methods employed and to algorithm comparison. Some suggestions are presented in order to facilitate future comparison of published results.

A. Fernandes (✉)
INOV—INESC Inovação, Rua Alves Redol, 9, 1000-029 Lisboa, Portugal
e-mail: arm.fernandes@gmail.com

V. Gomes · P. Melo-Pinto
CITAB-Centre for the Research and Technology of Agro-Environmental and Biological Sciences, Universidade de Trás-os-Montes e Alto Douro, 5000-801 Vila Real, Portugal

P. Melo-Pinto
Departamento de Engenharias, Escola de Ciências e Tecnologia, Universidade de Trás-os-Montes e Alto Douro, 5000-801 Vila Real, Portugal

C. Cruz Corona (ed.), *Soft Computing for Sustainability Science*,
Studies in Fuzziness and Soft Computing 358, DOI 10.1007/978-3-319-62359-7_5

1 Introduction

The continuous improvement of spectroscopic instruments has allowed the existence of small portable equipments that are very attractive for local field measurements, instead of remote, in various areas of application, such as agriculture and food quality assessment [24, 33, 34]. In recent years, the integration of spectroscopic techniques with soft computing and multivariate analysis algorithms has emerged as a promising method for non-destructive analysis. The cost effectiveness and environmental friendliness of this method makes it sustainable. The spectroscopic techniques measure the intensity of light from various wavelengths that come from a sample. This is called a spectrum and it varies depending on the chemical compounds present in the sample. The reason is that different compounds absorb and reflect light differently. Each spectrum may contain the measurement of hundreds of different wavelengths. The use of soft computing and multivariate analysis algorithms is necessary because of the large dimensionality (number of wavelength) of each spectrum and because each sample contains various chemical compounds whose spectra overlap. The advantages of using spectroscopy combined with soft computing and multivariate analysis algorithms over other conventional techniques can be summarized in the following points: (1) It is non-invasive and non-destructive; (2) It is chemical-free which reduces costs and makes measurements environmentally friendly and sustainable; (3) It allows the simultaneous analysis of several different chemical compounds present in a sample based on a single spectral analysis; (4) The equipment portability permits the creation of small and movable laboratories; (5) Once the models to extract the relevant information are built and validated, it is a fast and simple way of making an analysis.

This chapter will focus on the application of multivariate analysis and soft computing combined with non-destructive local spectroscopy to two viticulture subfields of increasing importance: (1) The measurement of enological parameters in samples composed of a small number of whole berries; and, (2) The identification of different varieties or clones of grapevine. Recent reviews did not address these matters in detail [8, 37], contrarily to what is intended with the present work. The enological parameter determination involves solving regression problems where the models output is continuous, and the variety or clone identification is a typical classification problem.

The determination of enological parameters in grapes is important for ripeness determination and harvest date definition. Nowadays, this is done using destructive wet chemistry methods. Most of the published works on enological parameter prediction by non-destructive spectroscopic analysis employ samples with a large number of grape berries which is easier to analyse than a small number of berries [12]. The use of whole berries instead of homogenates seems to make the problem even harder [5], but it has the advantage of making the technique non-destructive. Using a small quantity of whole berries can be interesting in terms of the selection of the best berries for producing high quality wines [29–31, 35]. The present review will focus on three enological parameters, sugar content, pH and anthocyanin content.

The identification of grapevine varieties and clones using non-destructive spectroscopic analysis is still in its childhood. Conventionally, it is done by ampelography [13] that separates the varieties or clones based on the plant characteristics, but this requires highly trained experts. The alternative is the use of costly and time consuming DNA analysis methods. The importance of identifying varieties and clones comes from the necessity to ensure truthness-to-type in plant nurseries, to ensure that certain varieties are not planted in certain apellation areas, and ultimately because grape price is variety dependent [21]. Proper clone identification can bring competitive advantages to growers because certain clones are more adaptable to certain regions than others and because some have been selected to produce more. A fast and simple clone and variety identification method is also useful for plant diversity conservation that is essential for sustainable development.

In the two subfields of the present review the soft computing algorithms used were discriminant analysis, neural networks isolated or associated in committees and support vector machines (SVM). This review analyses also multivariate analysis algorithms such as partial least squares (PLS) and multiple linear regression (MLR) in order to obtain a complete overview of the works published in the two subfields. Some comparisons between the performance of discriminant analysis, neural networks, SVM and PLS for the same problems will be presented.

2 Technology and Methods

2.1 Spectroscopy

Spectroscopy analyses the interaction between matter and electromagnetic radiation at various wavelengths. In this interaction, the radiation coming from the sample may have the same wavelength as the incident radiation, or it may change. For the works analysed in the present review only the former case is relevant, which excludes fluorescence. The radiation interacting with the sample may be totally or partially absorbed or reflected by the sample; the extent to which these phenomena happen is determined by the chemical compounds that form the sample, up to the depth that the radiation penetrates. The measurement of the intensity of radiation emitted by a sample as a function of the wavelength is called a spectrum. Since each chemical compound has a specific spectrum, by measuring the spectrum of a sample it is possible to know the chemical compounds that are present in the sample. However, this is difficult to do because each sample contains many chemical compounds whose spectra overlap to form a single spectrum. This is why it is necessary to use sophisticated soft computing and multivariate analysis algorithms to determine if a certain chemical compound contributed to a spectrum and in which quantity. Two good references on technology fundamentals are Sun [34] and Gowen et al. [18].

2.1.1 Signal Acquisition Modes

The present review will describe works whose spectroscopic measurements were done in reflectance, transmittance or interactance mode, with the last two being closely related [32]. These modes depend on how the sample is illuminated and how the electromagnetic radiation coming from the sample is captured. Reflectance, transmittance and interactance measure the percentage of electromagnetic radiation originating in the sample relative to the amount of incident radiation. In reflectance mode the electromagnetic radiation coming from the sample is mainly that reflected at the surface but also some that penetrated the sample and was still sent back to the receiver. In transmittance mode the electromagnetic radiation must cross the whole sample to emerge at 180° from the position where it penetrated in the sample. Interactance mode, can be seen as a generalization of transmittance where the angle between the electromagnetic radiation emitter and receiver, with the sample at the vertex of the angle, may be different from 180°. In both transmittance and interactance the receiver does not get the light reflected from the sample surface. Transmittance and interactance are usually harder to use than reflectance because the intensity of the signal originating in the sample is significantly lower for transmittance and interactance and also because there must usually be some form of contact with the sample in order to prevent any surface reflected electromagnetic radiation from reaching the receiver. In reflectance mode it is possible to do imaging without any contact with the samples.

2.1.2 Equipment

A fundamental characteristic of a spectrometer is the wavelength range that it measures [34]. For the present review, the two most important wavelength ranges are the visible from the 390 nm up to 770 nm and the near-infrared between 770 and 2500 nm. Below the visible region there is the near-ultraviolet region between 200 and 390 nm and beyond the near-infrared there is the mid-infrared between 2500 and 25000 nm. The wavelength range for each application is chosen so that it includes absorption bands of the chemical compounds that must be measured. It is also important to consider the number of wavelength bands measured on the whole equipment wavelength range since it is related with the spectral resolution of the equipment. Usually, the larger the number of bands is the better the spectral resolution is. Depending on the number of these bands, a few tens or hundreds, the equipment is called multispectral or hyperspectral, respectively. This review mentions various works that use equipments operating simultaneously in the visible and infrared between 400 and 1000 nm and present good results. This range is in-fact quite popular even though, the near infrared wavelength larger than 1000 nm contains important absorption bands. One possible reason for the popularity of the range between 400 and 1000 nm, besides the good results obtained, might be the tendency for the spectrometers operating at larger wavelengths to be more expensive.

It is important to distinguish between spectrometers with or without imaging capability. The former are able to resolve spatially the spectral information. These spectrometers usually get the spatial information for a line over a sample, so that each image produced has spectral information on one axis and spatial information on the other. To get the complete image of a sample it is usually necessary a mirror scanner or positioning table to gather information line by line. Imaging spectrometers also tend to be more expensive than those without imaging capability. Fourier transform infrared spectroscopy was not considered for this review.

2.2 Sampling Issues

In any work using soft computing or multivariate analysis the used dataset must be representative. This means that it should contain samples that represent all the characteristics that are possible to find. Since for validation purposes the dataset is usually split into subsets, it is good to have enough data so that each subset is itself representative of the whole. In practice, it is usually hard to know if a dataset is representative or not, so one of the most important strategies to obtain a representative set is by gathering large number of samples using a data collection process that is not biased. In practice, the difficulty and cost of collecting a large number of samples usually makes the datasets small.

In the subfields that are the focus of the present review sample variability may originate from various factors: (1) The variety or clone used, which can be hundreds of different ones; (2) The samples being collected in different location; and, (3) The year in which the samples were harvested. In each scientific work published it is very difficult to gather samples that comprehensively cover all possible sources of variability, consequently, the published scientific works usually narrow their scope regarding these topics by limiting the different varieties or clones, vintages and places where samples are collected. The information discussed in the following sections is summarised in Tables 1 and 2 that divide the works into enological parameter determination and variety or clone identification. Each row of the tables contains the information relative to one work. The columns contain informations such as the vintages used, the locations where the samples were collected, the number of varieties or clones employed in models, the total number of samples available and how they were divided per variety. These tables also mention the cases where multiple vintages, sample collection locations or varieties were employed simultaneously in a single model.

2.2.1 Varieties

The soft computing or multivariate analysis models for enological parameter determination were frequently created for single varieties [10, 12, 16, 17, 23]. However, there are some works that used different varieties in the same model [2, 4, 14, 26].

Table 1 Information regarding works on enological parameter determination with samples composed of a small number of grape berries. "CV" stands for cross-validation.

	Vintages	Sample harvesting location	Number of varieties per model/total number of samples per model/*number of samples from each variety*	Validation Methods	Sample split for validation
Fernandes et al. (2015) [12]	2012	One vineyard Pinhão, Portugal	One/240	n-fold CV with test set	87.5% n-fold CV 12.5% test
Herrera et al. (2003) [23]	2002	One vineyard Maipo Valley, Chile	One/150, 240, 260, 300 (total of 15 models)	Hold-out without test set	75% training 25% validation
Gomes et al. (2014) [17]	2012 and 2013 (One model does n-fold CV with one and tests with both)	One vineyard Pinhão, Portugal	One/84, 264 (two models)	n-fold CV with test set	60 or 210 samples for n-fold CV 24 or 54 samples for test
Fernandes et al. (2011) [10]	2009	One vineyard Vila Real, Portugal	One/46	Leave-one-out	
Gomes et al. (2014) [16]	2012	One vineyard Pinhão, Portugal	One/240	n-fold CV with test set	87.5% n-fold CV 12.5% test
Larrain et al. (2008) [26]	2003	Two vineyards Maipo Valley, Chile. Unclear if varieties come from both places.	One/135 to 740 (total of 14 models) Four/693/*205+218+135+135* Four/1633/*678+685+135+135+389* Four/1753/*724+740+144+145+409*	Hold-out without test set	75% training 25% validation

(continued)

Table 1 (continued)

	Vintages	Sample harvesting location	Number of varieties per model/total number of samples per model/ *number of samples from each variety*	Validation Methods	Sample split for validation
Geraudi et al. (2009) [14]	2008	France and Australia. Chardonnay is from both locations, but models seem to use only one location from this variety.	One and probably four/unclear	Leave-one-out	
Cao et al. (2010) [4]	Not clear	Bought in local fruit markets in China	Three/439/*115+127+197*	Hold-out without test set	66.7% training 33.3% validation
Arana et al. (2005) [2]	Not said. Probably one.	Three vineyards from Navarra, Spain. Models seem to use Chardonnay from two vineyards.	Probably: One/144, 288 (two models) Two/432/*144+288*	Leave-one-out	

Table 2 Information regarding works on identification of grapevine varieties and clones.

	Vintages	Sample collecting location	Number of varieties or clones per model/Number of samples per variety or clone	Validation Methods	Sample split for validation
Gutiérrez et al. (2015) [21]	2012	One vineyard, Navarra, Spain	10 varieties/20	Only mention cross-validation	
Cao et al. (2010) [4]	Not clear	Bought in local fruit markets in China	3 varieties/115, 127, 197	Hold-out without test set	66.7% training 33.3% validation
Arana et al. (2005) [2]	Not said. Probably one	One vineyard in Navarra, Spain.	2 varieties/144	Probably leave-one-out	
Gutiérrez et al. (2015) [19]	2012 and 2015 (6 variety model uses both)	A total of three vineyards in La Rioja and Navarra, Spain. Classifiers employ samples from one or three locations.	5, 6 and 20 varieties/20 or 24 in different classifiers	n-fold cross-validation	
Melo-Pinto et al. [27]	2011	A total of three vineyards from La Rioja and Navarra, Spain. Each clone was from only one location.	4 clones/25	Repeated hold-out without test sets	48% training 52% validation and vice-versa
Fernandes et al.(2015b) [11]	2011	Two vineyards from Navarra, Spain. Each clone was from only one location.	4 clones/25	Repeated hold-out without test sets	80% training 20% validation

(continued)

Table 2 (continued)

	Vintages	Sample collecting location	Number of varieties or clones per model/Number of samples per variety or clone	Validation Methods	Sample split for validation
Diago et al. (2013) [9]	2011	A total of three vineyards from La Rioja and Navarra, Spain. Two varieties came from two locations.	3 varieties/100	Repeated hold-out with and without test sets	28% training 72% validation or 28% training 28% validation and 44% test
Gutiérrez et al. (2016) [20]	2012	One vineyard, Navarra, Spain	10 varieties/20	n-fold cross-validation with test set	80% n-fold CV 20% test
Yang et al. (2012) [39]	2011 n	Local vineyard probably in Hangzhou, China	4 varieties/30	Hold-out without test set	67% training 33% validation
Lacar et al.(2001) [25]	2000	Single vineyard, Barossa valley, South Australia	4 varieties/120	No classification efficiency given	

Larrain et al. [26] used the largest number of different varieties, four, in a single model. When creating a model with different varieties the results are some sort of average of the models from individual varieties (as can be seen in Larrain et al. [26]), therefore, the values in terms of squared correlation coefficients (R^2) or root mean squared error (RMSE) may not be the best, but the model becomes more robust and has a more widespread application than the models using individual varieties.

For identification of varieties or clones, the largest number of different varieties, twenty, was used by Gutiérrez et al. [19]. For clones, a total of eight different clones, four for two different varieties, were separated in Melo-Pinto et al. [27]. The same four Cabernet Sauvignon clones were identified in the work of Fernandes et al. [11]. It is expected that the increase in number of different clones and varieties will bring a tendency for classification efficiency reduction. Nevertheless, testing different algorithms to find the most suitable for a certain problem as well as the increase in number samples might help to contradict this tendency.

2.2.2 Different Vintages

Even though the generalisation between vintages could be an important factor for the success of the models, in the present review, the samples employed were usually from the same vintage. Nevertheless, there were two exceptions. Gutiérrez et al. [19] has built a model for variety identification using 2012 and 2015 data and Gomes et al. [17] has built a model for sugar content prediction whose training was done with 2012 samples and the test employed 2013 samples.

2.2.3 Different Locations

Gathering samples from multiple locations also helps to increase the models' robustness, however, it is not yet a standard procedure. In addition, even though it might be stated in some works that samples were collected at different locations it is not always clear if the samples of a certain variety but from different locations were used in the same model. Arana et al. [2] seems to use samples of Cabernet Sauvignon harvested in two different locations to build a model for sugar content prediction. In two works it is clear that a single model employs samples of the same variety but from different locations: Gutiérrez et al. [19] created a model to separate six varieties collected in three different locations; Diago et al. [9] created a model that separates three varieties and whose samples of each one of two varieties came from two different locations.

2.2.4 Number of Samples

The total number of samples employed in the various works reviewed in the present chapter was highly variable. In enological parameter determination, for a single model, it may vary between 46 samples in Fernandes et al. [10] for anthocyanin determination and, 1753 in Larrain et al. [26] for sugar content determination. Larrain et al. [26] has also built models with 1633 and 693 samples for pH and anthocyanin determination, respectively. The works of Cao et al. [4] and Arana et al. [2] used approximately 430 samples, which can be considered an average number of samples.

When looking at the number of samples per variety the most common is to find between 100 and 300 [2, 4, 12, 16, 17, 23, 26] but the number can go down to 46 in Fernandes et al. [10] or up to 740 in Larrain et al. [26]. The number of samples should be analysed taking into consideration the validation method employed because of the way in which the patterns are split. For example, a model built in Larrain et al. [26] employed a total of 135 samples, but the use of hold-out validation method left 34 samples for validation, while the leave-one-out method in Fernandes et al. [10] allowed the use of all the 46 samples available for validation. A discussion on which validation method is more correct is beyond the purpose of the example.

For variety or clone identification the most important is the number of samples per variety or clone since each class should contain a number of samples that allowed creating a representative set. Cao et al. [4] used between 115 and 197 samples per variety, Arana et al. [2] used 144 and Diago et al. [9] 100. Various works such as Melo-Pinto et al. [27], Gutiérrez et al. [19], Gutiérrez et al. [21], Gutiérrez et al. [20], Fernandes et al. [11] employed 20 or 25 samples per variety or clone.

2.3 *Validation Methods*

The validation of all models is fundamental to make sure that they generalise well, i.e., that they are capable of providing accurate results for samples not used in the models training. This verification is particularly important when the models are built using a number of samples that is not several times, as rule of thumb 10 times [22], larger than the number of adjustable variables of the model, which is a situation that frequently happens with hyperspectral data. Due to the relevance of assessing proper generalization the present review will only analyse works that used validation. In the scientific works of this review two validation methods stand out: (1) Hold out; and (2) n-fold cross-validation. Tables 1 and 2 summarise the validation methods and the sample split for validation of the various works under review.

2.3.1 Hold-Out

In hold-out the samples are divided into a set for training, one for validation and, sometimes, another for test. The training set is used to create the model, the validation set to choose or adjust any parameters and the test set to analyse the model generalisation once all parameters are fixed. Frequently, the hold-out does not use a test set and so the generalisation is assessed in the validation set. Yang et al., Cao et al., Larrain et al. and Herrera et al. [4, 23, 26, 39] used hold-out without test set. The percentage of samples for training was 75% in Larrain et al. [26] and Herrera et al. [23] and 67% in Yang et al. [39] and Cao et al. [4] with the remnant samples being used for validation. A variation of the hold-out method is the repeated or Monte-Carlo hold-out where the training, validation and test procedures are executed multiple times with the samples being randomly assigned to the various sets in each time. The final R^2 or RMSE for validation or test are the average of the values from these parameters obtained in the multiple executions. Melo-Pinto et al. [27], Fernandes et al. [11] and Diago et al. [9] used repeated hold-out with just training and validation sets, but Diago et al. [9] also employed repeated hold-out with test set. In these three works the sample split differs. Melo-Pinto et al. [27] used approximately a 50% split for training and validation, Fernandes et al. [11] a 80/20% split and Diago et al. [9] a 28/72% split, respectively.

2.3.2 N-Fold Cross-Validation

In n-fold cross-validation the samples are divided into n folds and n-1 are used for training and one for validation. The training/validation process is repeated n times with a different fold being left out each time for validation. Fernandes et al. [10], Geraudie et al. [14] and Arana et al. [2] used leave-one-out, a situation in which the folds contained only one sample, and Gutiérrez et al. [19] employed 5-fold cross-validation. In Fernandes et al. [12], Gomes et al. [17], Gomes et al. [16] and Gutiérrez et al. [20] besides the n-fold cross-validation there was also a test set. The first three articles used 7-folds but Gomes et al. [17] also validated one model with 3-folds. In Fernandes et al. [12] and Gomes et al. [16] the percentage of samples for test was 12.5% while in Gomes et al. [17] it was 20 or 29% depending on the model. Gutierrez et al. [20] used 5-folds and the percentage of patterns left for test was 20%.

2.4 *Algorithms*

In the subfields of the present review the algorithms are essential to extract information from the high dimensionality data. The most used algorithm was Partial Least Squares (PLS), closely followed by neural networks and Support Vector Machines (SVM). PLS has its origin in the chemometrics community and has provided very good results due to being able to cope with a number of samples smaller than the number of variables which is normally prone to overfitting and also due to the ability to handle highly correlated variables. Neural networks and SVM have known successful and widespread application inside the machine learning community. Discriminant analysis was also used in the subfields reviewed but to less extent and Adaptive boosting of neural networks (Adaboost) and Multiple Linear Regression (MLR) were each used in only one article.

2.4.1 Brief Algorithm Description

PLS [38] creates new variables, called components, that correspond to the projection of the independent (X) and dependent (Y) variables into new directions that maximize the covariance between X and Y. It can do dimensionality reduction because only a few of these components are enough to explain most of the data variance. The best number of components is usually chosen to be that providing the smallest validation error. Besides dimensionality reduction, PLS calculates also the regression coefficients that transform the independent into the dependent variables. PLS is able to cope with a number of input variables larger than the number of measurements. The input variables can also be correlated.

Neural networks [22] are biologically inspired mathematical processors. They are composed of neurons connected by weights. The output of a neuron corresponds to adding the multiplication of its inputs by its weights and passing this result through a function called the activation function. The multilayer perceptron, which is the most widely used type of neural network, has an input and an output layer and at least one hidden layer. The hidden layer allows to create nonlinear functions if the neurons have nonlinear activation functions. Training is done by iteratively changing the weights in order to minimise the difference between the neural network output neurons outcome and a desired value provided by an expert for each input sample.

SVM [7] for classification find a hyperplane that maximizes the distance to the training points of any group of samples. In linearly separable problems this hyperplane is generated by a linear combination of some samples that are called the support vectors. When the problem is not linearly separable kernel functions are used to transform the input samples into a feature space where separation is easier. The transformed samples are afterwards combined. Obtaining the weights for the linear combination of the kernel transformed samples involves solving a quadratic optimization problem that minimises the weights norm which is relevant to have good generalization. This optimisation includes also a regularization parameter C that imposes a tradeoff between the flatness of the model and the number of wrongly classified samples. A parameter *nu* was introduced in SVM formulation because it is easier to interpret than C. *nu* is an upper bound on the fraction of misclassified samples. With LSSVM the quadratic optimization problem is transformed into a linear system of equations.

Discriminant analysis [1] finds linear, orthogonal, combinations of the input variables that best separate groups of samples.

Adaboost [22] is a method to create a committee of classification or regression models. These models are trained in sequence so that the training of the new model gives more emphasis to the samples that had larger error on the older model. With Adaboost it has been reported that, in experiments, the generalization error may continue to decrease as new models are added even when the training error has reached zero.

MLR [28] is an extension of the common least squares regression to multiple input and output variables. Contrarily to PLS, MLR may not work well when the input variables are correlated or when there are more variables than samples.

2.4.2 Algorithm Performance Comparison

Even though the two subfields do not yet have a large number of publications it was possible to find various algorithm comparisons. In Fernandes et al. [10] it was mentioned that neural networks were used instead of PLS because the former provided better results than the latter for models that determine anthocyanin content in grapes. The R^2 for a single neural network and PLS were 0.36 and 0.25, respectively. When associating four neural networks as an Adaboost, a type of committe machine, it was possible to raise the R^2 to 0.65. The mean absolute percentage error (MAPE) with one neural network was 18.6% while with the four from the Adaboost it came down to 13.4%. In Gomes et al. [16], neural networks were compared to PLS for sugar content determination in grapes, and no major difference was found in the performance of the two algorithms both in terms of R^2 and RMSE. The R^2 values were 0.929 and 0.924 for PLS and neural networks, respectively, while the RMSE was 0.939° and 0.955° Brix in the same order. In Melo–Pinto et al. [27], sixteen classifiers were built, eight using neural networks and eight using PLS. The leaves to separate belonged to eight different clones and, each classifier identified the leaves from a specific clone. The classification efficiencies varied between 78.3 and 100% for neural networks and between 90.8 and 100% for PLS, with a clear advantage for PLS. For Gutiérrez et al. [19] the PLS, neural networks and SVM were compared in the creation of variety classifiers. When analysing the differences between algorithms, the neural networks and SVM produced significantly better results than PLS. When comparing SVM and neural networks the difference was not statistically significant. In Yang et al. [39] linear discriminant analysis (LDA), SVM and neural networks were compared in the separation of grape seed varieties. In this work LDA was capable of 100% classification efficiency for the four varieties analysed, while for SVM and neural networks the classification efficiency was still 100% in two varieties but dropped down to 90% for the remaining two varieties.

In the present case, as in any other field of application, it is necessary to test various algorithms in order to find the one with the best performance.

3 Enological Parameter Determination in Samples with a Small Number of Grape Berries

The evaluation of the grapes maturation based on the evolution of enological parameters over time is one of the most important aspects to decide the optimal moment for harvest. During this process, the level of anthocyanins (phenolic compounds responsible for the grapes pigmentation) and sugars increase, while the acidity diminishes [3]. The models created using soft computing or multivariate analysis algorithms that allow to assess the grapes ripeness correspond to functions that transform a certain spectrum into a desired chemical attribute of the grape samples, in the present case, sugar content, pH or anthocyanin content. The models were created in a supervised way, meaning that the training algorithms receive, for each spectrum, the desired chemical attribute value measured by conventional methods. These are called regression problems since the models output is continuous. The spectra were collected in reflectance or interactance modes.

3.1 Number of Berries per Sample

The present revision focus is on the use of a small number of whole berries per sample. Using a small quantity of berries can be interesting in terms of the selection of the best berries for producing high quality wines [29–31, 35]. The analysis of a samples with a small number of whole berries can be automated in the near future using destemmers capable of extracting berries one-by-one from the bunches [15] and placing them over a conveyor belt that passes underneath a spectrometer [12]. Another interesting use of individual berries is the study of variations inside a grape bunch [6].

Arana et al., Cao et al., Fernandes et al., Larrain et al., Herrera et al. [2, 4, 10, 23, 26] used whole single berries in their works. This allowed reaching R^2 values of 0.93, 0.8 and 0.68 for sugar content, pH and anthocyanin content, respectively, and RMSE of 0.96° Brix, 0.09 and 0.18 $mg.g^{-1}$ in the same order. The R^2 and RMSE for each enological parameter are not from the same models. Cao et al. [4] reached an $R^2$of 0.96 for pH but the samples do not assume all the values between ~2.5 and ~4.5. Geraudie et al. [14] used between two and four whole berries and Fernandes et al. and Gomes et al. in two works [12, 16, 17] used 6 whole berries. In these four works the best R^2 values were 0.96, 0.73 and 0.95 for sugar content, pH and anthocyanin content, respectively, with corresponding RMSE values of 0.924° Brix, 0.18 and 14 $mg.L^{-1}$.

3.2 Difficulties of Using a Small Number of Whole Berries per Sample

Most of the published works on enological parameters predictions do it for a large number of grape berries, usually 50 or more. This is an easier problem than using a small number of berries since the sample variability diminishes with the increase of the number of berries due to the averaging out of the variations of the individual berries. The use of whole berries is also more difficult than using homogenates, also because in the latter case there is an averaging out of the berries variations. In fact, Cozzolino et al. [6] has shown the existence of a large variation in the spectra of individual berries and, suggests that this problem can be tackled by scanning the berries in several positions.

3.3 Articles Using Reflectance Mode

This section describes in detail the works that use reflectance mode for enological parameter determination in samples with a small number of grape berries. Table 3 summarizes the information for the various works. Each row of the table contains the results for a different work. The columns show the R^2 and RMSEP for thee enological parameters, sugar content, pH and anthocyanin content, and also the varieties, the wavelength range and mode of the spectroscopic measurement as well as the soft computing algorithm employed. The table contains the results for works using reflectance and interactance modes.

Arana et al. [2] developed a prediction model for sugar content using PLS regression. The authors made the spectroscopic measurements using a laboratory spectrometer operating at 800–2500 nm that was not portable. In spite of the non-portability of the equipment, that is an important characteristic of the works reviewed, this work was still included with the purpose of the overview being as complete as possible. The R^2 values obtained for sugar content were 0.70 and 0.58 for varieties Chardonnay and Viura, respectively, and the RMSE was 1.27 and 1.89° Brix. The model combining the two varieties presented a R^2 of 0.43 and RMSE of 2.16. The number of samples used was 288 for Chardonnay and 144 for Viura.

Cao et al. [4] used Least Squares Support Vector Machine method (LSSVM) to create a model from spectra collected with a portable spectrometer. The wavelengths for model creation were selected by genetic algorithm and corresponded to 418, 525, 556, 633 and 643 nm for sugar content and 446, 489, 504 and 561 nm for pH. For sugar content the R^2 was 0.82 and the RMSE 0.96° Brix, while for pH the R^2 and RMSE were 0.96 and 0.13, respectively. The models were created for a total of 439 samples of the varieties Manaizi, Mulage and Heiti, with 197, 127 and 115 samples per variety, in this order. The models created used the three varieties simultaneously.

In 2011, Fernandes et al. [10] tested the performance of committee machines of neural networks in the determination of anthocyanin content. A portable hyperspectral camera operating in the range 380-1028 nm was used. The results presented were 0.65 for R^2 and 88 mg.L^{-1} for RMSE. The work employed 46 samples of the Cabernet Sauvignon variety.

Fernandes et al. [12] also employed a portable hyperspectral camera in the range 380-1028 nm. Neural networks were used to create models for sugar content, pH and anthocyanin content. The results revealed R^2 values of 0.92, 0.73 and 0.95, and RMSE values of 0.95° Brix, 0.18 and 14 mg.L^{-1} for sugar content, pH and anthocyanin content, respectively. A total of 240 samples of the Touriga Franca variety were used.

Gomes et al. [17] used an equipment and wavelength range similar to that of Fernandes et al. [10, 12]. The variety employed was also Touriga Franca as in Fernandes et al. [12]. This work used samples collected in 2013 to evaluate neural network models for sugar content prediction created with 2012 or 2013 samples. This work was one of very few works that created models with grapes from one vintage and tested the models with grapes from another vintage. The test with 2013 samples for the

Table 3 Further information regarding works on enological parameter determination with samples composed of a small number of grape berries

	Sugar Content		pH		Anthocyanin content		Varieties	Wavelength range (nm)	Algorithm	Spectroscopy mode
	R^2	RMSEP (° Brix)	R^2	RMSEP	R^2	RMSEP				
Fernandes et al. (2015) [12]	0.92	0.95	0.73	0.18	0.95	14 $mg.L^{-1}$	Touriga Franca	380-1028	Neural Network	Reflectance
Herrera et al. (2003) [23]	0.74–0.94	1.06–1.35	-	-	-	-	Models with one of the varieties: Chardonnay, Carménère, Cabernet Sauvignon.	650–1100 and 750–1100	PLSR	Interactance
Gomes et al. (2014) [17]	0.91–0.96	1.03–1.17	-	-	-	-	Touriga Franca	380–1028	Neural Network	Reflectance
Fernandes et al. (2011) [10]	-	-	-	-	0.65	88 $mg.L^{-1}$	Cabernet Sauvignon	380–1028	Adaboost with neural networks	Reflectance
Gomes et al. (2014) [16]	0.939–0.955	0.924–0.929	-	-	-	-	Touriga Franca	380–1028	Neural Network and PLSR	Reflectance
Larrain et al. (2008) [26]	0.87–0.93	1.11–1.24	0.56–0.80	0.09–0.16	0.40–0.68	0.18–0.32 $mg.g^{-1}$	Models with one of the varieties and the four red: Chardonnay, Carménère, Cabernet Sauvignon, Merlot, Pinot Noir.	640–1300	PLSR	Interactance
Geraudi et al. (2009) [14]	0.78–0.95	0.92–1.63	-	-	0.83	0.08 $mg.g^{-1}$	Models with one of the varieties and probably the four red: Shiraz, Cabernet, Pinot Noir, Pinot Meunir, Chardonnay.	400–1100	MLR	Interactance

(continued)

Table 3 (continued)

	Sugar Content		pH		Anthocyanin content		Varieties	Wavelength range (nm)	Algorithm	Spectroscopy mode
	R^2	RMSEP (° Brix)	R^2	RMSEP	R^2	RMSEP				
Cao et al. (2010) [4]	0.82	0.96	0.96	0.13	-	-	Model with three varieties: Manaizi, Mulage, Heiti.	418, 525, 556, 633, 643 for sugar and 446, 489, 504, 561 for pH	LSSVM	Reflectance
Arana et al. (2005) [2]	0.58–0.70	1.27–1.89	-	-	-	-	Models with one and two of the varieties:Viura and Chardonnay.	800-2500	PLS	Reflectance

neural network created with samples from 2012 presented R^2 values of 0.91 and RMSE of 1.17° Brix. For a model trained and validated with 2013 samples the R^2 value was 0.96 and the RMSE was 1.03° Brix. The work involved 240 samples from 2012 and 84 from 2013.

Gomes et al. [16] used the same equipment, wavelength range and variety of Gomes et al. [17]. The models for sugar content were created with 240 samples using PLS and neural networks. The results in terms of R^2 were 0.929 and 0.924 for PLS and neural network, respectively, and the RMSE was 0.939° Brix and 0.955° Brix.

3.4 *Articles Using Interactance Mode*

This section describes in detail the works that use interactance mode for enological parameter determination in samples with a small number of grape berries. The information is summarized in Table 3.

Herrera et al. [23] developed PLS models for sugar content determination. Two different wavelength ranges were employed, from 650 to 1100 nm and from 750 to 1100 nm. The equipment used was a portable spectrometer. The R^2 obtained varied between 0.74 and 0.94 with the RMSE ranging between 1.06 and 1.35° Brix. The models were created for the following individual varieties: Cabernet sauvignon, Chardonnay and Carménére, with the total number of samples per variety ranging from 150 and 300. This article provided an interesting comparison of model results depending on the wavelength range used, the signal acquisition modes and the signal preprocessing.

Larrain et al. [26] used a portable spectrometer operating between 640 and 1300 nm and built PLS based models for sugar content, pH and anthocyanin. The results for sugar content prediction in models for individual varieties ranged between 0.87 and 0.93 for R^2 and between 1.1 and 1.2° Brix for RMSE. For pH, the results in terms of R^2 were between 0.56 and 0.80 and, in terms of RMSE were between 0.088 and 0.16. For anthocyanin content, the R^2 ranged between 0.4 and 0.68 and the RMSE between 0.18 and 0.32 $mg.g^{-1}$. This work presented also models for red varieties with R^2 values of 0.91, 0.74 and 0.62 and RMSE of 1.24° Brix, 0.15 and 0.30 $mg.g^{-1}$ for sugar content, pH and anthocyanin content, respectively. The varieties used were Cabernet Sauvignon, Carménère, Merlot, Pinot Noir and Chardonnay, with the number of samples per variety varying between 144 and 740 for sugar content, between 135 and 685 for pH and between 135 and 218 for anthocyanin. In red varieties models the number of samples was 1753, 1633 and 693 for sugar content, pH and anthocyanin content determination, in this order.

Geraudi et al. [14] used multiple linear regression to create models from spectra acquired with a portable spectrometer operating between 400 and 1000 nm. The models were produced for sugar and anthocyanin content measurement. For sugar content the R^2, depending on the variety, varied between 0.78 and 0.93 while the RMSE assumed values between 0.92 and 1.63. The varieties in question were Shiraz,

Cabernet (as named in the article), Pinot Noir, Pinot Meunier and Chardonnay. Two models were built with a set of red varieties, the R^2 obtained were 0.92 or 0.95 and the RMSE were 1.2 or 1.12° Brix. Two other models were also built for the white varieties, with R^2 values of 0.78 or 0.84 and RMSE values of 1.63 or 1.2° Brix. For anthocyanin, a model was created for the Shiraz variety with R^2 and RMSE values of 0.83 and 0.08 $mg.g^{-1}$, respectively. The number of samples was not clearly stated, however, it was mentioned that the samples for model creation were collected at a rate of three per day for each variety during two months.

From the various works concerning the creation of models for enological parameter measurement, including those in reflectance, transmittance or interactance mode, only Larrain et al. [26] and Geraudie et al. [14] have actually done measurements in the field with the grape berries in the vinetree. The grapes were harvested in all the remaining works. However, for Geraudie, the samples used in the model creation also seem to have been harvested.

4 Identification of Grapevine Varieties and Clones

The classifiers reported in this section were built in supervised way, meaning that it is necessary to know to which variety or clone the samples belong to in order to train the classifiers. These classifiers solve multiclass classification problems. Usually, in classification problems, only two classes need to be separated, while in the present subfield of application, the number of classes can be as many as the number of varieties or clones that must be identified. The encoding of the various labels into the training algorithms or the methods to separate the classes can be various but they were not always clearly described in the articles. In this subfield it was possible to find the two following labelling methods for the output of the classifier: (1) Sequential numbering of the classes (1, 2, 3) based on a priori information about the sorting of the average spectra from each class [9]; (2) Use of dummy variables composed of a sequence of zeros and ones, with each class being characterised by having the value one at a specific position of the sequence and zeros everywhere else. The different classes have ones at different positions of the sequence [11]. In Melo–Pinto et al. [27] it was used the one-versus-all approach, in which, there was a classifier that separated two classes by saying if a sample was from a certain clone or not. This has two disadvantages: (1) It requires a classifier per class, contrarily to the Diago et al. [9] and Fernandes et al. [11] approaches; and (2) The sets become unbalanced, since the class of the variety or clone to be identified will normally have a number of samples smaller than that of the remaining varieties or clones put together to constitute the other class. Table 4 compiles information relative to the various works. Each row of the table contains the results for a different work. The columns show the correct classification percentage of the samples from the different varieties or clones, the varieties or clones separated, and also the wavelength range and soft computing algorithm employed. All works in the table were done in reflectance mode.

Table 4 Further information regarding works on identification of grapevine varieties and clones. All measurements were done in reflectance mode

	Correct classification percentage (in the same order as varieties/clones)	Varieties/clones	Wavelength (nm)	Algorithm
Gutiérrez et al. (2015) [21]	95, 95, 95, 100, 90, 100, 95, 89.5, 90, 100	Cabernet Sauvignon, Caladoc, Carmenere, White Grenache, Pedro Ximenez, Pinot Noir, Tempranillo, Treixadura, Viognier and Viura	1600–2400	C-SVM
Cao et al. (2010) [4]	93.9%, 97.6% and 100%	Manaizi, Mulage and Heiti	636, 649, 693 and 732	LSSVM
Arana et al. (2005) [2]	95.8 and 98.6%	Viura and Chardonnay	800–2500	Canonical discriminant analysis
Gutiérrez et al. (2015) [19] (Model for 20 varieties)	100, 100, 95, 95, 90, 70, 70, 90, 90, 100, 50, 95, 80, 85, 75, 95, 95, 85, 90, 95 with neural network	Cabernet Franc, Cabernet Sauvignon, Caladoc, Carménère, Godello, Malvasia, Marselan, Pedro Ximénez, Pinot Noir, Touriga Nacional, Verdejo, Viognier, White Grenache, White Tempranillo, Viura, Grenache, Treixadura, Tempranillo, Syrah, Albariño	1600–2400	PLS, SVM and neural network
Melo-Pinto et al. [27] (Best result)	100, 98.5, 99.6, 100, 90.8, 95, 97.7, 100 with PLS	RJ24, RJ26, RJ43, RJ75 (Tempranillo clones) and CS 15, CS 169, CS 685 and CS R5 (Cabernet Sauvignon clones)	380–1020	PLS and neural network
Fernandes et al. (2015b) [11]	98.2%, 99.2%, 100% 97.8%,	CS 15, CS 169, CS 685 and CS R5 (Cabernet Sauvignon clones)	634–759	PLS
Diago et al. (2013) [9]	92.6%, 95.3% and 92.7%	Tempranillo, Grenache and Cabernet Sauvignon	380–1028	PLS

(continued)

Table 4 (continued)

	Correct classification percentage (in the same order as varieties/clones)	Varieties/clones	Wavelength (nm)	Algorithm
Gutiérrez et al. (2016) [20]	100, 100, 100, 100, 100, 75, 75, 100, 75, 100	Cabernet Sauvignon, Caladoc, Carmenere, White Grenache, Pedro Ximenez, Pinot Noir, Tempranillo, Treixadura, Viognier and Viura	1600 – 2400	nu-SVM
Yang et al. (2012) [39] (Best result)	100% with LDA	Rosario Bianco, Red globe, Muscat Kyoho and Fujiminori.	200-1100	Linear discriminant analysis, SVM and neural networks
Lacar et al. (2001) [25]	-	Cabernet Sauvignon, Merlot, Semillon and Shiraz	400-900	-

4.1 The Articles

Lacar et al. [25] employed a portable spectrometer to measure the spectra of leaves in the range 400–900 nm. They used ANOVA to understand if there were significant differences between the varieties Cabernet Sauvignon, Merlot, Semillon and Shiraz. With 120 samples per variety it was possible to conclude that there were statistically significant differences between mean values of different varieties at certain wavelengths. However, it was not presented the efficiency in identifying the different samples. The work presented the wavelengths 520, 550, 580, 615, 650, 720 and 755 nm as those with the greatest potential for variety identification.

Arana et al. [2] developed a classification model based on canonical discriminant analysis. The spectroscopic measurements were made using a laboratory spectrometer operating at 800–2500 nm that was not portable. As previously, in spite of the non-portability of the equipment, this work was still included for completeness of the present overview. The identification was based on the weight and sugar content of berries or in the spectra of the berries. The latter provided better results in the identification of two varieties, with a correct classification percentage of 95.8% for Viura and 98.6% for Chardonnay. The number of samples per variety was 144.

Cao et al. [4] used LSSVM to classify whole grape berry spectra collected with a portable spectrometer. The wavelengths employed in variety identification were selected by genetic algorithms and corresponded to 636, 649, 693 and 732 nm. The varieties identified were Manaizi, Mulage and Heiti with correct classification percentages of 93.9, 97.6 and 100%, respectively. The total number of samples per variety was 197, 127 and 115, respectively.

Yang et al. [39] used a portable spectrometer to obtain spectra for classification of grape seed varieties. The system operated between 200 and 1100 nm. The algorithms used to create the classifiers were linear discriminant analysis, SVM and neural networks. The first algorithm was the best of the three providing 100% classification efficiency for the four varieties analysed that were Rosario Bianco, Red globe, Muscat Kyoho and Fujiminori. Various spectra preprocessing methods such as Savitzky-Golay smoothing, multiplicative scatter correction, baseline offset correction, first and second order de-trending, first and second order derivatives and standard normal variate were tested but only the last was useful to improve the results. The total number of samples used was 120 with 30 per variety.

Diago et al. [9] used portable hyperspectral imaging of grapevine leaf discs in the wavelength region between 380–1028 nm to identify three varieties. The algorithm used for identification was PLS. The number of samples per variety was 100, with each variety containing samples from four different clones. The percentage of correct leaf classification was 92.6, 95.3 and 92.7% for Tempranillo, Grenache and Cabernet Sauvignon, respectively.

Gutiérrez et al. [19] have built PLS, SVM and neural network classifiers for the identification of varieties. The leaf measurements were done in the field, without removing the leaf from the vinetree, using a portable spectrometer that operated in the range 1600–2400 nm. The classifiers were built for leaves of 5 or 20 different varieties from the same location, or for leaves of 6 varieties from three different locations. For classification with data from the same location there were 20 leaves available per variety, while for the classification with data from multiple locations there were 24 leaves available per variety. The article concluded that when increasing from 5 to 20 the number of varieties identified the classification ability of PLS degrades considerably with respect to SVM or neural networks. The best results were obtained by neural networks with a classification percentage for 20 varieties ranging between 50 and 100% depending on the varieties. This percentage was 87.25% when considering all the varieties together. For 6 varieties, with the samples coming from multiple locations, the percentage of correctly classified samples with a neural network varied between 62.5 and 91.7% for the different varieties. With the varieties together, the correct classification percentage was 77.08%. The 20 varieties used were Cabernet Franc, Cabernet Sauvignon, Caladoc, Carménère, Godello, Malvasia, Marselan, Pedro Ximénez, Pinot Noir, Touriga Nacional, Verdejo, Viognier, White Grenache, White Tempranillo, Viura, Grenache, Treixadura, Tempranillo, Syrah, Albariño, with the last six being used to create the classifier with samples from multiple locations. The article also presented a study about the best type of spectrum preprocessing, concluding that it was the second order derivative and Savitsky-Golay filter and sometimes standard normal variate with detrending.

Gutiérrez et al. [21] and Gutiérrez et al. [20] used the same spectrometer operating in the same range as Gutiérrez et al. [19]. Leaf spectra were also acquired in the field. Both works employed SVM, but of different types, to identify ten different varieties, five red and five white: Cabernet Sauvignon, Caladoc, Carmenere, White Grenache, Pedro Ximenez, Pinot Noir, Tempranillo, Treixadura, Viognier and Viura. In Gutiérrez et al. [21] the C-SVM allowed to obtain correct classification percentages between 89.5 and 100% for the different varieties. With the varieties together this percentage was 95%. In Gutiérrez et al. [20] the nu-SVM reached slightly less good results with the classification efficiencies varying between 75 and 100% for the different varieties resulting in an overall value of correct classification of 92.5%. In both Gutiérrez et al. [21] and Gutiérrez et al. [20] the total number of samples used was 200 with 20 per variety.

Melo–Pinto et al. [27] is a Portuguese patent on the identification of clones using spectroscopy combined with multivariate analysis or artificial intelligence methods. The experimental results section shows 16 classifiers built with PLS or neural networks to identify 8 clones, RJ24, RJ26, RJ43, RJ75 from Tempranillo variety and CS 15, CS 169, CS 685 and CS R5 from Cabernet Sauvignon. The data was acquired using a portable hyperspectral camera operating between the 380 and 1020 nm. Each classifier identified only one clone against all others from a certain variety, being therefore a two class classifier. With PLS, the percentage of leaf discs correctly classified as belonging to a clone varied between 90.8 and 100% while for neural networks it varied between 78.3 and 100%. The percentage of leaf discs correctly classified as not belonging to the desired clone varied between 97.8 and 100% for PLS and between 96.4 and 100% for neural networks. The number of samples available per clone was 25.

Fernandes et al. [11] created a PLS classifier to identify leaf discs from different clones. PLS input was spectra in the range 634 to 759 nm acquired by portable hyperspectral imaging. The work presented a comparison of the impact of using various types of spectrum preprocessing methods in classifier efficiency, with the best method being a second derivative. The best classifier created was able to separate a total of 100 leaves into four clones of Cabernet Sauvignon, namely, CS 15, CS 169, CS 685 and CS R5 with correct classification percentage of 98.2, 99.2, 100 and 97.8%, respectively. The number of samples available per clone was 25.

The spectra for the studies in this section were collected in reflectance mode, and only Gutiérrez et al. [19–21] in three different works collected the data with the samples in the vinetree. Lacar et al. [25] did the measurements in the field but after removing the samples from the vinetree.

5 Suggestion for Information Summary

In order to facilitate the comparison between article results information about the critical aspects should be included.

For regression problems:

(1) Squared correlation coefficient; (2) Root mean and median squared error; (3) Mean and median absolute percentage error; (4) Mean absolute error;

For classification problems:

(5) Correct percentage of classification per variety or clone;

In general:

(6) Spectrometer portability; (7) Spectrometer imaging capability or absence of it; (8) Spectrometer wavelength range; (9) Varieties employed; (10) Vintages harvested; (11) Number of different locations where the samples were collected, (12) Total number of samples; (13) Number of berries per sample; (14) Validation method; (15) Split of samples between training, validation and test sets; (16) The algorithms used for model creation; and, finally, (17) The place where the spectral measurements were done, in the field or in the laboratory and under which conditions.

Parameters in points 1–5 are relevant to have a more accurate characterisation of the results. By having more parameters to compare it is possible to get a better understanding on how two works compare to each other. Points 6–8 and 17 characterise the technology employed. With more sophisticated equipment or under less demanding experimental conditions it is reasonable to expect better results. Points 9–12 characterise the diversity of the samples employed to build a model. Increasing the number of varieties, vintages, locations or samples in each model should lead to more robust models. Point 13, the number of berries per sample, has impact on the results because the smaller it is the larger the sample variability is. Points 14 and 15 characterise the validation method and are fundamental to understand the reliability of the created model. Point 16 allows to know which algorithm should be used when creating new models.

6 Final Comments

The authors acknowledge that gathering large number of samples is costly and the resulting models may exhibit a bad performance leading to possible waste of resources. It is the authors' opinion that, in face of the good results obtained so far, an increase in the number of samples has a limited risk and might even help to improve results due to the datasets becoming more representative. Nevertheless, there is always the risk that increasing the number of samples might increase sample variability to a point that model results become worse. But it would also be fundamental to have this information. The expansion of the models validity by including for each model data from more varieties, vintages and locations should also be a priority. If this is not done there is a risk that the subfields of the present review might be considered of limited interest and, consequently, might tend to die out. Creating models capable of handling data collected under field conditions, which is yet rare, can also increase the interest in these subfields.

Regarding validation methods it was sometimes hard to understand what has been done due to the use of the same designation for different meanings. Consequently, training set should be reserved for the samples used to train the models, validation set for the samples that were used to select parameters and test set for model generalisation evaluation. This generalisation evaluation implies that all model parameters have been previously defined. Even though this is standard designation, in practice, it was common to find people designating validation results as test, disregarding the fact that the samples were employed in parameter choice. In addition, it was not always straightforward to understand if the calibration set was the training set alone or the training and validation sets.

The validation set results from hold-out or k-fold cross-validation are widely used and accepted to assess the generalization of the created models. However, these results might still be biased, which is why it would be interesting to use, in the future, more test sets containing a large enough number of samples for these sets to be representative of all possible samples. Alternatively, it should be considered the use of nested n-fold cross-validation [36] that allows to use all available data as a test set.

The current review on enological parameter determination in samples with a small number of grape berries used only squared correlation coefficient and root mean squared error for analysis since these were the standard parameters reported. However, the values of these parameters might be influenced by outliers. A more robust comparison between works could have been done if the value distribution within sample points for parameters such as the error, absolute error and absolute percentage error had been provided separately for training, validation and testing sets. A simple way to show these parameters distribution is by using boxplots with whiskers where at least the 5, 25, 50, 75, and 95th percentiles are visible. Providing the values for percentiles in tables would also facilitate comparisons.

The authors would like to finish this chapter by saying that making overviews of other people's work is always very hard and apologising for any wrong information that the chapter might contain. In addition, we would like to ask readers to send us an email reporting new articles in the review subfields and any mistakes that might be found.

Acknowledgements This work is supported by: European Investment Funds by FEDER/COMPETE/POCI – Operacional Competitiveness and Internacionalization Programme, under Project POCI-01-0145-FEDER-006958, National Funds by Portuguese Foundation for Science and Technology (FCT), under the project UID/AGR/04033/2013 and Project VITINOV - PA 52306 - financed by "Fundo Europeu Agrícola de Desenvolvimento Rural" (FEADER) and by the Portuguese State through "Medida 4.1. - Cooperação para a Inovação do programa PRODER - Programa de Desenvolvimento Rural". Armando Fernandes acknowledges a post doctoral grant with number SFRH/BPD/108060/2015 from Portuguese Foundation for Science and Technology (FCT), financed by the Social European Fund and the Portuguese Ministry of Education and Science.

References

1. Alpaydin, E.: Introduction to Machine Learning, 2nd edn. MIT Press, Cambridge (2010)
2. Arana, I., Jaren, C., Arazuri, S.: Maturity, variety and origin determination in white grapes (Vitis vinifera L.) using near infrared reflectance technology. J. Infrared Spectrosc. **13**, 349–357 (2005)
3. Bisson, L.: In Search of Optimal Grape Maturity. Pract, Winery Vineyard (2001)
4. Cao, F., Wu, D., He, Y.: Soluble solids content and pH prediction and varieties discrimination of grapes based on visible-near infrared spectroscopy. Comput. Electron. Agric. **71**, S15–S18 (2010). doi:10.1016/j.compag.2009.05.011
5. Cozzolino, D., Esler, M., Dambergs, R., et al.: Prediction of colour and pH in grapes using a diode array spectrophotometer (400–1100 nm). J. Infrared Spectrosc. **12**, 105–111 (2004). doi:10.1255/jnirs.414
6. Cozzolino, D., Cynkar, W., Janik, L., et al.: Prediction of total anthocyanins in individual grape berries using visible and near infrared spectroscopy. In: Proceedings Twelfth Australian Wine Industry Technical Conference, pp 24–29 (2004)
7. Cristianini, N.: An Introduction To Support Vector Machines: And Other Kernel-based Learning Methods. Cambridge University Press, New York (2000)
8. Dambergs, R., Gishen, M., Cozzolino, D.: A review of the state of the art, limitations, and perspectives of infrared spectroscopy for the analysis of wine grapes, must, and grapevine tissue. Appl. Spectrosc. Rev. **50**, 261–278 (2015). doi:10.1080/05704928.2014.966380
9. Diago, M.P., Fernandes, A.M., Millan, B., et al.: Identification of grapevine varieties using leaf spectroscopy and partial least squares. Comput. Electron. Agric. **99**, 7–13 (2013). doi:10.1016/j.compag.2013.08.021
10. Fernandes, A.M., Oliveira, P., Moura, J.P., et al.: Determination of anthocyanin concentration in whole grape skins using hyperspectral imaging and adaptive boosting neural networks. J. Food Eng. **105**, 216–226 (2011). doi:10.1016/j.jfoodeng.2011.02.018
11. Fernandes, A.M., Melo-Pinto, P., Millan, B., et al.: Automatic discrimination of grapevine (Vitis vinifera L.) clones using leaf hyperspectral imaging and partial least squares. J. Agric. Sci. **153**, 455–465 (2015). doi:10.1017/S0021859614000252

12. Fernandes, A.M., Franco, C., Mendes-Ferreira, A., et al.: Brix, pH and anthocyanin content determination in whole port wine grape berries by hyperspectral imaging and neural networks. Comput. Electron Agric. **115**, 88–96 (2015)
13. Galet, P.: A Practical Ampelography: Grapevine Identification. Comstock, New York (1979)
14. Geraudie, V., Roger, J.M., Ferrandis, J.L., et al.: A Revolutionary Device for Predicting Grape Maturity Based on NIR Spectrometry (2009)
15. Goldfarb, A.: Don't Call 'Em Crushers. Wines Vines (2008)
16. Gomes, V.M., Fernandes, A.M., Faia, A., Melo-Pinto, P.: Comparison of different approaches for the Prediction of Sugar Content in Whole Port Wine Grape Berries using Hyperspectral Imaging. In: ENBIS 14 - 14th Annual Conference of the European Network for Business and Industrial Statistics (2014)
17. Gomes, V.M., Fernandes, A.M., Faia, A., Melo-Pinto, P.: Determination of sugar content in whole Port Wine grape berries combining hyperspectral imaging with neural networks methodologies. In: CIES 2014 IEEE Symposium on Computational Intelligence Engineering Solution, pp 188–193. IEEE, New York (2014)
18. Gowen, A.A., O'Donnell, C.P., Cullen, P.J., et al.: Hyperspectral imaging - an emerging process analytical tool for food quality and safety control. Trends Food Sci. Technol. **18**, 590–598 (2007). doi:10.1016/j.tifs.2007.06.001
19. Gutiérrez, S., Tardaguila, J., Fernández-Novales, J., Diago, M.P.: Support vector machine and artificial neural network models for the classification of grapevine varieties using a portable NIR spectrophotometer. PloS One **10**, e0143197 (2015)
20. Gutiérrez, S., Tardaguila, J., Fernández-Novales, J., Diago, M.P.: Data mining and NIR spectroscopy in viticulture: applications for plant phenotyping under field conditions. Sensors **16**, 236 (2016)
21. Gutiérrez, S., Tardaguila, J., Fernández-Novales, J., Diago, M.P.: Data mining and non-invasive proximal sensing for precision viticulture. In: Proceedings 2nd International Electronic Conference Sensors Application (2015). doi:10.3390/ecsa-2-S2003
22. Haykin, S.: Neural Networks: A Comprehensive Foundation. Prentice Hall, USA (1999)
23. Herrera, J., Guesalaga, A., Agosin, E.: Shortwave-near infrared spectroscopy for non-destructive determination of maturity of wine grapes. Meas. Sci. Technol. **14**, 689 (2003)
24. Huang, H., Liu, L., Ngadi, M.O.: Recent developments in hyperspectral imaging for assessment of food quality and safety. Sensors **14**, 7248–7276 (2014)
25. Lacar, F.M., Lewis, M.M., Grierson, I.T.: Use of hyperspectral reflectance for discrimination between grape varieties. In: IEEE 2001 International IEEE Geoscience and Remote Sensing Symposium IGARSS01, pp. 2878–2880 (2001)
26. Larrain, M., Guesalaga, A.R., Agosin, E.: A multipurpose portable instrument for determining ripeness in wine grapes using NIR spectroscopy. IEEE Trans. Instrum. Meas. **57**, 294–302 (2008). doi:10.1109/TIM.2007.910098
27. Melo-Pinto, P., Fernandes, A.M., Tardaguila, J., et al.: Processo de Identificação Ultrarrápida, Destrutiva ou Não Destrutiva e Amiga do Ambiente de Clones de Plantas Utilizando Espectroscopia, Análise Multivariada ou Métodos de Inteligência Artificial. Patent PT106253 (2012)
28. Montgomery, D.C.: Introduction to Linear Regression Analysis, 5th edn. Wiley, Hoboken (2012)
29. Noguerol-Pato, R., González-Barreiro, C., Cancho-Grande, B., et al.: Floral, spicy and herbaceous active odorants in Gran Negro grapes from shoulders and tips into the cluster, and comparison with Brancellao and Mouratón varieties. Food Chem. **135**, 2771–2782 (2012). doi:10.1016/j.foodchem.2012.06.104
30. Noguerol-Pato, R., González-Barreiro, C., Cancho-Grande, B., et al.: Aroma potential of Brancellao grapes from different cluster positions. Food Chem. **132**, 112–124 (2012). doi:10.1016/j.foodchem.2011.10.042
31. Noguerol-Pato, R., González-Barreiro, C., Simal-Gándara, J., et al.: Active odorants in Mouratón grapes from shoulders and tips into the bunch. Food Chem. **133**, 1362–1372 (2012). doi:10.1016/j.foodchem.2012.01.113

32. Sun, D.-W.: Computer Vision Technology for Food Quality Evaluation. Elsevier, Amsterdam (2008)
33. Sun, D.-W.: Modern Techniques for Food Authentication, 1st edn. Elsevier, Amsterdam (2008)
34. Sun, D.-W.: Hyperspectral Imaging for Food Quality Analysis and Control, 1st edn. Elsevier, London (2010)
35. Tarter, M.E., Keuter, S.E.: Effect of rachis position on size and maturity of Cabernet Sauvignon berries. Am. J. Enol. Vitic. **56**, 86–89 (2005)
36. Varma, S., Simon, R.: Bias in error estimation when using cross-validation for model selection. BMC Bioinform. **7**, 91 (2006)
37. Wang, H., Peng, J., Xie, C., et al.: Fruit quality evaluation using spectroscopy technology: a review. Sensors **15**, 11889–11927 (2015)
38. Wold, S., Sjöström, M., Eriksson, L.: PLS-regression: a basic tool of chemometrics. Chemom. Intell. Lab. Syst. **58**, 109–130 (2001)
39. Yang, H.Q., Luo, W.Q., Wang, W.J.: Nondestructive discrimination of grape seed varieties using UV-VIS-NIR spectroscopy and chemometrics. In: Applied Mechanics Material, pp 89–94. Trans Tech, Switzerland (2012)

Consumer Segmentation Through Multi-instance Clustering Time-Series Energy Data from Smart Meters

Alejandro Gómez-Boix, Leticia Arco and Ann Nowé

Abstract With the rollout of smart metering infrastructure at large scale, demand-response programs may now be tailored based on consumption and production patterns mined from sensed data. In previous works, groups of similar energy consumption profiles were obtained. But, discovering typical consumption profiles is not enough, it is also important to reveal various preferences, behaviors and characteristics of individual consumers. However, the current approaches cannot determine clusters of similar consumer or prosumer households. To tackle this issue, we propose to model the consumer clustering problem as a multi-instance clustering problem and we apply a multi-instance clustering algorithm to solve it. We model a consumer as a bag and each bag consists of instances, where each instance will represent a day or a month of consumption. Internal indices were used for evaluating our clustering process. The obtained results are general applicable, and will be useful in a general business analytics context.

1 Introduction

On the way towards a low-carbon future, electricity networks are considered as enablers and one of the critical areas to be studied under the Strategic Energy Technologies Plan. The first European Electricity Grid Initiative (EEGI) Roadmap 2010–2018 was approved by the European Commission and the Member States alongside the creation of EEGI in June 2010 [22]. The EEGI Roadmap defines the

A. Gómez-Boix (✉) · L. Arco (✉)
Department of Computer Science, Central University of Las Villas,
Santa Clara, Cuba
e-mail: aggboix@gmail.com

L. Arco
e-mail: leticiaa@uclv.edu.cu

A. Nowé
COMO Lab, Vrije Universiteit Brussel, Brussel, Belgium

C. Cruz Corona (ed.), *Soft Computing for Sustainability Science*,
Studies in Fuzziness and Soft Computing 358, DOI 10.1007/978-3-319-62359-7_6

research, development and demonstration challenges that both European transmission and distribution system operators should address in the next years with the aim to face the requirements linked to the evolution of power systems and to respond to different external factors. For this reason, smart-grid projects are receiving a lot of attention [11, 25, 27, 36]. New perspectives emerge for energy management. A large amount of smart-meters and sensors are being deployed and they results in a new data deluge we will have to face. With the rollout of smart metering infrastructure at scale, demand-response programs may now be tailored based on users' consumption and production patterns as mined from sensed data.

In [5], a general methodology and a specific two-level clustering approach were introduced for obtaining groups of similar energy consumption profiles. Thus, characteristic load and production profiles per time period can be determined by the user (a time period could be a week, a month or a year). This research was conducted as part of the SCANERGY project [36] which took into account the changing role of dwellings, neighborhoods and cities, concerning the electricity consumption, production, distribution and storage. Discovering consumption profiles is not enough, it is also important reveal various preferences, behaviors and characteristics of individual consumers. Nevertheless, this approach cannot determine clusters of similar consumer or prosumer[1] households. To tackle this issue, in this chapter we will model the consumer clustering problem as a multi-instance clustering problem and we will apply a multi-instance clustering algorithm to solve it. We will model a consumer as a bag and each bag consists of some instances, where each instance will represent a day or a month of consumption. Thus, we will discover groups of energy consumers according to the similarity of their daily and monthly consumption.

In next section, we describe the problem to be solved in more detail. In Sect. 3 we will survey previous work in multi-instance objects. In Sect. 4 we introduce a general approach for applying multi-instance clustering algorithms based on distances between bags. Section 5 contains the results of our approach on smart meter data and Sect. 6 concludes with a summary and directions for further works.

2 Cluster Analyses of Smart Meter Data

Smart meter data are time series; which makes the analysis quite complex. For that reason, cluster analysis of consumption data has been explored in some papers [6, 8, 23, 26, 33, 34, 41, 50], not so much the clustering of production data. From now on we will refer to consumption data clustering approaches; however, all proposals are applicable to production data clustering as well. While some authors have been working on grouping consumers considering the similarity among time series models, such as ARMA and ARIMA [10]; others have been focusing on grouping

[1] It is called prosumer to those consumers who have installed solar panels and therefore, also they produce energy that can consume or put on the power grid; thus they produced and consumed, hence the name prosumer.

consumers considering the time series as feature vectors. In literature four approaches are proposed to cope with the feature vector definition:

1. Consider features as interval consumption measurements (e.g., each 15 min) [13, 24].
2. Only use global features (e.g., mean and standard deviation of an overall day) for characterizing each consumer [33, 45].
3. Extend the time series data by additional global features or other external measures [6].
4. Create local patterns for characterizing the time series [16, 34].

The first one follows a raw-data-based approach, the last two follow a feature-based approach and the third one considers an extension of the raw data including other features.

The definition of a distance measure between time series is necessary in all four approaches [28], deciding on the appropriate distance measure depends on the clustering objective, which can be similarity in time, similarity in shape or similarity in change [54]. In this research we are interested in time series clustering where the main clustering objective is similarity in time, because we need to cluster together consumers with similar energy consumption series, which vary in a similar way at each time interval. For this reason, in the first approach it is necessary to define a distance measure based on the specific characteristics of time series data. Secondly, the arithmetic means of the single time segments are the starting point for the formation of global consumer behavior, but global features only do not properly represent the customers' behavior. Thus, the second approach is not enough to segment the customers, and make groups of households with similar consumption patterns and determine on the fly the cluster membership of a given load curve. In the third approach, the dimensionality of the time series is increased and it could be difficult to manage different kind of features, global and local in the same clustering process. Finally, the last approach could be useful for detecting clusters with similar load profiles, but it could be depending on the homogeneity of the data from the global feature point of view.

As we pointed out, the above approaches have some advantages and disadvantages. Thus, some authors prefer to develop hybrid approaches for time series clustering in order to solve the above disadvantages [1, 32, 35, 37, 42, 54]. There are several reasons for developing hybrid time series clustering models [32]. For instance, we might obtain very different clustering results for the same time series dataset when different time granules are considered. For time series clustering, dimensionality reduction methods are often applied to reduce data dimension before clustering. Consequently, the information of subsequence may be overlooked. Therefore this might result in different clustering results after considering the subsequence information. Some conventional clustering methods require prior information and domain knowledge; others do not require prior information but are to computationally expensive to be applied on very large data sets. The combination of clustering methods can mitigate the disadvantages of some and enhance the benefits of others. For some applications, the clustering objective might not be that apparent. The selection of the time series representation and the similarity measure depends on the clustering objective. Thus, different clustering approaches are required.

Some hybrid clustering methods are proposed in the area of clustering analysis of smart metering data [2, 23, 45, 46]. Most of them apply Self-Organizing Maps (SOM) [29] in the first level and k-means [38] or hierarchical clustering algorithms [46] in the second level. SOM is used to obtain a reduction of the dimension of the initial dataset and k-means is used to group the weight vectors of the units of SOM and the final clusters are obtained [23, 45, 46]. Another approach applies k-means first and uses spectral clustering to segment a collection into classes of similar statistical properties [2]. These hybrid approaches only exploit the combination of clustering methods in order to mitigate the disadvantages of ones and enhance the benefits of other. However, they do not exploit other important reasons for developing hybrid time series clustering models.

A two-level clustering methodology for smart metering data was proposed in [5]. The proposed methodology was used through the application of the two-level clustering approach to Belgian energy consumption and production data. Daily consumption and production profiles were obtained, considering global features such as total daily consumption or production, and local features such as hourly consumption or production, respectively. Whereas, prototypical consumption and production profiles were discovered considering global features such as total yearly consumption or production, and local features such as daily consumption or production, respectively. Thus, the proposed methodology allows obtaining different clustering results using different time granules. Moreover it allows considering different clustering objectives. In the first level, only some general statistics about the data are required; while in the second level, the main objective is similarity in time for identifying the consumption or production profiles. The approach also allows for dimensionality reduction via feature extraction in the first level; doing so the clustering algorithm becomes more efficient. It extracts a set of measures from the original time series; as such it is possible to obtain good results using the Euclidean distance, whereas this measure cannot handle the original time series directly. It uses a finite set of statistical measures to capture the global and local nature of the time series; thus, the computational efficiency of the clustering algorithms can be improved and use of more advanced clustering algorithms becomes possible. Finally, it allows obtaining clusters with homogeneous consumption or production levels in the first level, and clusters with the same profile in the second level. Nevertheless, this approach cannot reveal various preferences, behaviors and characteristics of individual consumers considering their daily consumption profiles.

To the best of our knowledge, all approaches obtained are still in a research phase, especially when it comes to clustering methods for obtaining groups of consumers considering the consumption profile; because of if we represent a consumer considering a large consumption series, the dimensionality will be high and it will influence the distance measure results between consumer, and consequently, the quality of the applied clustering method. For that reason, in this paper, we propose to apply the multi-instance clustering techniques for solving this problem.

3 Multi-instance Clustering Approaches

In modern data mining applications, the complexity of analyzed data objects is increasing rapidly. Molecules are analyzed more precisely and with respect to all of their possible spatial conformations [17]. Earth observation satellites are able to take images with higher resolutions and in a variety of spectra which was not possible some years before. Data mining started to analyze complete websites instead of single documents [31]. All of these application domains are examples for which the complexity demands a richer object representation than single feature vectors. Thus, for these application domains, an object is often described as a set of feature vectors or a multi-instance (MI) object. For example, a molecule can be represented by a set of feature vectors where each vector describes one spatial conformation or a website can be analyzed as a set of word vectors corresponding to its HTML documents. With the rollout of smart metering infrastructure at scale, it is necessary to discover the users' consumption and production patterns. Thus, consumers should be analyzed more precisely and with respect to all of their daily or monthly consumption profiles. For that reason, a consumer (or prosumer) can be represented by a set of feature vectors where each vector describes a consumption profile for a specific period.

As a result the research community started to develop techniques for multi-instance learning that where capable to analyze multi-instance objects. In [17] was proposed the notion of multi-instance learning, where the training set is composed of many bags each containing many instances. There are many multi-instance learning algorithm for supervised classification and prediction problems [9, 14, 17, 52, 56]. Unsupervised multi-instance learning may help the solution of the consumer clustering considering their energy consumption profiles. However, unsupervised multi-instance learning algorithms where bags are without labels have not been extensively studied and consumer clustering has not been treated with the multi-instances clustering approach.

Note that at the first glance, multi-instance clustering may be regarded as merely a simple extension of traditional clustering task where the objects to be clustered are now sets of instances instead of single instances. However, the task of clustering multi-instance bags has its own characteristics. Specifically, in order to cluster objects described by sets of instances, the most intuitive strategy is to let the instances contained in one set contribute equally to the clustering process. While for multi-instance clustering, this kind of strategy may not be appropriate since the instances comprising the bag usually exhibit different functionalities.

Therefore, although in multi-instance clustering the labels of the bags are missing, the bags should not be regarded as simple collections of independent instances while the idiosyncrasies and relationships of the instances in the bags should be carefully investigated.

The first multi-instance clustering algorithm is named BAMIC (BAg-level Multi-Instance Clustering) [55]. Concretely, BAMIC tries to partition the unlabeled training bags into k disjoint groups of bags, where several forms of Hausdorff metric [3] is utilized to measure distances between bags with which the popular k-Medoids algorithm is adapted to fulfill the clustering task. Experimental results show that BAMIC could effectively reveal the inherent structure of the multi-instance data set. In [55], the authors regard bags as atomic data item and use some distance metric to measure the distances between bags. Then, they adapt the k-medoids algorithm to cluster bags. Their method is efficient in some applications. In [53], the authors solve the multi-instance clustering problem in a different way. They formulate a novel Maximum Margin Multiple Instance Clustering (M3IC) problem based on Maximum Margin Clustering (MMC) [51]. The new formulation aims at finding desired hyperplanes that maximize the margin differences on at least one instance per bag in an unsupervised way. But the formulation of M3IC is a nonconvex optimization problem, and they could not solve it directly. Therefore, the original M3IC problem was relaxed and a new method named M3IC-MBM was proposed, which is a combination of Constrained Concave-Convex Procedure (CCCP) and Cutting Plane methods, to solve the relaxed optimization task.

BAMIC uses a distance measure for comparing bags. A problem of this approach is that the selection of a meaningful distance measure has an important impact of the resulting clustering. For that reason, other multi-instance clustering approach has been proposed in [30]. In [30] the authors introduced an algorithm for clustering multi-instance objects that optimizes probability distributions to describe the data set. Part of this work is based on expectation maximization (EM) clustering for ordinary feature vectors using Gaussians. This approach models instances as members of concepts in some underlying feature space. Each concept is modeled by a statistical process in this feature space. A multi-instance object is considered as the result of selecting several times a concept and generating an instance with the corresponding process. Clusters of multi-instance objects are described as multinomial distributions over the concepts. In other words, different clusters are described by having different probabilities for the underlying concepts. The proposed method is formalized in three steps. In the first step a mixture model describing concepts in the instance space is derived. The second step finds a good initialization for the target distribution by subsuming each multi-instance object by a so-called confidence summary vector and afterwards clustering these vectors using the k-means method. In the final step, a final EM clustering step optimizing the distribution for each cluster of multi-instance objects is employed.

Although, as claimed by [44], defining distances between bags in an unsupervised way may not reflect their actual content differences, and the calculation of the distances between bags is quite time consuming, since it always needs to calculate all the pairwise distances between instances in different bags, this approach has interesting properties for clustering consumers considering their consumption profiles. As our objective is clustering consumers by similarity in time [54]; it is good to compare bags by comparing their instances. The distance measures used by BAMIC compare bags by considering their individual consumptions, thus contributing to cluster with respect to similarity in time.

4 Multi-instance Clustering Approach for Smart Metering Data

We introduce a general approach for applying multi-instance clustering algorithms based on distances between bags, in order to grouping consumers from smart metering data. The general schema of the proposed methodology is shown in Fig. 1. General schema of the proposed methodology. The proposed approach consists of the following stages:

- **Stage 1**: **Data gathering**. Read time series; e.g., the time intervals of interest for data representation are typical 1 min, 15 min or 1 h in the context of smart metering applications [15].

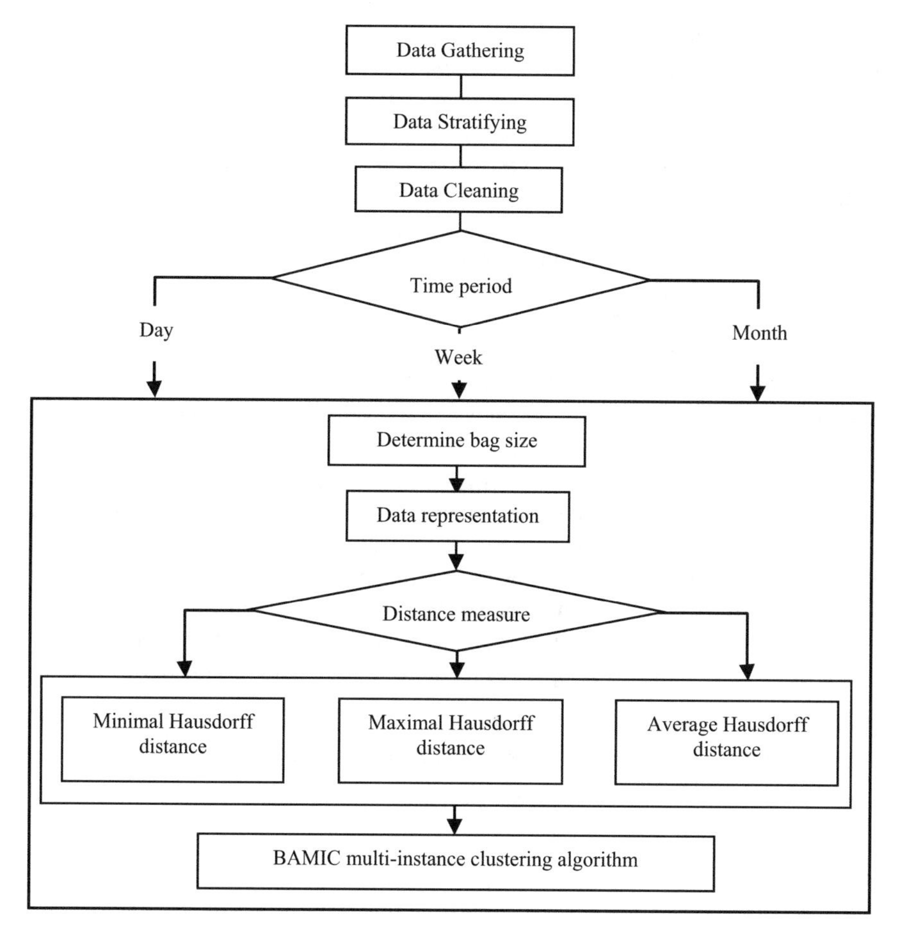

Fig. 1 General schema of the proposed methodology

- **Stage 2**: **Data stratifying**. Segment the data, which separates the raw data sets in-to more homogeneous subsets, in order to sustain scalability; e.g., data can be stratified using a split between weekend and weekdays, or between summer and winter months when we are working with smart metering data [13, 24].
- **Stage 3**: **Data cleaning**. Detect and remove errors and inconsistencies from data in order to improve the data quality. In the smart metering domain some strategies can discard data sets showing more than one hour of recording gaps [24]; check for inconsistencies in the data and remove outliers [23]; detect missing values and replace them using regression techniques [23]; remove special days (e.g., public holidays) [13]; remove non-continuous data [39].
- **Stage 4**: **Determine bag size**. The size of the bags and instances depends on the time period that we need to consider and the granularity to take in to account in the cluster analysis. For instance, it is possible to define bags which include the energy consumption data of a summer month where each instance represents a whole week of this month.
- **Stage 5**: **Data representation**. Represent data considering the time period and the bag granularity. We suggest using raw-data-based representation because our clustering objective is similarity in time. For example, each bag might represent a consumption month and each instance represents the daily consumption in this month. For this example, each bag will have 30 instances and each instance will be represented by 96 features (if we consider 15 min interval consumption during a day). It is useful to apply dimensionality reduction and normalization methods over data representation.
- **Stage 6**: **Determine distance measure**. It is necessary to specify the distance metric to be used to calculate distances between bags, which could take the form of maximal, minimal or average Hausdorff distance considering the approaches proposed in [55]. Our approach does not impose any restriction about the distance metric to be applied. The selection of the distance metric depends on the clustering objective, which can be: similarity in time, in shape or in change [54]. The objective of this paper is clustering taking into account similarity in time, due to it is necessary to cluster consumers with similar consumption profiles. However, it is possible to apply other distances for comparing bags.
- **Stage 7**: **Apply multi-instance clustering**. We suggest applying BAMIC algorithm because it is an algorithm based on distances between bags; for that reason, it is more appropriate to compare the time series that make up the bags.

If we follow this approach, it is possible to obtain clusters of consumers with similar consumption profiles. This approach is applicable to grouping prosumer either considering consumption or production profiles.

5 A Case of Study on Belgian Data

In Belgium the authority on energy policy is shared between the federal and the regional administrations. The responsible authority for the smart metering roll-out in Flanders is the regional energy regulator, VREG, while there are two Distribution System Operators (DSO): Eandis and Infrax [21, 47].

There are few studies on profile identification and consumer segmentation in Belgium [4, 5, 20]. The oldest study [20] was based on consumption data containing hourly consumption values from substations within the Belgian grid. The typical daily profile for each consumer is first identified, and, after that, the k-means algorithm is applied for capturing the different profiles. A large dataset of over 1300 load profiles of residential customers forms the basis for modeling in the study accomplished in [4]. Each load profile is a sequence of measured data, with a resolution of 15 min, over duration of one year. A multiway spectral clustering without the use of pre-modeling steps was used to detect consumer profiles. And, as we mention before, in [5], a general methodology and a specific two-level clustering approach were introduced for obtaining groups of similar energy consumption profiles. But, discovering consumption profiles is not enough to reveal preferences, behaviors and characteristics of individual consumers.

In the research presented in this paper, we use real-life data, provided by Eandis.[2] Houses in Belgium are connected to the electricity grid of the DSO and they are organized in neighborhoods of different sizes. All houses in a given neighborhood are connected via the low-voltage grid to one substation of the DSO. The dataset is comprised of aggregated 15 min intervals of electricity consumption and production of 2928 homes from 44 substations in Belgium.

Our objective is to apply our multi-instance clustering approach for detecting clusters of consumers considering their behaviors in order to contribute to future decision making problems in this field.

5.1 Consumer Data

The structure of a consumer data is illustrated in Fig. 2. Each row represents a consumption day of one consumer, respectively, and each numbered column represents a 15 min consumption or production interval, respectively. Each consumer has 365 consumption days, consumptions could be selected by an interval of several days, and it can be a trimester, a month or a week. Consumption values are no negative real values, including zero, which indicates that there is no consumption.

The consumption's behavior changes depending on weekends or weekdays. Furthermore, the consumption changes depending of the season of the year, it brings

[2]http://www.eandis.be.

Id	Day	Month	Year	1	2	3	...	96
3186808	1	11	2013	0.025	0.008	0.015	...	0.014
3186808	2	11	2013	0.007	0.019	0.023	...	0.012
3186808	3	11	2013	0.007	0.022	0.023	...	0.009
3186808	4	11	2013	0.01	0.014	0.011	...	0.014
3186808	5	11	2013	0.01	0.007	0.014	...	0.021
3186808	6	11	2013	0.02	0.008	0.015	...	0.023
3186808	7	11	2013	0.008	0.015	0.016	...	0.034
3186808	8	11	2013	0.02	0.017	0.012	...	0.01
3186808	9	11	2013	0.014	0.019	0.019	...	0.014
3186808	10	11	2013	0.013	0.006	0.014	...	0.014
3186808	11	11	2013	0.004	0.014	0.009	...	0.009
3186808	12	11	2013	0.01	0.021	0.016	...	0.015

Fig. 2 Example of consumer data from a specific substation

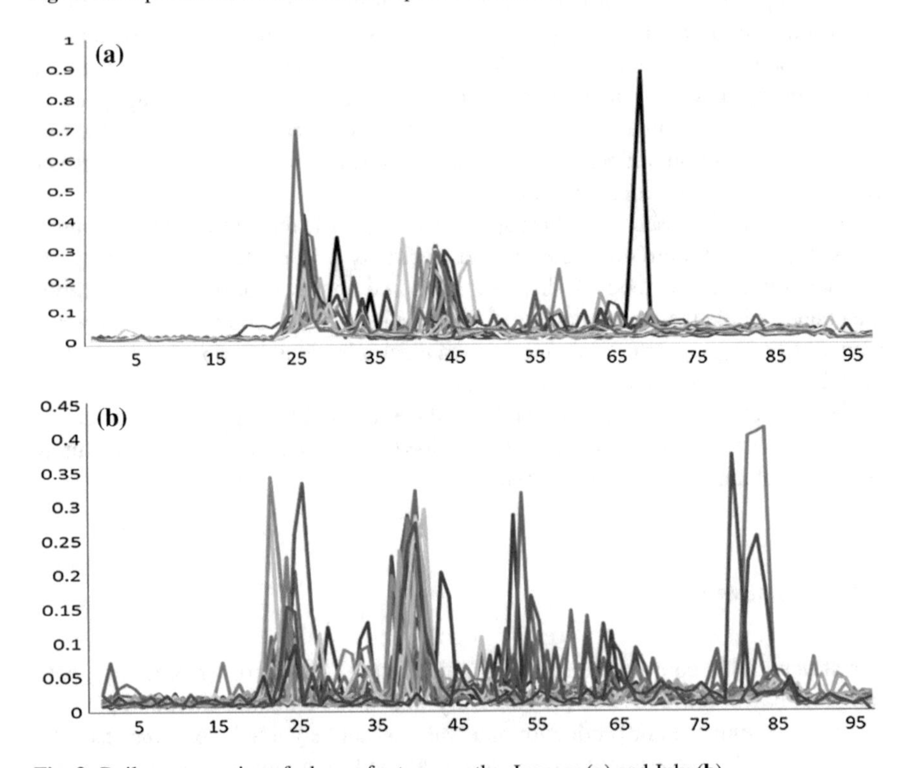

Fig. 3 Daily consumption of a house for two months: January (**a**) and July (**b**)

about that the consumption of the same house behaves differently in winter than in summer. Figure 3 shows the daily consumption of a particular house for two months of different seasons: January and July, respectively. Since there is a lot of variability over the different days it is not possible to identify an overall consumer profile. Thus, we are interested in the detection of consumption profiles, these profiles coincide with the consumer's profiles.

5.2 Experimental Results

We applied the multi-instance clustering approach to the dataset with the electricity consumption of 2928 homes from 44 substations in Belgium, considering the period between November 1st of 2013 and October 31st of 2014.

Discovering clusters of houses according to their behavior is the main objective, taking into account all consumptions of a house. Besides, we want to discover the most typical consumers considering their monthly consumption behavior; for this purpose, an experiment was designed.

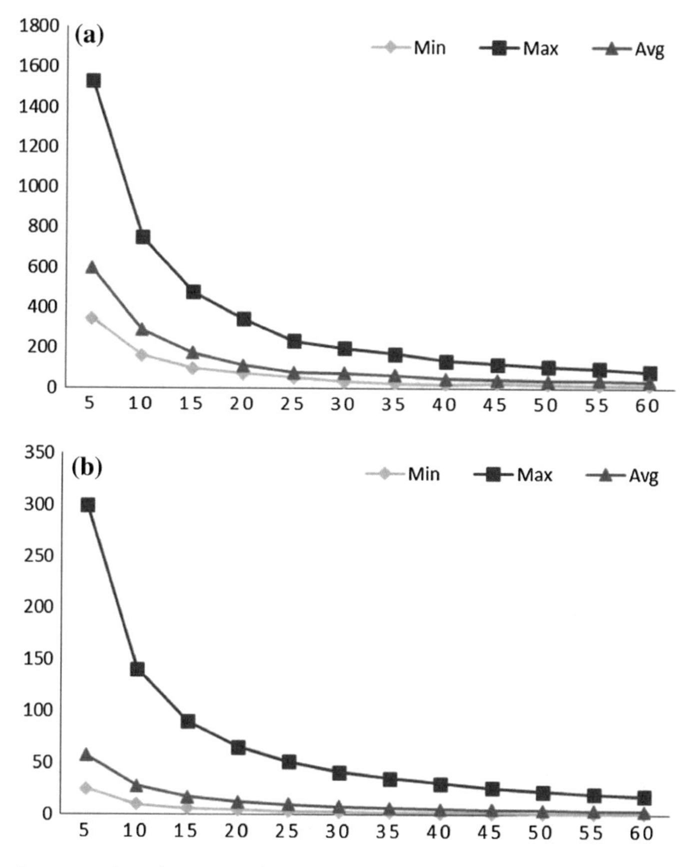

Fig. 4 Performance clustering comparison among forms of Hausdorff metric based on Ball-Hall index, **a** January and **b** July

5.2.1 House's Consumption Profiles

We prepared two datasets by selecting daily consumption of 400 houses, for the period of a month: one dataset with the consumption in January and another with the consumption in July. Thus, our datasets are composed by sets of houses, and each house consists of a set of consumption days. Daily consumption consists of time series with 96 energy consumption intervals.

BAMIC clustering algorithm was applied to both datasets. Three distance functions were used for comparing bags [55]: minimal Hausdorff distance, maximal Hausdorff distance and average Hausdorff distance. The number of clusters was fixed at intervals of five clusters, from five to 60 clusters.

For the evaluation of clustering solutions, we will use the usual clustering validity indices. There are two kinds of validity indices: external indices and internal indices.

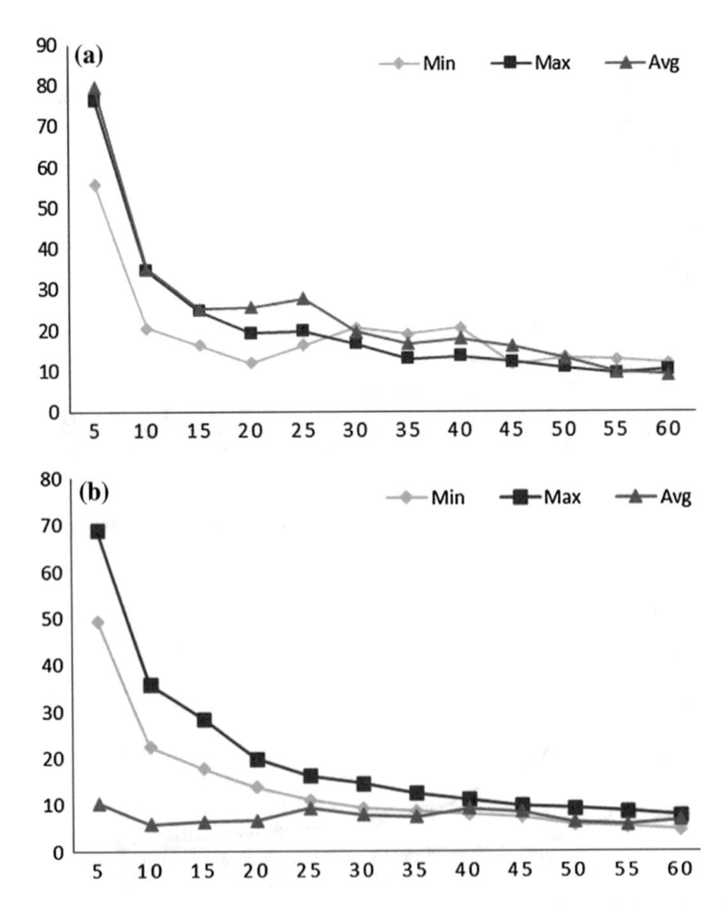

Fig. 5 Performance clustering comparison among forms of Hausdorff metric based on Calinski-Harabasz index, **a** January and **b** July

An external index is a measure of agreement between two partitions where the first partition is the a priori known clustering structure, and the second results from the clustering procedure [18]. Internal indices are used to measure the goodness of a clustering structure without external information [49]. For internal indices, we evaluate the results using quantities and features inherent in the data set. The optimal number of clusters is usually determined based on an internal validity index. The internal indices used for evaluate our clustering process are shown below:

- Ball-Hall index: The mean dispersion of a cluster is the mean of the squared distances of the points of the cluster with respect to their barycenter. The Ball-Hall index [7] is the mean, through all the clusters, of their mean dispersion. The largest difference between levels is used to indicate the optimal solution [40].
- Calinski-Harabasz index: The measures of between-cluster isolation and within-cluster coherence; its maximum value determines the optimal number of clusters [12].

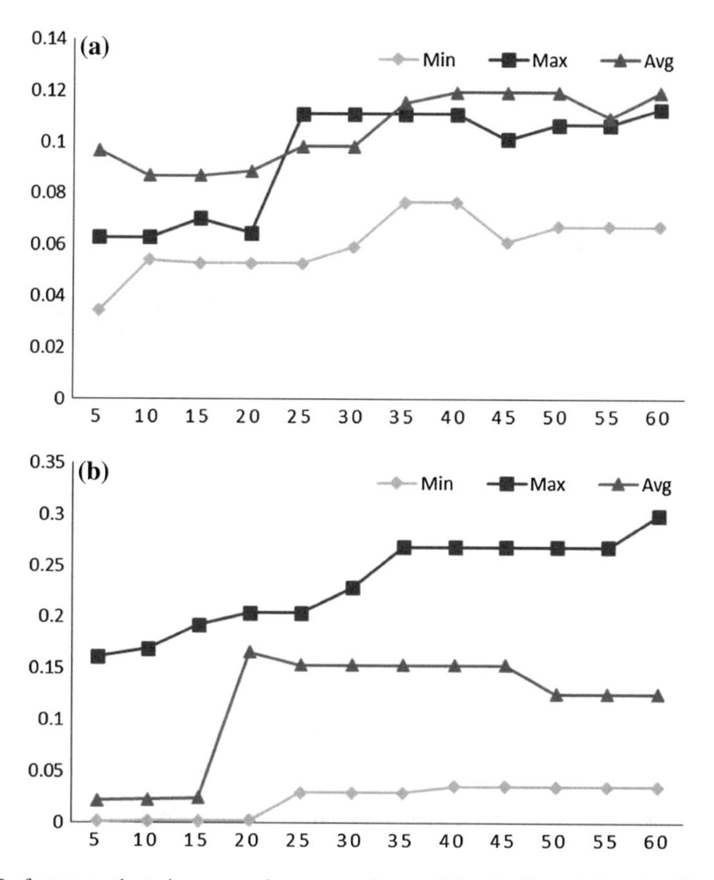

Fig. 6 Performance clustering comparison among forms of Hausdorff metric based on Dunn index, **a** January and **b** July

- Dunn index: A measure that maximizes the inter-cluster distances while minimizing the intra-cluster distances; its large values indicate the presence of compact and well-separated clusters, so the number of clusters that maximizes the index is taken as the optimal number of clusters [19].
- PBM index: The PBM index (acronym constituted of the initials of the names of its authors, Pakhira, Bandyopadhyay and Maulik) is calculated using the distances between the points and their barycenters and the distances between the barycenters themselves. Its maximum value determines the optimal number of clusters [43].
- Silhouette index: The Silhouette coefficient combines both cohesion and separation. An overall measure of the goodness of a clustering can be obtained by computing the average Silhouette coefficient of all points. The coefficient assumes the best results when it is close to one [48].

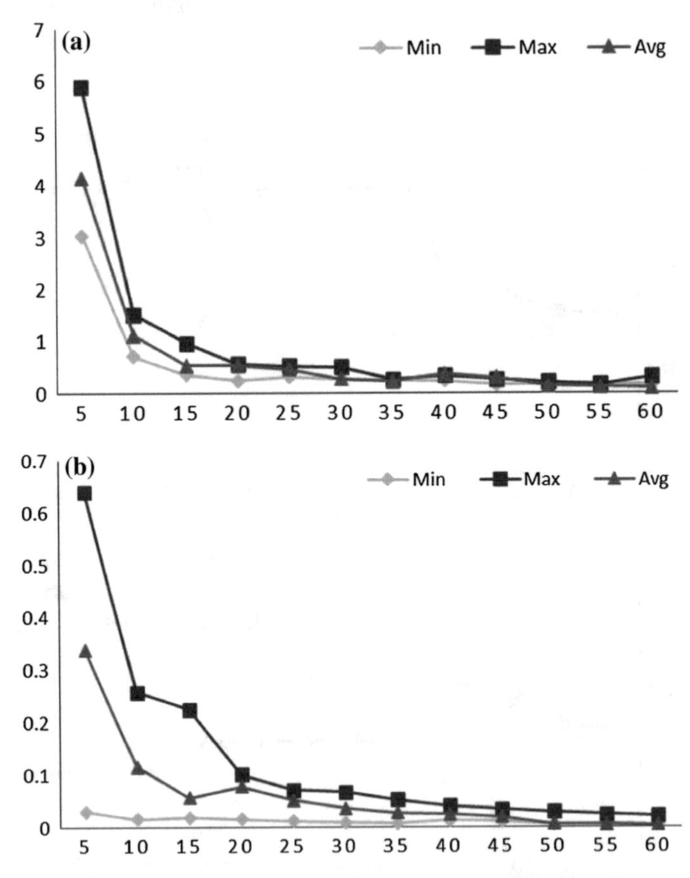

Fig. 7 Performance clustering comparison among forms of Hausdorff metric based on PBM index, **a** January and **b** July

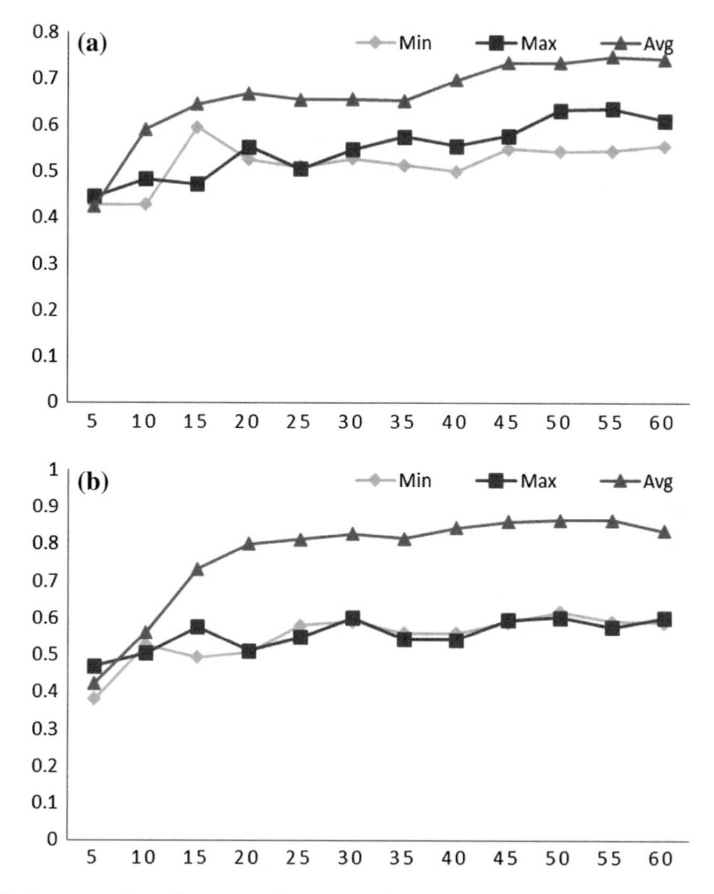

Fig. 8 Performance clustering comparison among forms of Hausdorff metric based on Silhouette index, **a** January and **b** July

Hereinafter, the graphics used to compare the internal indices that evaluate our clustering show the values of BAMIC algorithm with the three forms of Hausdorff metric: minimal Hausdorff distance ("Min"), maximal Hausdorff distance ("Max") and average Hausdorff distance ("Avg"). The x-axis contains the number of clusters and y-axis contains the analyzed validity index.

Figure 4 shows the results for the Ball-Hall index. Our objective is to maximize the value of Ball-Hall index. Both in January and in July, the best values are obtained for maximal Hausdorff distance function, as the best value for the number of clusters k is the one corresponding to the greatest difference between two successive slopes, it is achieved if $k = 15$, but for our purpose, the stability is more important and this is reached from $k = 30$ onwards.

Figure 5 shows the results of the Calinski-Harabasz index, maximum values determine the optimal number of clusters. In January the average Hausdorff distance function has the best performance up to 25 clusters. In July, the maximal Hausdorff

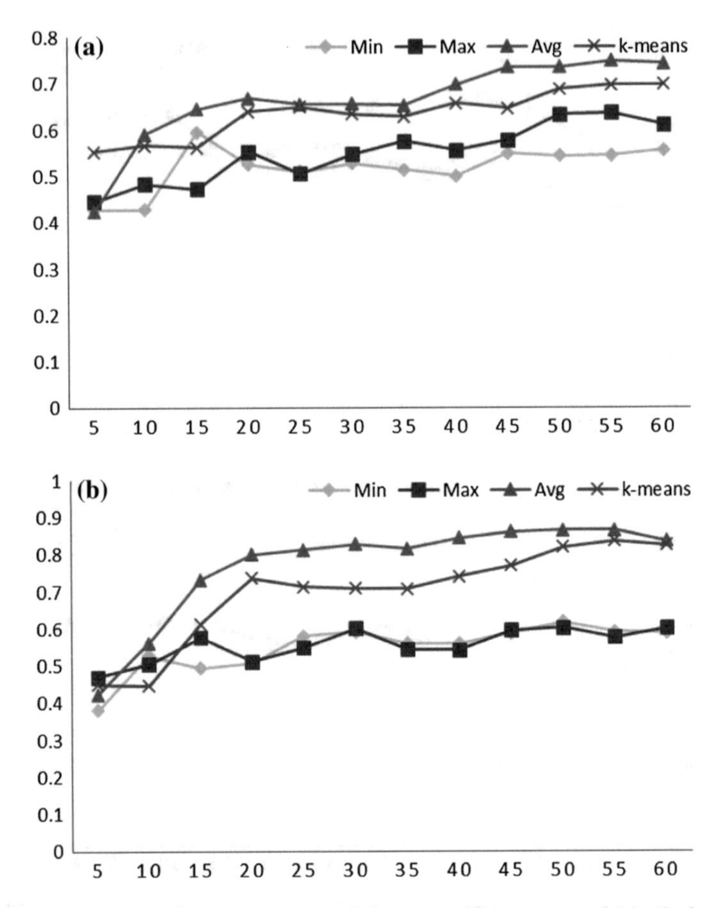

Fig. 9 Performance clustering comparison between multi-instance and classical clustering approaches based on Silhouette index, **a** January and **b** July

distance function has the best performance. In both cases from 30 clusters on, all three distance functions have a very similar behavior.

Figure 6 shows the results for the Dunn index, maximum values determine the presence of compact and well-separated clusters. In January the behavior is not clear, the best results are for average Hausdorff distance function from 40 to 50 clusters.

In July, the maximal Hausdorff distance function has the best performance, the best values are reached from $k = 35$ onwards.

Figure 7 shows the results for the PBM index, maximum values determine the optimal number of clusters. In both cases the maximal Hausdorff distance function has the best performance up to 15 clusters, but from 20 clusters on, all three distance functions have a very similar behavior.

Figures 8 and 6 shows the results for the Silhouette index, maximum values determine the optimal number of clusters. In both months, January and July, average

Hausdorff distance function obtains best results. In both cases best results are from 45 to 55 clusters.

Best values were obtained for the maximal and average Hausdorff distances, average Hausdorff distance obtained best results for the Calinski-Harabasz and Dunn indices in January, and Silhouette index in both months. Although the optimal number of clusters could not be determined exactly, Ball-Hall and Calisnki-Harabanz indices achieve stability in the interval from 30 clusters onwards; similarly PBM index achieves stability from 20 clusters on. The number of clusters with Dunn index is not clear, in the month of January best results are obtained from 40 to 50 clusters, in the month of July best results are obtained for 35 clusters onward. On the other hand, Silhouette index shows stability, in both cases from 45 to 55 clusters. So in this latter case we cannot draw conclusions on the optimal number of clusters.

In order to compare the proposed multi-instance approach with the classical clustering approach, another experiment was made. In this experiment a consumer was represented as a single consumption time series. The series is composed by all the individual consumptions, resulting in 2976 consumptions, for a month period. K-means algorithm was used for discovering the consumer clusters. For most of validity indices, Hausdorff distance metrics have the best results. Figure 9 illustrates that the proposed multi-instance approach obtains best results with respect to the classical clustering approach, comparing them taking into account the Silhouette index.

6 Conclusions and Future Work

The proposed approach based on multi-instance clustering was applied on real-life data of Belgian energy consumption. Clusters of daily consumption profiles were obtained, considering a one month period of consumption data. Months of January and July were selected, due to the variability of consumption behaviors in different seasons of the year.

The proposed approach has the following advantages: It allows obtaining different clustering results using different distance functions to compare sets of elements. Moreover, it allows considering different clustering objectives. Five internal validity indices were used to measure the quality of clustering results and to evaluate our clustering solutions, in most cases, maximal Hausdorff distance metric obtained best results, the optimal number of clusters it is not determined correctness, due to it depends on measured property by internal index applied and the desired granularity of clustering. For instance, Calinski-Harabasz and PBM indices reach best results from five to 15 clusters, however Ball-Hall index reaches best results at 30 clusters and Dunn index, from 40 to 50 clusters. On the other hand, Silhouette index shows stability, in both months from 45 to 55 clusters. Because the decision about the numbers of clusters to obtain depends on the objective of clustering; an analysis of what measures the validity index is required.

One may choose a validity index to estimate an optimal number of clusters, where the optimal clustering solution is found from a series of clustering solutions under different number of clusters. However, finding the best clustering solution for a clustering task depends on not only a validity index but also the appropriate clustering procedure. The clustering method applied has as objective similarity in time, for this reason in further works, it required performing studies having as objective similarity in change or similarity in shape. On the other hand, the proposed methodology uses three different distance functions for comparing bags, these measures cannot reflect differences of bags with respect to their own content, making necessary to find distance functions for bag comparison that reflect structural differences.

References

1. Aghabozorgi, S., Ying Wah, T., Herawan, T., Jalab, H.A., Shaygan, M.A., Jalali, A.: A hybrid algorithm for clustering of time series data based on affinity search technique. Sci. World J. **2014**, 562194 (2014)
2. Albert, A., Rajagopal, R.: Smart meter driven segmentation: what your consumption says about you. IEEE Trans. Power Syst. **28**(4), 4019–4030 (2013)
3. Alkhansari, M.G., Huang, T.S.: Fractal-based image and video coding. Video Coding, pp. 265–303. Springer, Berlin (1996)
4. Alzate, C., Espinoza, M., De Moor, B., Suykens, J.A.K.: Identifying customer profiles in power load time series using spectral clustering. In: Artificial Neural Networks - ICANN, 2009, vol. 5769, pp. 315–324. LNCS (2009)
5. Arco, L., Casas, G., Nowé, A.: Two-level clustering methodology for smart metering data. In Fifth International Workshop on Knowledge Discovery, Knowledge Management and Decision Making, (2015)
6. Ardakanian, O., Koochakzadeh, N., Singh, R.P., Golab, L., Keshav, S.: Computing Electricity Consumption Profiles from Household Smart Meter Data. In: EDBT Workshop on Energy Data Management, pp. 140–147 (2014)
7. Ball, G.H., Hall, D.J.: ISODATA, a novel method of data analysis and pattern classification. DTIC Document (1965)
8. Binh, P.T.T., Ha, N.H., Tuan, T.C., Khoa, L.D.: Determination of representative load curve based on Fuzzy K-Means. In: The 4th International Conference on Power Engineering and Optimization, no. Lc, pp. 281–286 (2010)
9. Blockeel, H., Page, D., Srinivasan, A.: Multi-instance tree learning. In: Proceedings of the 22nd international Conference on Machine Learning, pp. 57–64 (2005)
10. Brockwell, P.J., Davis, R.A.: Introduction to Time Series and Forecasting. Springer Texts in Statistics, 2nd edn. Springer, New York (2002)
11. Brunner, H., De Nigris, M., Gallo, A.D., Herold, I., Hribernik, W., Karg, L., Koivuranta, K., Papič, I., Lopes, J.P., Verboven, P.: Mapping & gap analysis of current European smart grids projects, p. 55, April 2012
12. Caliński, T., Harabasz, J.: A dendrite method for cluster analysis. Commun. Stat. Methods **3**(1), 1–27 (1974)
13. Cao, H.A., Beckel, C., Staake, T.: Are domestic load profiles stable over time? An attempt to identify target households for demand side management campaigns. In: 39th Annual Conference of the IEEE Industrial Electronics Society (IECON 2013), pp. 4733–4738 (2013)
14. Chen, Y., Bi, J., Wang, J.Z.: MILES: multiple-instance learning via embedded instance selection. IEEE Trans. Pattern Anal. Mach. Intell. **28**(12), 1931–1947 (2006)
15. Chicco, G.: Overview and performance assessment of the clustering methods for electrical load pattern grouping. Energy **42**(1), 68–80 (2012)
16. Dent, I., Aickelin, U., Rodden, T.: Application of a clustering framework to UK domestic electricity data. In: UKCI 2011, 161–166 (2011)

17. Dietterich, T.G., Lathrop, R.H., Lozano-Perez, T.: Solving the multiple instance problem with axis-parallel rectangles. Artif. Intell. **1**(89), 31–71 (1997)
18. Dudoit, S., Fridlyand, J.: A prediction-based resampling method for estimating the number of clusters in a dataset. Genome Biol. **3**(7), 1–21 (2002)
19. Dunn, J.C.: Well-separated clusters and optimal fuzzy partitions. J. Cybern. **4**(1), 95–104 (1974)
20. Espinoza, M., Joye, C., Belmans, R., DeMoor, B.: Short-term load forecasting, profile identification, and customer segmentation: a methodology based on periodic time series. IEEE Trans. Power Syst. **20**(3), 1622–1630 (2005)
21. European Commission.: Benchmarking smart metering deployment in the EU-27 with a focus on electricity (2014)
22. European Commission: Cost-benefit analyses & state of play smart metering deployment in the EU-27, pp. 1–8 (2014)
23. Figueiredo, V., Rodrigues, F., Vale, Z., Gouveia, J.B.: An electric energy consumer characterization framework based on data mining techniques. IEEE Trans. Power Syst. **20**(2), 596–602 (2005)
24. Flath, C., Nicolay, D., Conte, T., Van Dinther, C., Filipova-Neumann, L.: Cluster analysis of smart metering data: an implementation in practice. Bus. Inf. Syst. Eng. **4**, 31–39 (2012)
25. Giordano, V., Gangale, F., Jrc-ie, G.F., Sánchez, M., Dg, J., Onyeji, I., Colta, A., Papaioannou, I., Mengolini, A., Alecu, C., Ojala, T., Maschio, I.: Smart grid projects in Europe?: lessons learned and current developments. Europe **2011**, 1–118 (2011)
26. Hossain, J., Kabir, A.N.M.E., Rahman, M., Kabir, B., Islam, R.: Determination of typical load profile of consumers using fuzzy C-means clustering algorithm. Int. J. Soft Comput. Eng. **1**(5), 169–173 (2011)
27. Hübner, M., Prüggler, N.: Smart Grids Initiatives in Europe - Country Snapshots and Country Fact Sheets. Framework, (2011)
28. Iglesias, F., Kastner, W.: Analysis of similarity measures in times series clustering for the discovery of building energy patterns. Energies **6**, 579–597 (2013)
29. Kohonen, T.: Self-organized formation of topologically correct feature maps. Biol. Cybern. **43**(1), 59–69 (1982)
30. Kriegel, H., Pryakhin, A., Schubert, M.: An EM-Approach for Clustering Multi-Instance Objects (2006)
31. Kriegel, H.-P., Schubert, M.: Classification of websites as sets of feature vectors. In: Databases and Applications, pp. 127–132 (2004)
32. Lai, C.-P., Chung, P.-C., Tseng, V.S.: A novel two-level clustering method for time series data analysis. Expert Syst. Appl. **37**(9), 6319–6326 (2010)
33. Lavin, A., Klabjan, D.: Clustering time - series energy data from smart meters. Energy Effic. **8**(4), 681–689 (2014)
34. Lee, T.E., Haben, S.A., Grindrod, P.: Modelling the electricity consumption of small to medium enterprises. In: The 18th European Conference on Mathematics for Industry Conference, pp. 1–7 (2014)
35. Liao, T.W.: A clustering procedure for exploratory mining of vector time series. Pattern Recognit. **40**, 2550–2562 (2007)
36. Losa, I., De Nigris, M., Van, T.: Analysis of the on-going Research and demonstration efforts on smart grids in Europe. 22nd International Conference on Electricity Distribution, pp. 10–13, June 2013
37. Lumpur, K.: Incremental clustering of time-series by fuzzy clustering. J. Inf. Sci. Eng. **688**, 671–688 (2012)
38. MacQueen, J.: Some methods for classification and analysis of multivariate observations. In: Proceedings of the 5th Berkeley Symposium Mathematical Statistics and Probability, vol. 1, pp. 281–297 (1967)
39. McLoughlin, F., Duffy, A., Conlon, M.: Analysing domestic electricity smart metering data using self organising maps. In: Integration of Renewables into the Distribution Grid, CIRED. Workshop, 2012, 1–4 (2012)

40. Milligan, G.W., Cooper, M.C.: An examination of procedures for determining the number of clusters in a data set. Psychometrika **50**(2), 159–179 (1985)
41. Mutanen, A., Ruska, M., Repo, S., Järventausta, P.: Customer classification and load profiling method for distribution systems. IEEE Trans. Power Deliv. **26**, 1755–1763 (2011)
42. Oates, T., Firoiu, L., Cohen, P.R.: Clustering time series with hidden Markov models and dynamic time warping. In: Proceedings of the IJCAI-99 Workshop on Neural, Symbolic and Reinforcement Learning Methods for Sequence Learning, pp. 17–21 (1999)
43. Pakhira, M.K., Bandyopadhyay, S., Maulik, U.: Validity index for crisp and fuzzy clusters. Pattern Recognit. **37**(3), 487–501 (2004)
44. Rahmani, R., Goldman, S.A.: MISSL: multiple-instance semi-supervised learning. In: Proceedings of the 23rd International Conference on Machine Learning, pp. 705–712 (2006)
45. Räsänen, T., Kolehmainen, M.: Feature-based clustering for electricity use time series data. In: 9th International Conference, ICANNGA: 2009, vol. 5495, 401–412. LNCS, (2009)
46. Räsänen, T., Voukantsis, D., Niska, H., Karatzas, K., Kolehmainen, M.: Data-based method for creating electricity use load profiles using large amount of customer-specific hourly measured electricity use data. Appl. Energy **87**, 3538–3545 (2010)
47. Renner, S., Heinemann, C.: European smart metering landscape report. Imprint **2**, 1–168 (2011)
48. Rousseeuw, P.J.: Silhouettes: a graphical aid to the interpretation and validation of cluster analysis. J. Comput. Appl. Math. **20**, 53–65 (1987)
49. Thalamuthu, A., Mukhopadhyay, I., Zheng, X., Tseng, G.C.: Evaluation and comparison of gene clustering methods in microarray analysis. Bioinformatics **22**(19), 2405–2412 (2006)
50. Wijaya, T.K., Ganu, T., Chakraborty, D., Aberer, K., Seetharam, D.P.: Consumer segmentation and knowledge extraction from smart meter and survey data. In: SDM 2014, (2014)
51. Xu, L., Neufeld, J., Larson, B., Schuurmans, D.: Maximum margin clustering. In: Advances in Neural Information Processing Systems, pp. 1537–1544 (2004)
52. Zhang, Q., Goldman, S.A.: EM-DD: An improved multiple-instance learning technique. In: Advances in Neural Information Processing Systems, pp. 1073–1080 (2001)
53. Zhang, D., Wang, F., Si, L., Li, T., Lafayette, W., Science, C., Lafayette, W., Sciences, I.: M3IC: Maximum Margin Multiple Instance Clustering. In: International Joint Conference on Artifical Intelligence, pp. 1339–1344 (2003)
54. Zhang, X., Liu, J., Du, Y., Lv, T.: A novel clustering method on time series data. Expert Syst. Appl. **38**, 11891–11900 (2011)
55. Zhang, M.L., Zhou, Z.H.: Multi-instance clustering with applications to multi-instance prediction. Appl. Intell. **31**(1), 47–68 (2009)
56. Zhou, Z.-H., Sun, Y.-Y., Li, Y.-F.: Multi-instance learning by treating instances as non-iid samples. In: Proceedings of the 26th Annual International Conference on Machine Learning, pp. 1249–1256 (2009)

A Multicriteria Group Decision Model for Ranking Technology Packages in Agriculture

Juan Carlos Leyva López, Pavel Anselmo Álvarez Carrillo and Omar Ahumada Valenzuela

Abstract The problem of ranking a set of technology packages that are best suited for growing crops, is developed with a multicriteria group decision model. The group decision model is based on ELECTRE GD, a group decision method for multicriteria ranking problems, strongly based on ELECTRE III, developed to work on those cases where there is great divergence among the decision-makers. We use a practical case study to show our approach, where a group of decision-makers evaluates among the available technology packages to an agricultural company, in order to select the most appropriate alternative. The proposed model generates an agreed collective solution that aids those decision-makers with different interests, to reach (through an iterative process) an agreement on how to rank the technology packages. The proposed procedure is also based on a preference disaggregation approach for reaching agreement between individuals. To support the proposal of a temporary collective solution, individual inter-criteria parameters are inferred concerning individual and global preference for outranking methods in a feedback process.

Keywords Technology packages · Multicriteria decision analysis · Preference disaggregation analysis · Outranking methods · Multiobjective genetic algorithms · Group decision-making

---

J.C.L. López · P.A. Álvarez Carrillo (✉)
Department of Economic and Management Sciences, University of Occident,
80020 Culiacan, Sinaloa, Mexico
e-mail: pavel.alvarez@udo.mx

O.A. Valenzuela
CONACYT Research Fellow, Management Sciences Doctorate Program,
University of Occident, 80020 Culiacan, Sinaloa, Mexico

J.C.L. López
Autonomous University of Sinaloa, 80010 Culiacan, Sinaloa, Mexico
e-mail: juan.leyva@udo.mx

O.A. Valenzuela
e-mail: omar.ahumada@udo.mx

C. Cruz Corona (ed.), *Soft Computing for Sustainability Science*,
Studies in Fuzziness and Soft Computing 358, DOI 10.1007/978-3-319-62359-7_7

1 Introduction

Crop management is a very important activity for improving profitability of growers, particularly for those that have only one cropping season per year, and it has lately gather a lot of attention from the research community [1]. One aspect often overlooked on these models, is the need for qualifying technology packages, to achieve the goals of agricultural companies in terms of economic, productivity and the environmental criteria.

Given the rapid development of technology, it is harder for farmers to select among the different alternatives of technology packages, in particular when there is more than one decision-maker (DM) evaluating the alternatives.

According to [14], group decision-making is understood as the reduction of different individual preferences in a given set for a single collective preference. Group decision problems involve different DMs. They may diverge in their perception of the problem and have different interests, but all are responsible for the well-being of the organization and share responsibility for the decision implemented. When a situation of decision-making involves multiple actors, each with different values, the final decision is generally the result of an interaction between these individual preferences, in this case some group decision-making approach can be useful [16]. Nevertheless, this process is not conflict-free, and there may be some differences among the DMs, caused by a large number of factors, such as different ethical or ideological beliefs, different specific objectives or different roles within the organization [21, 28].

A previous work presents similar issues [26], since it describes an approach used with multicriteria decision support for aggregating group preferences. Under this method, DMs may exchange opinions and relevant information, but a group consensus is only needed to define a potential set of alternatives. Then, each member defines his/her own criterion, the appropriate evaluations, parameters (weights, thresholds, *etc.*) and a multicriteria method is used to obtain their personal ranking. Thereafter, each criterion of the decision-maker is considered separately, and the information contained in his/her individual preference is aggregated into a final collective order, using this self-same multicriteria decision approach or another one. A similar process can be found in [25].

This chapter presents a multicriteria group decision model for ranking technology packages based on the outranking methods ELECTRE III [28] and ELECTRE GD [21]. The model is based on individual solutions generated by decision-makers with ELECTRE III and a global solution generated with ELECTRE GD representing every decision-maker. This model is appropriate for those cases where there is great divergence among the DMs. However, most group aiding approaches based on Multicriteria Decision Analysis (MCDA) lacks adequate support in parameter redefinition in the feedback step of the agreement group stage. Usually, in this stage, the DM is asked to provide their own different values for weights or other inter-criteria parameters using tools that are not in the group context, e.g., sensibility analysis only generates results related to the individual preferences, since it was not designed for a group context, making the results obtained from this, or similar approaches for parameter recommendations, inappropriate in the context of group preference.

The feedback process is an important step in the multicriteria group decision making methodology, since it uses a procedure based on the aggregation of individual results for supporting DMs to reach a collective result, and usually the DMs must face the situation where the individual results are divergent. This situation is a typical situation of conflict because there is a level of disagreement of the DMs with the collective solution.

If DMs cannot negotiate among themselves, then the conflicts may be solved individually, i.e., DMs should consider being flexible in the definition of some preferences or, as appropriate, redefine some of their parameters again. In this chapter, we illustrate this scenario with a practical application on prioritizing technology packages in order to identify which of them have to be selected. A group of DMs from an agricultural company was supported to reach a consensus solution, modelling it as an instance of the group-ranking problem using a multicriteria group decision aiding methodology based on the outranking approach. In this context, DMs with different interests generated their own ranking of technology packages. Later, a collective ranking of technology packages was generated using ELECTRE GD. Naturally, disagreements were found between the individual rankings and the collective solution.

Due the complexity of the task, a group decision aiding procedure was used to support the DMs to prioritize technology packages regarding the collective and individual prioritization. Operationally, the feedback process was based on an indirect procedure called preference disaggregation analysis (PDA) [13] for a group decision aid approach applied to ranking of a set of technology packages. The approach uses a model for inferring inter-criteria parameters, where the DMs generate individual rankings that exhibit less disagreement with the collective solution [4]. For solving the problem of ranking technology packages, the DMs, with the support of a facilitator-analyst, used a web-based group decision support system named SADGAGE[1] [18]. The system is based on a multicriteria group decision aiding agreement model and has implemented the methods ELECTRE III and ELECTRE GD. The model includes a process for reaching agreement. In this process, the feedback mechanism generates advice on how the DMs should change their preferences to reach a ranking of alternatives with a high degree of agreement. This advice is based on an inferring inter-criteria parameter tool, which identifies particular inter-criteria parameters that contribute more to the agreement level.

2 Agricultural Technological Packages

Since the green revolution of the 1960s farmers have significantly increased the amount of food available for human consumption, such increment, has not only been based in an increment in the land dedicated to agriculture, but mainly from the

[1]The system SADGAGE is hosted at http://mcdss.udo.mx/sadgage.

development of new technology packages. However, such rate of improvement has slowed down in recent years, which is troubling, given that growers need to double crop production over the next 35 years, just to ensure that there is enough safe and nutritious food to feed a rapidly crowding planet [11].

Agricultural technologies are often presented as a package of interrelated technologies for example, high yielding seeds, fertilizer, herbicides, and chemicals. Technologies in fact consist of sub-components that are available individually or jointly as a package [2]. Accordingly, one major focus in the literature in recent years has been the investigation of the decision-making process characterizing choice of the optimal combinations of the components of a technological package over time [9], but empirical evidence suggest that for the most part, farmers choose to adopt inputs sequentially, adopting initially only a part of the package and subsequently adding components over time [17]. Previous studies showed that farmers adopt pieces of the package at first disregarding the advantage of the interaction between components of technological packages [5].

Incremental adoption of technology packages is reasonable, in terms of the uncertain outcome of the new technology. One example shown by authors is the increase of yield by adopting a new seed, may be further improved by applying appropriate levels of fertilizer and irrigation or modifying plant density [17]. The divisibility of the technology package can create a particular adoption dynamic. In the case described in [2], they develop a Bayesian model of adoption dynamics that demonstrates how uncertainty with a technological package with known risk can lead to a sequential adoption pattern in which farmers adopt a single component first.

It seems there is evidence that farmers prefer to have sequential adoption of technologies, even when there is a high interaction between the individual components, due to the risk in terms of investment and the learning curve required by some of the technologies [5]. Adoption studies of technologies for farmers have considered single innovations in isolation ignoring the process of adoption among a set of components of technological package [8], but new developments such as Genetic Modified seed varieties often work in tandem with herbicides that improve yields considerably [30].

An appropriate model to select the most suitable technological package given their particular benefits, both in terms of yield, but more importantly in their return on investment (given the often high cost of new technological improvements) is an important problem that needs to be tackled, to have better decisions and quicker adoption of those technologies that will improve farmers competitiveness profitability [9].

3 Multicriteria Methods

Relevant approaches in multicriteria decision analysis are Analytic Hierarchy Process (AHP) and Attribute (MAUT). AHP was developed by [29] and has been used in complex decision-making problems when the evaluation criteria can be hierarchized in subcriteria. The decision-making process in AHP include four stages; hierarchical structure of the problem, data input, estimating relative importance of evaluation criteria, evaluation of alternatives by combining the relative importance. The strengths of AHP are that intuitive appeal to the decision-makers, decomposes a decision problem into its constituent parts and builds hierarchies of criteria, capture both subjective and objective evaluation measures.

On the other hand, MAUT is a multicriteria decision-making approach that consists in multiplying the score of each alternative and for a criterion, by the weight assigned to that criterion. The values found are added, the selected alternative is the ones that gets highest value from this summation [24]. The strength of MAUT is that it accounts for uncertainty. Some illustrative examples about AHP and MAUT can be found in [24].

Considering the importance of multicriteria methods in dealing with human reality, several approaches have been proposed. All of them replace the search for the optimal solution, to favor solutions with the greatest commitment. They are used to support decision making by individuals and organizations. In this chapter, the multicriteria methods used are ELECTRE III and ELECTRE GD. The first is needed to generate individual solutions (valued outranking relation and ranking of alternatives). The second uses individual solution as input to generate global solution (collective ranking). Both methods are based in outranking relations, for individual and collective solutions, respectively. The shared strengths in these methods have, are the flexibility to express preferences, since transitivity is not required, and that it can handle the incomparability between alternatives.

3.1 ELECTRE III

The ELECTRE III method is based on a pair-wise comparison of the alternatives, leading to valued preference degrees (see Roy 1996, [28]). We briefly show the essence of the ELECTRE III method. Let $A = \{a_1, a_2, ..., a_m\}$ be the set of decision alternatives or potential actions, and assume that there exist defined criteria $g_k,\ k = 1, 2, ..., r$. For each pair of alternatives $(a_i, a_j) \in AxA$, there are a concordance index $C(a_i, a_j)$ and a discordance index $d_k(a_i, a_j)$. In the concordance index, we have, in a manner of speaking, a measure of the extent to which we are in harmony with the assertion that a_i is at least as good as a_j, whilst in the discordance index we take in account the discordance associated with this assertion. The model building phase combines these two measures to produce a measure of the degree of outranking, that is, a credibility index $\sigma(a_i, a_j)$ $(0 \leq \sigma(a_i, a_j) \leq 1)$ which assesses the strength of the assertion that "a_i is at least as good as a_j, $a_i S a_j$." The credibility degree for each pair $(a_i, a_j) \in AxA$ is defined as follows:

$$\sigma(a_i, a_j) = \begin{cases} C(a_i, a_j),\ if\ K(a_i, a_j) = \phi \\ C(a_i, a_j) \bullet \prod_{k \in K(a_i, a_j)} \frac{1 - d_k(a_i, a_j)}{1 - C(a_i, a_j)}\ K(a_i, a_j) \neq \phi \end{cases} \quad (1)$$

where $K(a_i, a_j)$is the set of criteria such that $d_k(a_i, a_j) > C(a_i, a_j)$.

This formula assumes that if the strength of the concordance exceeds that of the discordance, then the concordance value should not be modified. Otherwise, we are forced to question the assertion that $a_i S a_j$ and modify $C(a_i, a_j)$ according to the above equation. If the discordance is 1 for any $(a_i, a_j) \in AxA$ and any criterion k, then we have no confidence that $a_i S a_j$; therefore, $\sigma(a_i, a_j) = 0$. Therefore, we have constructed a valued outranking relation S_A^σ defined on AxA; this means that we associate with each ordered pair $(a_i, a_j) \in AxA$ a real number $\sigma(a_i, a_j)$ $(0 \leq \sigma(a_i, a_j) \leq 1)$ that reflects the degree of strength of the arguments favouring the crisp outranking $a_i S a_j$.

A second phase, the exploitation of S_A^σ is carried out using the distillation method to obtain a ranking of the alternatives. As an outranking method, the ELECTRE III method may suffer from a pair-wise rank reversal effect [23]. The distillation method does not have a mechanism or way to detect and minimize this kind of irregularity. Our approach for exploitation a valued outranking relation S_A^σ is to use a multi-objective evolutionary algorithm-based heuristic method [20].

3.2 ELECTRE GD

It is a method strongly based on the ELECTRE III multicriteria outranking methodology to assist a group of DMs with different value systems to achieve a consensus on a set of possible alternatives. The preference aggregation procedure starts with N individual rankings and N corresponding valued preference functions, and uses the natural heuristic provided by ELECTRE methodology for obtaining a valued outranking relation representing the collective preference. The heuristic used in ELECTRE GD [21] is based on majority rules combined with concessions to significant minorities. Figure 1 shows a group decision process integrating individual result to reach a collective solution.

The ELECTRE GD method uses as inputs the fuzzy outranking relation and rankings from every DM. The method integrates the solution from individuals and generates a collective preferential model. The obtained model, is presented as a fuzzy outranking relation between alternatives that can be exploited for obtaining a ranking of alternatives. Another well-known method is GDSS PROMETHEE developed by [22] for group decision-making in outranking approach.

4 Multicriteria Group Decision Model

This section presents a group decision model, based on the ELECTRE methodology, targeting cases in which a great divergence occurs between the decision-makers. The procedure presented in Fig. 1 illustrates the case for using the model presented in this section. Because reaching a collective solution is an iterative procedure where the

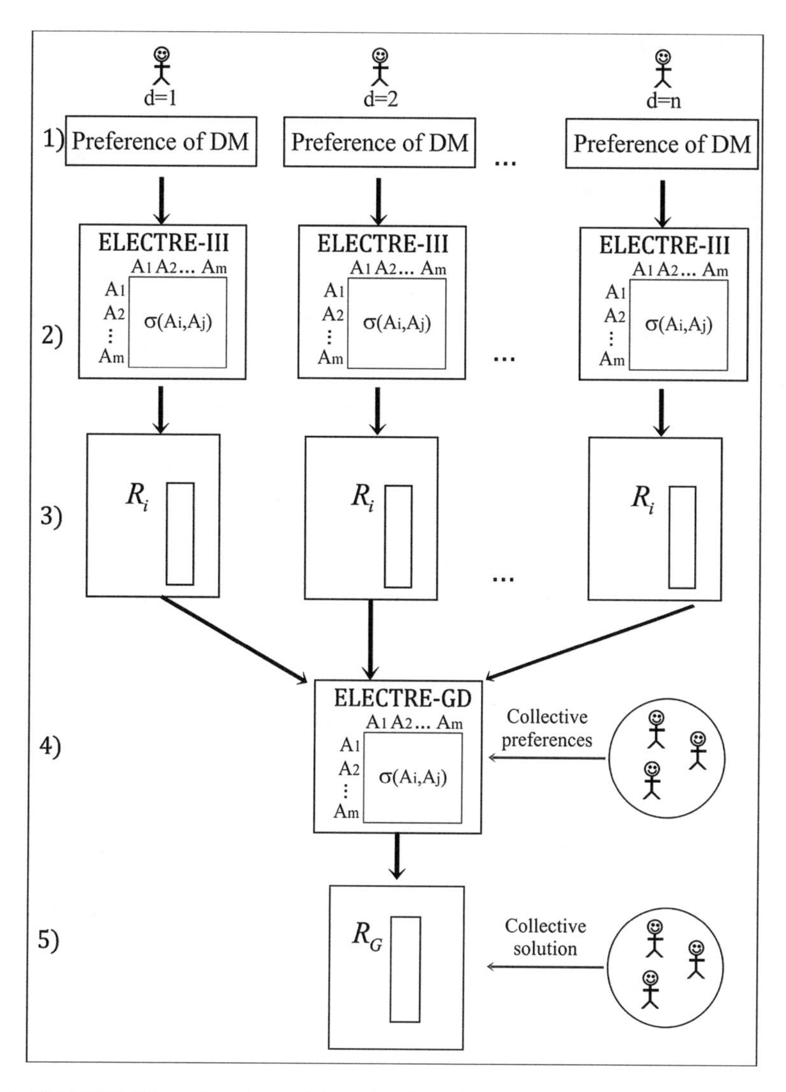

Fig. 1 ELECTRE GD method for reaching collective solution

DM needs to generate again the individual solution in successive iterations. In this situation, the DM needs to be supported in some stages of the process. For example, a particular stage is when DM is back at stage 1 at Fig. 1, to change parameters of the method (ELECTRE III or PROMETHEE) to generate again the individual solution. Because set inter-criteria parameters is not an easy task for DMs, the model proposed generates parameters in a group context without requiring more effort to the DM.

4.1 Description of the Model

In the model proposed in this chapter, each DM determines his/her parameters; at the same time, they should determine the criteria and evaluate the alternatives in a collective way. Each DM is responsible for the establishment of weights and thresholds according to his/her own value judgment.

Group multicriteria decision analysis (GMCDA) processes usually require a facilitator-analyst to formulate the problem initially; moreover, successful group decision aiding requires appropriate coordination processes for incorporating diverse individual views into an aggregated final decision. A coordination mode refers to a series of procedures and aggregation methods, which incorporate the group and individual member activities and facilitate them to reach agreement of a high quality group decision [3]. In such an environment, each participant can sometimes work individually and/or collaborate with the rest of the group at other times.

The coordination methods are associated to the degree to which the individual activities are coupled. On the one hand, if the group is loosely coupled, i.e. there is great divergence among the decision-makers; the members may use the parallel coordination mode. In this type of coordination, individuals work in parallel without interacting with others during most of the group decision-aiding process. On the other hand, if the group is tightly coupled, in which members have to interact with others to get consensus in the middle stages during the entire process. In an extreme case, the members work in sequential coordination mode [3]. This paper describes a consensus reaching process for loosely coupled groups using a parallel coordination mode.

Alternative coordination modes are described by [27]. The parallel, pooled, concurrent, sequential and reciprocal are other coordination methods. The authors generated a comparison called contingency relation between coordination modes in terms of synchronization needs, modularity and flexibility. Regarding the need of interaction between the DMs, parallel coordination requires the lowest synchronization and reciprocal requires the highest. In this sense, we are working with the most flexible method.

4.1.1 Parallel Coordination Mode

Parallel coordination means everyone in a group works independently through most steps during the decision aiding process. The procedure and respective aggregation methods are described systematically as follows:

I. Preliminary stage: structuring the decision problem

The preliminary stage is a phase of knowledge acquisition and problem structuring. A facilitator has first to be appointed. On one hand, the facilitator has to be familiar with the ELECTRE methodology and, on the other hand, he or she needs to have a reasonable knowledge of the actual group decision problem and its context. The following steps can be considered as potentially occurring:

Step 1. First contact of facilitator and DMs. Each DM is encouraged to express his own preferences in order to enrich the understanding of the facilitator with respect to the decision process.
Step 2. Problem description. The DMs meet. The facilitator comments on the available infrastructure and gives an overall description of the problem. A group consensus is needed to define the decision problem.
Step 3. Define the possible evaluation criteria. Each DM works alone and defines his own consistent family of criteria.
Step 4. Alternative generation. The DM work alone during this step.
Step 5. Choose a stable set of alternatives. A group consensus is needed to define the stable set of alternatives.
Step 6. Comments on the alternatives. This step ends the preliminary stage and the next evaluation stages can start.

II. Individual evaluation stage

Step 7. The proposed alternatives are evaluated by each criterion.
Step 8. Define weights and thresholds of the criteria.
Step 9. The individual ELECTRE analysis. ELECTRE-III is applied to construct a valued outranking relation and subsequently a multiobjective evolutionary algorithm (MOEA) [20] is applied to exploit it and derive a ranking in decreasing order of preferences.

During the stage II, each DM works individually, with the possible assistance of the facilitator. At the end of this stage, everybody has a good personal view of the decision problem. Everybody has idea on how to decide. More precisely, each DM has a ranking of the alternatives in decreasing order of preference.

III. Group evaluation stage

The purpose is now to focus on group decision support in order to take into account the specific points from the perspective of the different DMs.

Step 10. Global evaluation. At the end of the individual evaluation stage, the facilitator collects the rankings and valued preference relations coming from the DMs and with this information the ELECTRE-GD is applied to construct a valued outranking relation and again the MOEA procedure is applied to exploit it. As a result, it recommends a complete ranking of the alternatives from the best to the worst ones.

At the end of the step 10, a global evaluation is obtained for the group. The ELECTRE-GD_MOEA methodology proposes a best compromise. If the group agrees with the result of the global analysis, the best compromise can be adopted and the session can be closed. On the other hand, if for some reason some DMs do not agree on this compromise, the conflicts have to be tackled. This process is repeated until individuals' preferences become sufficiently similar. Normally, a consensus model is necessary to address these conflicts.

4.1.2 The Multiobjective Evolutionary Algorithm for Deriving Final Ranking

The exploitation process deals with the outranking relation in order to clarify the decision through a partial or total preordering reflecting some of the irreducible indifferences and incomparabilities [10]. The individual and collective fuzzy outranking relations were generated by the ELECTRE III and ELECTRE GD, respectively. The MOEA implemented for exploiting those fuzzy outranking relation used in our process was proposed by [20]. This MOEA optimizes $f,\ u,\ \lambda$ objective functions. f is the number of incomparabilities between pairs of actions into the individual in the sense of the crisp relation S_A^{λ}. u is the number of preferences between actions into the individual which are not "well-ordered" in the sense of S_A^{λ}. $\lambda(0 \leq \lambda \leq 1)$ is related with the credibility level of a crisp outranking relation. The result of this MOEA is a consistent ranking with the preferential model (fuzzy outranking relation).

5 Consensus Model Based on Consensus Measures

When the consensus model is based on consensus measures, the goal of achieving a consensus on the ranking of alternatives can be viewed as a dynamic process in which a facilitator, via an exchange of information and rational arguments, attempts to persuade individuals to adapt their preferences. This dynamic procedure is iterative and interactive and in each round a degree of the existing consensus and the proximity of individual orders to the collective temporary order are measured. The facilitator uses the degree of consensus to control the process. This process is repeated until the group becomes closer to a maximum consensus or a predefined number of round has been completed. A collective order is qualified as a consensus only when the group agreement level with respect to this order reaches a certain pre-established threshold (consensus threshold). The use of consensus threshold is a common practice in-group decision-making process. Even when the way to set up the threshold value is not clear, we can use some background information. In [7] the 0.75, 0.80 and 0.85 values are used for consensus threshold. In [12, 15] used 0.85 for the same threshold. However, the value can be related with the ability of the group to reach a consensus. It implies that in groups with divergent thinking a high threshold is difficult to reach, so lower threshold should be defined. In groups with convergent thinking a higher threshold should be defined in order to reach better consensus. In this case, a consensus threshold between 0.75 and 0.85 can be an appropriate value as a goal to be reached by the group.

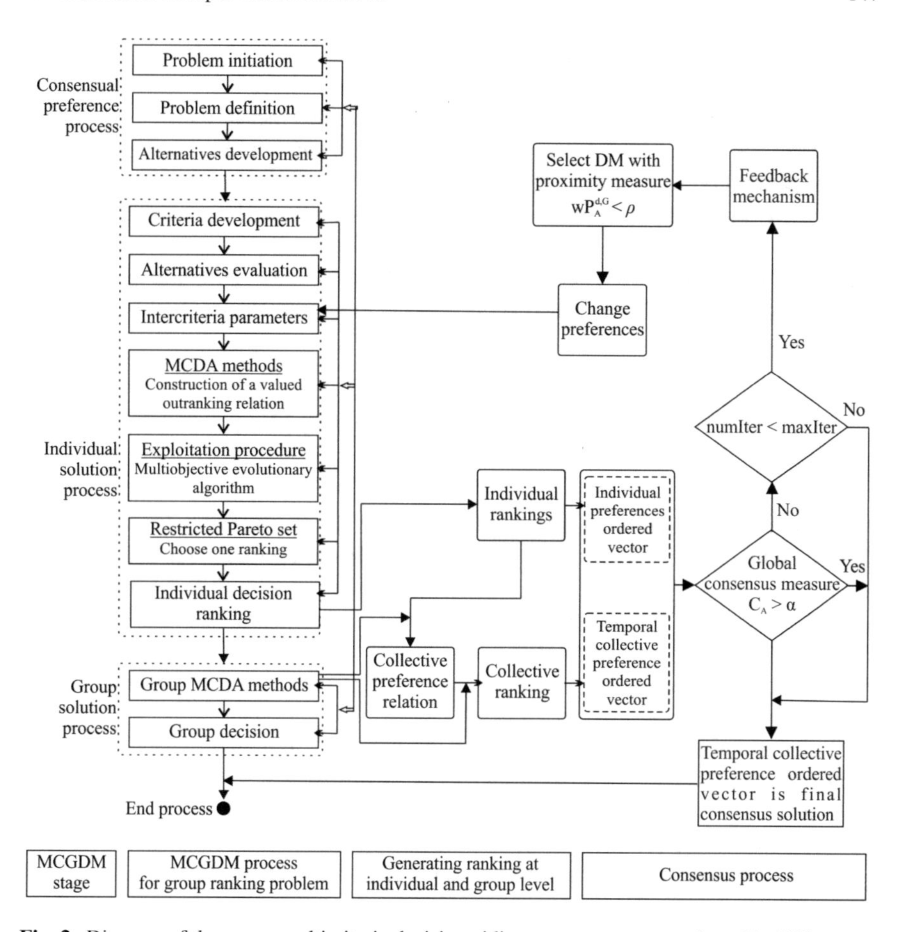

Fig. 2 Diagram of the group multicriteria decision aiding consensus process based in [18]

5.1 Consensus Measures

In each round of the consensus process, we calculate two consensus parameters, a *consensus measure* and a *proximity measure*. The first parameter guides the consensus process, and the second parameter supports the group discussion phase of the consensus process. A consensus threshold α is fixed in advance ($\alpha \in [0, 1]$, $\alpha > 0.5$). When the consensus measure reaches this threshold, the consensus process is finished and the solution is obtained. If that scenario does not occur, then the preferences of some decision-makers should be modified in a group discussion session in which the facilitator uses a proximity measure to propose a feedback process. In Fig. 2, we present the consensus model used in this group multicriteria ranking problem, it is described in further detail in the following subsections.

5.1.1 Consensus and Proximity Measures

The consensus model contemplates the position-weighted measure $wP_A^{d,G}$ [19]. This index measures the agreement level between an individual opinion and a collective opinion when both are expressed by rankings of a set of alternatives. It constitutes a weighted version of the Kendall's ranks correlation index. The index accounts for the position weights of the alternatives in an individual order to quantify the agreement level of the individual order with respect to a collective temporary order. The definition of the position-weighted version of the Kendall distance ($K_{\bar{p}}$) is presented as follows:

$$K_{\bar{p}}(O^k, O^G) = \sum_{i=1}^{m-1} \sum_{j>i} \max\left\{\bar{p}_i(O^k), \bar{p}_j(O^k)\right\} \bar{K}_{i,j}(O^k, O^G) \quad (2)$$

where $\bar{p}_i(O^k)$ (or $\bar{p}_j(O^k)$) is the position weight of the alternative a_i (or the alternative a_j) in the individual order O^k. $\bar{K}_{i,j}(O^k, O^G)$ is the Kendall's ranks correlation index.

The index $wP_A^{d,G}$ operates with the equivalent representation of a ranking as a list of ranks $O^d = [o^d(1), ..., o^d(m)]$ showing the position of alternative j in the ranking [6], converting the difference of the ranking into numerical data, and representing an ordered vector of alternatives from best to worst.

Each consensus parameter requires the use of $wP_A^{d,G}$ to obtain the level of agreement between the individual solution of DM d, $O^d = [o^d(1), ..., o^d(m)]$, and the collective solution $O^G = [o^G(1), ..., o^G(m)]$. Practical applications of this proximity index can be found in [4, 18, 19].

We define consensus indicators by comparing the positions' weighted rank of alternatives in two preferences vectors, using the following procedure:

1. To use a multicriteria ranking method to obtain individual rankings R_d for each DM; then, to use a multicriteria group ranking method to obtain collective ranking of alternatives R_G.
2. To calculate ordered vector of alternatives $\{O^d;\ d = 1, 2, ..., n\}$ (from individual rankings R_d) and collective ordered solution O^G (from the collective ranking R_G), where n is the number of DMs in the group.
3. To calculate the *proximity measure* of the *d-th* individual solution of the decision-maker to the collective temporary solution, $wP_A^{d,G}$.
4. To calculate the *consensus measure* C_A of all of the DMs using the following expression:

$$C_A = \sum_{d=1}^{n} \frac{wP_A^{d,G}}{n}$$

where n is the number of DMs in the group. C_A, is calculated by aggregating the above consensus degrees for each DM.

5.2 Consensus Control Mechanism

We use the *consensus measure* (C_A) as a control parameter: When C_A exceeds a minimum satisfaction threshold value $\alpha \in [0, 1]$, the consensus-reaching process stops and the temporal collective preference ordered vector is the final consensus solution. This is the end of the decision process. Otherwise, the feedback mechanism is activated. Additionally, a parameter to control the maximum number of consensus rounds, *maxIter*, is used (independently of the current C_A value) to avoid stagnation.

5.3 Feedback Mechanism

A feedback mechanism is applied when the consensus level is not satisfactory and it is finished when a satisfactory consensus level is reached. In the first case, the consensus measure C_A has not reached the consensus level required, then, some appropriate decision-makers' rankings should be modified.

The feedback mechanism consists of two sub-modules: Identification of the preferences (numerical values) that need to be modified and generation of advice. For the first one, we use a two-step process to identify first the DMs and then, the alternatives values or particular intercriteria parameter values that contribute less to the consensus level.

The procedure for generating an agreed collective solution is focused in the feedback process stage, proposing new alternative values or inter-criteria parameters for helping each DM to generate new individual solutions regarding individual and group preferences [4]. The rules of this feedback process are as follow:

"If proximity of the *d-th* individual solution of the decision-maker to the collective temporary solution $wP_A^{d,G}$ is less than a predefined threshold $\rho \in [0, 1]$ then the DM should change his or her preferences, where the DM freely decides whether to change his/her preferences, and then a new preferential model and ranking are generated based on the DM changing his/her preferences. It will be carried out as follows":

Step 1. Identification of decision-makers: Identify members d whose proximity measure $wP_A^{d,G}$ is less than predetermined threshold ρ; then, these members should change their preferences.

Step 2.

Step 2.a. Identification of alternatives: The set of alternatives that the above DMs should consider to change their evaluations on a particular criterion k is the one that reduces pairwise disagreements between the *d-th* order and the collective temporary order, which is provided by the solution of a multiobjective optimization problem [4].

Step 2.b. Identification of intercriteria parameters:

The set of intercriteria parameters that one DM should consider to change his or her determination on a particular criterion k is the one that reduces pairwise disagreements between the *d-th* order and the collective temporary order to increase the value of proximity measure $wP_A^{d,G}$, which is provided by the solution of a multiobjective optimization problem [4].

6 A Practical Application

In this section, we explain a real instance of a group multicriteria ranking problem with aims of ranking a set of technology packages regarding that it is an important factor that have a crucial role in crop management. We follow the parallel coordination mode for solving this problem.

I. Preliminary stage-structuring the decision problem

Step 2. Problem description.

As mentioned before, we are applying the group decision methodology to a practical case study in a small agricultural company that deals with the selection of a technology package, which is a major tactical decision implemented every season. The selection of the right technology package could reduce unnecessary costs and/or increase crop yields, thus improving the profitability of the company.

However, there are usually differences among the DMs over their preferred technologies, given that the technology package should consider many aspects, such as finances, land, labor, biological requirements, the goals of the DMs and those of the company.

In order to develop a decision that could be supported by all the DMs, we participated as a consulting team and developed an instance of a group multi-criteria ranking problem for a set of technology packages for crop management. We followed the parallel coordination mode for solving this problem.

Based on a technical study the company suggests eight technology packages as decision alternatives. The company cannot establish which technology package is the most profitable option; however, they can identify possible consequences for each one of the technology packages.

Four of the enterprise's top managers were asked to participate as DMs in a multicriteria group decision aiding process to make a ranking of technology packages proposed, in order to select one of them based on this order.

It is at this stage that the true expectations and objectives regarding the corn crop are determined. Executives from the company define formally the required decision criteria to be used in ranking technology packages. First, the group of DMs representing the company is presented: Director of the supplier company (DM1), representative of the financial department (DM2), representative of production department DM3 and representative of operational (logistic) department (DM4). DM1 is representing the supplier company; DM2, DM3 and DM4 are representing the agricultural company. However, DM2 represents financial interest and DM3 and DM4 represent interests related with the production and operation activities in the agricultural company. They expressed their points of view in meetings and listed the factors they perceived important to consider in the ranking process.

In order to make an adequate evaluation of the alternatives, taking in account the point of view of every DM, the decision process was supported by the multicriteria group decision support system (MCGDSS) SADGAGE, which has implemented the consensus model proposed in this chapter. The described problem was treated using SADGAGE and the consensus threshold was defined as $\alpha = 0.80$, based on previous practical applications by [4, 7, 12]. This threshold regards to the minimum consensus level approved by the group members, and, $\rho = 0.82$, the minimum proximity level to the temporary collaborative solution approved by the group members.

Step 3. Define the possible evaluation criteria.

After being designated by their organization and in possession of the scope of the project, the DMs hold a meeting with the facilitator-analyst in order to identify the criteria to be used in the selection process. We now show, the criteria definition for technology package for corn production. The consultants propose two main evaluation dimensions for packages of corn production. Six criteria are described in the following analytical way.

Cost of rent (C1): The package includes land for rent, because the agricultural company requires to rent land for planting the crops. This criterion is the cost for using the land and is measured in Mexican pesos in the northwest area, which are considered relatively good farmlands in the country.

Crop yields (C2): It is related with the type of seed and some elements of the ground preparation and planting maintenance. The crop yield is measured in quantity of tons per hectare harvested.

Success probability (C3): Each type of seed presents differences characteristics related to environments elements like weather resistance and plague types. The supplier presents a probability to harvests expected crop and obtains the expected benefit with the planned package. The measuring scale is 0–100.

Return on investment (C4): It is the package return on investment, which farms can obtain when the crop is sold. It is a common practice when the crop is gathered the farmers obtain around 100% or more of the investment in a technological package for corn production. For this criterion the seed cost is in consideration of the return on investment.

Preparation and maintenance (C5): This is an indicator composed by five types of activities in the land. This criterion concerns ground preparation, fertilizing, weed control, plague control and irrigation. This criterion is measured in Mexican pesos.

Harvest cost (C6): It is the process of harvesting the crop and it included as a part of the investment.

Steps 4 and 5 regarding the generation of alternatives and choosing a stable set of alternatives, respectively. However, the consulting firm proposes to the agricultural company a set of eight technology packages as decision alternatives based in supplier information and products. The eight alternatives suggested by the firm and endorsed by the group of DMs are coded in the following form:

A: Technology package 1
B: Technology package 2
C: Technology package 3
D: Technology package 4
E: Technology package 5
F: Technology package 6
G: Technology package 7
H: Technology package 8

This multicriteria decision process is well structured, because the activity is a typical situation of selecting a technology package, and the seed suppliers supported this process with all the technical information about the corn varieties based on the characteristics of the land and the requirements of the company.

The decision process is structured based on the agenda showed in Sect. 4.1.1. We can see it as an interactive and iterative procedure. The realization of the steps listed at Sect. 4.1.1 can be seen as a first iteration, if the consensus is not reached. A second iteration must follow with the DMs who result in a low proximity level compared to the group of temporary solutions. Then we have to encourage those DMs to modify their data and/or inter-criteria parameters to reach better proximity level and then achieve a consensus for the final solution. In this example, we are going to see that the group does not achieve consensus at the first iteration, hence some DMs must to modify their data or parameters to proceed with the second iteration, which constitutes a feedback mechanism.

II. Individual evaluation stage

Step 6. The proposed alternatives are evaluated by each criterion. At this point, the DMs have carried out the preliminary stage. For step 6, the DM has the information given by the consulting firm, the seed suppliers and the own information generated by the company. The information is regarded to each technology package for each criterion, so the performance of alternatives is showed in the Table 1.

The performances of the alternatives are expressed numerically. The measurements for criteria C1, C5 and C6 are in Mexican pesos. Criterion C2 is measured in tonnes harvested per hectare. Criterion C3 as a probability value, is measured in 0–100 scale. Criterion C4 is the return on investment; the values are in the range 100–120%.

Steps 7 and 8. Define weights and thresholds of the criteria. The decision of DMs assign weights of relative importance to each criterion. Table 2 presents the selection

Table 1 Performance of the alternatives

	C1	C2	C3	C4	C5	C6
A	9,500	11.89	65.25	119.07	12,248.03	2,161.49
B	4,500	6.10	95.15	100.73	7,989.43	1,737.35
C	6,500	9.00	78.48	109.90	9,938.79	2,007.95
D	7,000	10.20	75.74	113.94	10,501.56	2,060.87
E	5,000	6.82	92.37	102.10	9,173.64	1,752.72
F	7,500	10.84	73.04	115.75	10,887.03	2,116.94
G	6,000	7.70	89.28	105.56	9,300.12	1,930.59
H	8,500	11.14	69.84	117.73	11,359.21	2,150.92

Table 2 Parameters: weights, indifference and preference thresholds in iteration 1

	DM1				DM2				DM3				DM4			
	Dir	*w*	*q*	*p*	Dir	*w*	*q*	*p*	Dir	*w*	*q*	*p*	Dir	*w*	*q*	*p*
C1	Min	0.16	300	500	Min	0.17	400	600	Min	0.2	200	500	Min	0.19	300	500
C2	Max	0.2	0.5	1	Max	0.22	0.5	1	Max	0.17	0.4	0.9	Max	0.17	0.6	1.1
C3	Max	0.18	5	15	Max	0.13	5	15	Max	0.19	5	10	Max	0.19	6	12
C4	Max	0.22	1	2	Max	0.19	2	3	Max	0.18	1	2	Max	0.18	1.7	2.8
C5	Min	0.16	1200	2200	Min	0.17	1500	2000	Min	0.2	1000	2000	Min	0.19	1500	2500
C6	Min	0.08	600	1000	Min	0.12	200	400	Min	0.06	500	950	Min	0.08	1000	2000

Table 3 Individuals rankings

Position	DM1	DM2	DM3	DM4
1	H	A	G	H
2	A	H	H	G
3	D	F	B	E
4	F	D	A	D
5	C	C	F	A
6	G	G	E	F
7	E	B	D	B
8	B	E	C	C
λ	0.580	0.590	0.649	0.610

λ: Credibility index used in the generation of individual ranking

criteria, preference directions of the criteria, weights and thresholds regarding every criterion selected by the DMs.

First Iteration

Step 9. The individual ELECTRE analysis. According to the additional information pointed out before, for each DM we applied ELECTRE III to construct a valued outranking relation. Afterwards, we used the evolutionary algorithm presented in [20] to exploit the outranking relation and derive a final ranking of the alternatives in a decreasing order of preferences. Those methods are embedded in SADGAGE system. Table 3 shows the individual rankings resulted from the exploitation of the valued preference relations. DM1 and DM2 present similar ranking between them. Alternatives A and H are showed in first two positions, alternatives D and F appear in following positions. DM3 and DM4 present rankings less similar to those of DM1 and DM2, because it seems just G and H are consistent, they appear in the first two positions of the ranking, the rest of the positions for the alternatives, present more difference.

III. Group evaluation stage.

Step 10. Global evaluation. The activities developed at this point by the DMs permitted to construct the credibility matrix and the individual ranking for every

Table 4 Individuals and group ranking in iteration 1

Position	DM1	DM2	DM3	DM4	Collective
1	H	A	G	H	H
2	A	H	H	G	D
3	D	F	B	E	A
4	F	D	A	D	G
5	C	C	F	A	F
6	G	G	E	F	B
7	E	B	D	B	C
8	B	E	C	C	E
λ	0.580	0.590	0.649	0.610	
Dis	5	6	7	8	
Proximity	0.821	0.825	0.802	0.735	
Consensus level (C_A) 0. 796					

Dis (disagreements): the number of differences between two rankings viewed in pairwise format

DM. Those outcomes are used as the input data to the ELECTRE-GD method. This method integrates the individual rankings and generates a group ranking, representing the collective preference in the form of valued preference relation. Afterward, the resultant preference relation is exploited by the MOEA to derive a collective temporary ranking as showed in the Table 4.

First Stage in the Consensus Reaching Process

At the end of the step for modelling group preferences, a comparison of the individual rankings with the collective temporary ranking is carried out. We can realize that the individual solutions are different from collective solution. This information can be asserted with the data showed in Table 4. Here, we have the proximity level of each DMs from the collective solution. We realize the DM3 and DM4 have a proximity level less than $\rho = 0.82$, a minimum predefined threshold we arbitrarily established, to indicate the individual solution is close to the collective solution. However, the consensus index is less than $\alpha = 0.8$, which is the predefined threshold for the consensus index. Thus the current index value (0.796), indicates the consensus level has not been reached and the DMs share significant differences in their individual solutions. In this situation, the facilitator should interact with the DMs in a second iterative process of feedback. The interaction procedure aims to modify their individual preference, in order to obtain a better consensus solution where the similarities of the individual solutions are significantly similar to the collective solution. It means that some DMs must change their preferences by going through a modifying parameters stage.

Second Iteration - Feedback Process

To improve the consensus level, a greater similarity value is required for some DMs. For our technological package selection problem, the DMs with low proximity level

($\rho = 0.82$) were required to interact in a preference modifying procedure, to obtain a new solution, which does not have major differences that make it harder to accept. In Table 4, DM1 and DM2 present 0.821 and 0.825 proximity levels, higher than required. DM3 and DM4 present 0.802 and 0.735 proximity levels, lower than required. Then, DM3 and DM4 were involved to interact in a preference modifying procedure.

For the feedback stage of the consensus process the DM3 and DM4 are the k members with $wP_A^{k,G} < \rho$ in iteration 1. To support this stage we use an inferring parameter model developed by [4] that use as input a set of similar preference between individual and group solutions. The model concerns number of parameters' changes ($f1$), number of disagreements ($f2$) and agrees remain for the first ranking ($f3$); those objectives are evaluated in the new parameters proposal. Regarding the feedback stage of the consensus process, where the DM3 and DM4 need to change their individual preference (inter-criteria parameters), the output of the inferring parameter tool is shown in Tables 5 and 6, respectively. The new parameters suggested for the DM3 and DM4 implies some changes in the weight, indifference and preference thresholds.

In Tables 5 and 6, columns 2, 3 and 4 present the evaluation of the $f1$, $f2$, $f3$, respectively. Column 5 is the epsilon value used for the new proposal of parameters. The epsilon value is used to specify the neighborhood in which the parameter may vary. Column 6 presents the proposed parameters and column 7 the ranking generated. The last column is the proximity index that new ranking presents with the collective ranking of iteration 1.

In iteration 2, the new inter-criteria parameters suggested by the inferring model at Tables 5 and 6 are shown to DM3 and DM4, respectively to support the modifying parameters stage. DM3 supported proposals 2 and 6 (see Table 5) because alternatives G, H and A remain in the higher positions of the ranking. The rest of proposals can obtain acceptable proximity levels, however the alternative G is shown in lower positions of the ranking. Then, other proposal cannot be supported by the DM3 because G is one of the most preferred. In iteration 2, DM3 selected proposal 2 by the inferring model because it does not involve large changes in the most preferred alternatives from the top positions of the ranking.

The DM4 supported more proposals 2 and 7 (see Table 6) because the alternative H remains in the first position and G remains in high ranking positions. The rest of the proposals show H in higher positions and G in very low positions. In iteration 2, DM4 selected proposal 2 from the inferring model because it does not involve large changes in the top ranking positions, and presents H in the first position, G and D in high positions. Those three alternatives appear in four first position of previous ranking (iteration 1). For DM4, proposal 2 does not involve a large deviation from the preferred alternatives in the top ranking positions.

Second Iteration and Consensus Collective Solution

The proposal selected by DM3 and DM4 were used to generate new preference models and rankings for the DM3 and DM4, DM1 and DM2 remain in their previous ranking (see Table 7). With the rankings shown in Table 7, a new collective ranking

Table 5 Inter-criteria parameters proposals for improving the DM3' agreement

No.	f_1	f_2	f_3	ε	Proposal	Ranking	Proximity
1	7	8/28	−12/18	0.50	w1 = 0.108, w2 = 0.172, q2 = 0.297, q3 = 7.5, w4 = 0.27, q4 = 0.597, q5 = 1180	H≻A≻F≻G≻D≻C≻B≻E	0.887
***2**	6	9/28	−12/18	0.50	w1 = 0.108, w2 = 0.172, q2 = 0.297056, q3 = 7.5, w4 = 0.27, q4 = 0.597751	H≻G≻A≻F≻D≻B≻E≻C	**0.873**
3	9	6/28	−13/18	0.75	w1 = 0.076, w2 = 0.244, q2 = 0.245584, q3 = 8.75, p3 = 17.5, w4 = 0.31, q4 = 1.75, w5 = 0.08, w6 = 0.1	H≻A≻F≻D≻G≻C≻B≻E	0.892
4	7	8/28	−12/18	0.75	w1=0.077, w2 = 0.205, q2 = 0.245, q3 = 4.25, p3 = 11.5, w4 = 0.268, q4 = 1.75	H≻A≻F≻D≻G≻C≻B≻E	0.892
5	7	8/28	−12/18	0.75	w1=0.077, w2 = 0.205, q2 = 0.245, q3 = 4.25, p3 = 13.5, w4 = 0.268, q4 = 1.75	H≻A≻F≻D≻G≻C≻B≻E	0.892
6	6	5/28	−14/18	1.00	w1 = 0.112, w2 = 0.258, p2 = 1.8, w4 = 0.36, w5 = 0.02, q5 = 1242.42	H≻A≻G≻F≻D≻C≻B≻E	**0.888**
7	5	6/28	−14/18	1.00	w1 = 0.112, w2 = 0.258, w4 = 0.36, w5 = 0.02, q5 = 691.421	H≻A≻F≻G≻D≻C≻E≻B	0.810
8	3	10/28	−14/18	1.00	w4 = 0.36, w5 = 0.02, q5 = 691.421	H≻F≻A≻G≻D≻B≻C≻E	0.916

*: proposal selected by the DM3

is generated (see column 6). With the new temporal solution, the proximity level for DM1 reduces to 0.825 and DM2 increases to 0.911. The individual solutions from DM3 and DM4 reflect an improved proximity level: 0.881 and 0.807, respectively. The number of disagreement for every DM was reduced to 4, 3, 4 and 7 for DM1, DM2, DM3 and DM4, respectively. This new proximity between rankings generates a better consensus level($C_A = 0.856$) greater than the required ($C_A > \alpha$). In this particular application, the procedure finished with the consensus level obtained by iteration 2.

Table 6 Inter-criteria parameters proposals for improving the DM4' agreement

No.	f_1	f_2	f_3	ε	Proposal	Ranking	Proximity
1	7	8/28	−15/21	0.50	w1 = 0.192, q2 = 0.645, w3 = 0.153, w4 = 0.266, p4 = 3.265, w5 = 0.177, w6 = 0.042	A≻H≻D≻F≻C≻G≻B≻E	0.871
***2**	7	10/28	−16/21	0.50	w1 = 0.192, q2 = 0.645, w3 = 0.216, w4 = 0.266, p4 = 3.265, w5 = 0.114, w6 = 0.042,	H≻A≻G≻D≻F≻E≻B≻C	**0.835**
3	9	6/28	−15/21	0.75	w1 = 0.074, w2 = 0.282, q2 = 0.600261, p2 = 0.925, w3 = 0.192, w4 = 0.301, q4 = 0.425, p4 = 1.7, w5 = 0.0708	A≻H≻F≻G≻D≻C≻E≻B	0.807
4	7	8/28	−15/21	0.75	w1 = 0.146, q2 = 0.600261, p2 = 0.629, q3 = 4, q4 = 1.425, p4 = 2.775, w5 = 0.234	A≻D≻G≻H≻B≻E≻F≻C	0.767
5	6	9/28	−14/21	0.75	w1 = 0.146, q2 = 0.600261, p2 = 0.629018, q4 = 2.425, p4 = 2.77532, w5 = 0.234	G≻H≻A≻D≻F≻E≻B≻C	0.808
6	7	8/28	−15/21	1.00	w1 = 0.192, q2 = 0.645, w3 = 0.153, w4 = 0.266, p4 = 2.265, w5 = 0.177, w6 = 0.042	A≻H≻F≻G≻D≻C≻B≻E	0.871
7	7	10/28	−16/21	1.00	w1 = 0.192, q2 = 0.645, w3 = 0.216, w4 = 0.266, p4 = 3.265, w5 = 0.114, w6 = 0.042	H≻A≻F≻G≻D≻E≻B≻C	**0.808**

*: proposal selected by the DM4

The second collective ranking is now the temporary group solution. The interactive process we proposed, generated a new ranking. Table 8, shows an evaluation of the proximity of the individual ranking with the collective ranking and the value of the consensus index reached until now.

The proximity index shows the similarity between individual and collective solution for every DM. The proximity index takes in account the number of disagreements (rank reversals) between rankings. In iteration 1, the DM1, DM2, DM3 and DM4 present 5, 6, 7 and 8 disagreements with collective ranking, respectively. As the col-

Table 7 Individuals and group ranking in iteration 2

Position	DM1	DM2	*DM3	*DM4	+Collective
1	H	A	H	H	H
2	A	H	G	A	A
3	D	F	A	G	F
4	F	D	F	D	D
5	C	C	D	F	G
6	G	G	E	E	B
7	E	B	B	B	C
8	B	E	C	C	E
λ	0.579	0.589	0.560	0.559	
Dis	4	3	4	5	
Proximity	0.825	0.911	0.881	0.807	
Consensus level (C_A) 0.856					

Dis (disagreements): the number of differences between two rankings viewed in pairwise format
* DM with new ranking
+ New collective ranking

Table 8 Proximity level of the DMs' individual rankings compared with the collective temporary solution

	Proximity
DM2	0.911
DM3	0.881
DM1	0.825
DM4	0.807
Consensus index	0.856

lective ranking was changed for the iteration 2, the proximity index was updated for every DM. If the proximity index improves it implies disagreements were decreased and/or rank reversals present are from on the lower part of the ranking. Since the position-weighted version of the Kendall distance used for computing the proximity index, takes penalizes less those rank reversal occurred in bottom position compared with rank reversal occurred in top positions. In iteration 2, the DM1, DM2, DM3 and DM4 present 4, 3, 4 and 5 disagreements with collective ranking, respectively. It shows a reduction in disagreements between individual solutions and the related collective solution. The new collective solution, in correspondence with the iteration 1, presents higher consensus. In this case the DM can see their individual preferences are reflected better in the last collective solution than in previous collective solutions.

Here we can objectively see a significant improvement in the individual solutions. In addition, we can see that the predefined threshold value of 0.8 for the consensus index was crossed (it reached the value of 0.856).

With this interactive and iterative process, we have shown a procedure, which guides the facilitator objectively in a consensus process, and a discussion process with a consensus measures and a proximity measure, respectively. This practical application illustrates that the multicriteria group decision process can be guided by those two measures.

Results Analysis

The developed procedure, together with the proposed model, support the DMs to obtain a collective solution (iteration 2), which reflect their individual preference better than the first collective solution. In the first collective solution in Table 4, the DM3 and DM4 presented more disagreement with the collective solution than with the final solution. In this situation, the collective solution did not reflect their individual preferences, or at least not as much as for DM1 and DM2. The collective solution in iteration 1 (Table 4) presents alternatives H, D, A and G in the high ranking positions. That solution represented DM1 and DM2 individual ranking. DM3 and DM4 presented lower proximity levels because in their individual ranking top positions were occupied by H and G. However, the next alternatives like B and E were in low level of collective ranking. However, in iteration 2 (Table 7) the new individual results for DM3 and DM4 improve the alternatives A and F, retaining the preferred alternatives H and G in high level ranking positions. This solution improves the proximity level of DM1, DM2, DM3 and DM4, increasing the consensus level of the collective solution (0.856). In different situation, the proximity level obtained for each DM in iterations can be reduced for some DMs and increment for other. The goal is get equilibrium in the agreement context, and this has been reflected in the consensus level. The results must include a collective solution reflecting the preference of every DM.

7 Conclusions

Given the complexity of crop management, which has been increasing over the years in medium-sized agricultural companies, a growing emphasis has been given to the development of new forms for selecting technology packages, with the aim of adapting to the environmental changes, preventing attack by pest and diseases, and minimizing conflicts between the different actors involved. Additionally and habitually, the goals of the agricultural company are generally centered on maximizing the harvest at the least possible cost. Therefore, this chapter has presented a proposal for a multicriteria group decision model for ranking technology packages, by means of using an outranking approach based on ELECTRE methodology. The chapter presents a practical case and it is used as a reference to the numerical application. It is important to point out that the objectives of this study were achieved since the model presented can be used for problems where there is divergence about the preferences of the DMs. However, the model can be applied in different context for group decision-making problem when reaching consensus is required. In this sense, the model has been applied in problems of commercial location, location projects for water supply, franchise business decisions, but could also be applied to any related group decision-making problems.

In general, decisions related to the selection of technology packages involve a group of people, of an agricultural organization, in order to represent the interest of the company. Some studies found in the literature about technology packages do not tackle the question of group decision-making and consider decision-making only as if a single person were responsible for this decision, which does not correspond to a real situation. Thus the model developed, in the context applied, indicates the best technology package according to the different evaluation criteria.

Derived from the application of the proposed model some further research direction could arose about systematized procedure without further interaction (the minimum interaction) with DM and linguistic assessment for performance alternatives. Those research and the actual proposal can be applied for sustainable problems, agricultural process as technological packages and agrifood products.

References

1. Ahumada, O., Villalobos, J.R.: Application of planning models in the agri-food supply chain: a review. Eur. J. Oper. Res. **196**(1), 1–20 (2009)
2. Aldana, U., Foltz, J.D., Barham, B.L., Useche, P.: Sequential adoption of package technologies: the dynamics of stacked trait corn adoption. Am. J. Agric. Econ. **93**(1), 130–143 (2011). doi:10.1093/ajae/aaq112
3. Álvarez Carrillo, P.A., Leyva López, J.C., Sánchez Castañeda, MdlD: An empirical study of the consequences of coordination modes on supporting multicriteria group decision aid methodologies. J. Decis. Syst. **24**(4), 383–405 (2015). doi:10.1080/12460125.2015.1080583
4. Alvarez, P.A., Morais, D.C., Leyva, J.C., de Almeida, A.T.: A Multi-objective genetic algorithm for inferring inter-criteria parameters for water supply consensus. In: Gaspar-Cunha, A., et al. (ed.) Evolutionary Multi-Criterion Optimization, vol. Springer-Verlag's Lecture Notes in Computer Science (LNCS) series, pp. 218–233. EMO 2015, Part II, LNCS 9019, Guimarães, Portugal (2015)
5. Byerlee, D., de Polanco, E.H.: Farmers' stepwise adoption of technological packages: evidence from the Mexican Altiplano. Am. J. Agric. Econ. **68**(3), 519 (1986). doi:10.2307/1241537
6. Chiclana, F., Herrera, F., Herrera-Viedma, E.: Integrating three representation models in fuzzy multipurpose decision making based on fuzzy preference relations. Fuzzy Sets Syst. **97**(1), 33–48 (1998). doi:10.1016/s0165-0114(96)00339-9
7. Chiclana, F., Mata, F., Martinez, L., Herrera-Viedma, E., Alonso, S.: Integration of a consistency control module within a consensus model. Int. J. Uncertain. Fuzziness Knowl. Based Syst. **16**(1), 35–53 (2008). doi:10.1142/s0218488508005236
8. Feder, G., Just, R.E., Zilberman, D.: Adoption of agricultural innovations in developing countries: a survey. Econ. Dev. Cult. Chang. **33**(2), 255–298 (1985)
9. Feder, G., Umali, D.L.: The adoption of agricultural innovations: a review. Technol. Forecast. Soc. Chang. **43**(3), 215–239 (1993)
10. Fodor, J., Roubens, M.: Fuzzy Preference Modeling and Multicriteria Decision Support. Kluwer, Dordrecht (1994)
11. Grose, T.K.: The next green revolution. ASEE Prism **25**(4), 28–31 (2015)
12. Herrera-Viedma, E., Herrera, F., Chiclana, F.: A consensus model for multiperson decision making with different preference structures. IEEE Trans. Syst. Man Cybern. Part A Syst. Hum. **32**(3), 394–402 (2002). doi:10.1109/tsmca.2002.802821
13. Jacquet-Lagrèze, E., Siskos, Y.: Méthodes de Décision Multicritère. Editions Hommes et Techniques, Paris (1983)

14. Jelassi, T., Kersten, G., Zionts, S.: An Introduction to Group Decision and Negotiation Support. 537-568 (1990). doi:10.1007/978-3-642-75935-2_23
15. Kacprzyk, J., Zadrożny, S.: Supporting consensus reaching processes under fuzzy preferences and a fuzzy majority via linguistic summaries. Preferences and Decisions. Studies in Fuzziness and Soft Computing, pp. 261–279. Springer, Berlin (2010). doi:10.1007/978-3-642-15976-3_15
16. Khan, M.S., Quaddus, M.: Group decision support using fuzzy cognitive maps for causal reasoning. Group Decis. Negot. **13**(5), 463–480 (2004). doi:10.1023/B:GRUP.0000045748.89201.f3
17. Leathers, H.D., Smale, M.: A Bayesian approach to explaining sequential adoption of components of a technological package. Am. J. Agric. Econ. **73**(3), 734 (1991)
18. Leyva López, J.C., Álvarez Carrillo, P.A., Gastélum Chavira, D.A., Solano Noriega, J.J.: A web-based group decision support system for multicriteria ranking problems. Int. J. Oper. Res. 1–36. doi:10.1007/s12351-016-0234-0
19. Leyva López, J.C., Alvarez Carrillo, P.A.: Accentuating the rank positions in an agreement index with reference to a consensus order. Int. Trans. Oper. Res. **22**(6), 969–995 (2015). doi:10.1111/itor.12146
20. Leyva, J.C., Aguilera, M.A.: A multiobjective evolutionary algorithm for deriving final ranking from a fuzzy outranking relation. In: Coello Coello, C.A., Zitzler, E., Hernández Aguirre, A. (eds.) Evolutionary Multi-Criterion Optimization. Third International Conference, EMO 2005, Lecture Notes in Computer Science, vol. 3410, pp. 235–249. Springer, México (2005)
21. Leyva, J.C., Fernández, E.: A new method for group decision support based on ELECTRE III methodology. Eur. J. Oper. Res. **148**(1), 14–27 (2003). doi:10.1016/s0377-2217(02)00273-4
22. Macharis, C., Brans, J.P., Mareschal, B.: The GDSS PROMETHEE procedure - a PROMETHEE-GAIA based procedure for group decision support. J. Decis. Syst. **7**, 283–307 (1998)
23. Mareschal, B., De Smet, Y., Nemery, P.: Rank reversal in the PROMETHEE II method: some new results. 959–963 (2008). doi:10.1109/ieem.2008.4738012
24. Munier, N.: A Strategy for Using Multicriteria Analysis in Decision-Making. Springer, Netherlands (2011)
25. Palomares, I., Estrella, F.J., Martínez, L., Herrera, F.: Consensus under a fuzzy context: taxonomy, analysis framework AFRYCA and experimental case of study. Inf. Fusion **20**, 252–271 (2014). doi:10.1016/j.inffus.2014.03.002
26. Parreiras, R.O., Ekel, P.Y., Morais, D.C.: Fuzzy set based consensus schemes for multicriteria group decision making applied to strategic planning. Group Decis. Negot. **21**(2), 153–183 (2011). doi:10.1007/s10726-011-9231-0
27. Rana, A.R., Turoff, M., Hiltz, S.R.: Task and Technology Interaction (TTI): a theory of technological support for group tasks. In: Proceedings of The Thirtieth Annual Hawwaii International Conference on System Sciences, pp. 66–75. HI: IEEE Computer Society Press, Maui (1997)
28. Roy, B.: Multicriteria Methodology for Decision Aiding. Kluwer Academic Publishers, The Netherlands (1996)
29. Saaty, T.: The Analytic Hierarchy Process. McGraw-Hill, New York (1980)
30. Scandizzo, P.L., Savastano, S.: The adoption and diffusion of GM crops in United States: a real option approach. (2010)

Fuzzy Degree of Geographic Appropriateness for Social Impact Investing

Vicente Liern and Blanca Pérez-Gladish

Abstract Impact investing is an investment practice that is characterized by the explicit intentionality of attaining a social impact and the requisite of report and measure this impact in a transparent way. The investment decision making process has two main stages. In the first stage, filters are applied regarding four critical issues: target geography, impact theme, asset class and target return category. In this phase, the set of possible investment alternatives are determined based on their appropriateness for impact investment in terms of those four essential aspects. In a second stage, efficient portfolios are obtained taking into account financial criteria (maximizing expected return, minimizing risk) and trying to maximize the social impact of the portfolio of investments. In this chapter, we will focus on the establishment of the target geography for the impact investment proposing a fuzzy indicator of the appropriateness of a geographic area in terms of impact investment. This indicator will be based on Soft Computing techniques which are an attractive tool given the imprecise, ambiguous and uncertain nature of data related to social impact investment.

1 Introduction

The Global Impact Investing Network (GIIN) defines Impact Investments as investments made into companies, organizations, and funds with the intention to generate social and environmental impact alongside a financial return [2, 3, 7]. In the last years, there has been an important growth of impact investing. Impact investing provides capital in key sectors such as sustainable agriculture, clean technology, microfinance, housing, healthcare, and education, among other. Governments therefore, play a key role in supporting impact investment by facilitating as much as possible investments in their countries. In this context, the easy of doing business in a country

V. Liern (✉)
Universitat de València, Valencia, Spain
e-mail: vicente.liern@uv.es

B. Pérez-Gladish
Universidad de Oviedo, Asturias, Spain
e-mail: bperez@uniovi.es

C. Cruz Corona (ed.), *Soft Computing for Sustainability Science*,
Studies in Fuzziness and Soft Computing 358, DOI 10.1007/978-3-319-62359-7_8

is the first question a potential investor will look at. What defines an impact investor? The impact investor has the explicit and prior goal of positively influencing society through his investments. However, these investors also aim at obtaining a financial return. Nevertheless, they are more flexible than socially responsible investors are and target financial returns that range from below market to risk-adjusted market rate. They invest across all asset classes including cash equivalents, fixed income, venture capital and private equity. Among investors interested in impact investment, we can find financial institutions, pension funds, private foundations, insurance companies, development finance institutions, specialized financial institutions, large-scale family offices, fund managers and individual investors. These investors prefer to invest mainly directly into companies rather than making indirect investments [7].

In this work, we are mainly concerned about one of the main challenges to the growth of impact investing: the inadequate impact measurement practice. As reported by GIIN and J.P. Morgan in their last survey, 98% of respondents feel that standardized impact metrics are at least somewhat important to the development of the industry. The use of metrics aligned with such standards is also very important as 80% of respondents reported using metrics that align with some international standards.

Strategies in impact investing generally initially focus on three critical aspects: geography, industry sector and social impact. In this chapter, we will focus on the geographic appropriateness of impact investments proposing a fuzzy aggregate measure of the adequateness of a geographic area for impact investment in terms of the ease of doing business. We are interested in determining the universe of investments, opportunity set, for the impact investment portfolio selection problem. Time and resources constraint an investor. Knowing how to identify regions, economies and industries that are investable and poised for success is critical for being an effective investor. Weighting both, social and financial mission viability early on in the investment decision making process is critical for selecting the right investments for due diligence.

The measurement of the appropriateness degree of a geographic area (in terms of impact investment) is a priority objective of researchers on Social Responsibility. We will proposed a fuzzy measure based on Soft Computing techniques which are an attractive tool given the imprecise, ambiguous and uncertain nature of data related to social impact investment.

2 Geographic Appropriateness for Impact Investing

The starting point for impact investing is sourcing and screening. Sourcing and screening consists of selecting potential investments. The methods employed and the decisions made at this stage lead to a variety of paths that often have an influential role for decisions at later stages of the investment process. Sourcing strategies usually focus on four critical aspects: geography, industry sector or impact theme, asset class and target return class [1].

In this work, we will focus on the attainment of a global indicator of geographical appropriateness for impact investing. We will take into account two different groups of indicators. On the one hand, we will measure how ease is to do business in the target geography for impact investing. On the other hand, we will measure the human development level of the target geography as for an impact investor, the developmental need is a priority and should be reflected in the analysis [1]. We will work with 189 countries $\{C_i\}_{i=1}^{189}$. The initial list of countries is displayed in alphabetic order.

In this paper, in order to measure the ease of doing business we will use the Doing Business database from the World Bank Group (http://www.doingbusiness.org/). Doing Business provides ten quantitative sets of indicators on regulation for doing business in 189 countries referred to the following topics: dealing with construction permits, getting electricity, registering property, getting credit, protecting minority investors, paying taxes, trading across borders, enforcing contracts and resolving insolvency. The choice of these sets of indicators has been done based on economic research and firm-level data from the World Bank Enterprise Surveys. Table 1 displays a description of the topics.

Table 1 Description of country composite indicators of ease of doing business

	Indicators	Description
I_1	Starting a business	Procedures, time, cost and paid-in minimum capital to start a limited liability company
I_2	Dealing with construction permits	Procedures, time and cost to complete all formalities to build a warehouse and the quality control and safety mechanisms in the construction permitting system
I_3	Getting electricity	Procedures, time and cost to get connected to the electrical grid, the reliability of the electricity supply and the cost of electricity consumption
I_4	Registering property	Procedures, time and cost to transfer a property and the quality of the land administration system
I_5	Getting credit	Movable collateral laws and credit information systems
I_6	Protecting minority investors	Minority shareholders rights in related-party transactions and in corporate governance
I_7	Paying taxes	Payments, time and total tax rate for a firm to comply with all tax regulations
I_8	Trading across borders	Time and cost to export the product of comparative advantage and import auto parts
I_9	Enforcing contracts	Time and cost to resolve a commercial dispute and the quality of judicial processes
I_{10}	Resolving insolvency	Time, cost, outcome and recovery rate for a commercial insolvency and the strength of the legal framework for insolvency

Source http://www.doingbusiness.org

Table 2 Overall country ranking based on distance to frontier scores (DTF)

Rank	Country	DTF overall HDI
1	C_{149} = Singapore	87.34
2	C_{121} = New Zealand	86.75
3	C_{45} = Denmark	84.26
4	C_{88} = Korea, Rep.	83.91
5	C_{72} = Hong Kong SAR, China	82.87
6	C_{179} = United Kingdom	82.18
7	C_{180} = United States	82.15
8	C_{125} = Norway	81.53
9	C_{163} = Sweden	81.40
10	C_{57} = Finland	80.95

Source http://www.doingbusiness.org

Doing Business provides an aggregate measure for each data set which is called distance to frontier score (DTF) measuring the distance of each country to the frontier, which represents the best performance observed on each of the indicators across all countries in the Doing Business sample since 2005 or the third year in which data were collected for the indicator. The distance to frontier (ranging from 0 to 100, where 0 indicates the worst performance and 100 the frontier) is calculated for each indicator and then averaged across all indicators to compute the aggregate distance to frontier score. Doing Business weights all indicators equally. Table 2 displays the ranking of the top ten countries based on the overall distance to frontier score.

As acknowledge by Doing Business, the distance to frontier scores vary, often substantially, across indicators, indicating that strong performance by one country in one topic can coexist with weak performance in another (http://www.doingbusiness.org).

Let us consider for example the case of Singapore. It ranks first with respect to the DTF overall score but it performs badly when we consider trading across borders. Another interesting example is United States, performing quite well in DTF overall terms but very badly with respect to most of the individual indicators (exceptions are dealing with construction permits and resolving insolvency).

The level of human development will be measure using the Human Development Index (HDI), a statistic created by the United Nations Development Programme (UNDP). This index is a summary measure of average achievement in key dimensions of human development: a long and healthy life, being knowledgeable and have a decent standard of living. The health dimension, I_{health}, is assessed by life expectancy at birth; the education dimension, $I_{\text{education}}$, is measured by calculating the arithmetic mean of the mean of years of schooling for adults aged 25 years and more and expected years of schooling for children of school entering age (United Nations calculates the arithmetic mean of the two indicators). The standard of living dimension, income dimension, I_{income}, is measured by gross national income per capita. The HDI uses the logarithm of income, to reflect the diminishing importance of income with

Table 3 Description of indicators for the human development index

Indicators	Description
I_{11}	Life expectancy at birth (years)
I_{12}	Years of schooling for children of school entering age (years)
I_{13}	Gross national income (GNI) per capita. The HDI uses the logarithm of income, to reflect the diminishing importance of income with increasing GNI. The estimation of GNI per capita is done in 2011 purchasing power parity (2011 PPP $)

Source http://hdr.undp.org/en/content/human-development-index-hdi

increasing GNI, Gross National Income. Minimum and maximum values (goalspots) are set by United Nations in order to transform the indicators expressed in different units into dimension indexes, DI, taking values between zero and one.

$$I_{\text{health}} = \frac{\text{actual value} - \text{min value}}{\text{max value} - \text{min value}}.$$

The justification for setting these values is mainly based on historical evidence. The minimum value set for the health dimension is 20, as historical evidence shows that no country in the 20th century had a life expectancy of less than 20 years. The maximum value is set at 85 years [11, 13, 14]. For the education dimension, the minimum is set at 0 for both, expected and mean years of schooling, and the maximum is set at 18 and 15, respectively as societies can subsist without formal education, justifying the education minimum of 0 years. The maximum for mean years of schooling, 15, is the projected maximum of this indicator for 2025. The maximum for expected years of schooling, 18, is equivalent to achieving a masters degree in most countries. The low minimum value for gross national income (GNI) per capita, $100, is justified by the considerable amount of unmeasured subsistence and nonmarket production in economies close to the minimum, which is not captured in the official data. The maximum is set at $75,000 per capita. Kahneman and Deaton [9] have shown that there is a virtually no gain in human development and well-being from annual income beyond $75,000. Assuming an annual growth rate of 5%, only three countries are projected to exceed the $75,000 ceiling in the next four years. Table 3 shows a description of the four indicators used for the calculation of the three dimension indices. The scores for the three HDI dimension indices are finally aggregated into a composite index using geometric mean (http://hdr.undp.org/en/content/human-development-index-hdi).

$$HDI = (I_{\text{health}} \cdot I_{\text{education}} \cdot I_{\text{income}})^{1/3}.$$

Table 4 displays an example of the ranking provided by the UNDP based on the HDI. Notice that not all the countries provide data about human development. We

Table 4 2015 United Nations' Country Ranking based on the HDI

Rank	Country	Overall HDI
1	C_{125} = Norway	0.944
2	C_8 = Australia	0.935
3	C_{164} = Switzerland	0.930
4	C_{45} = Denmark	0.923
5	C_{120} = Netherlands	0.922
6	C_{62} = Germany	0.916
7	C_{79} = Ireland	0.916
8	C_{179} = United Kingdom	0.915
9	C_{121} = New Zealand	0.913
10	C_{31} = Canada	0.913

Source http://www.hdr.undp.org.

have handled the list of 189 countries from the World Bank and we have assigned a zero score when a human development indicator was not available for the country.

In the next section, we will present an aggregating method to obtain an overall fuzzy measure of the appropriateness of a geographic area in terms of impact investing, taking into account the variability and different nature of the scores in each topic for both, the ease to do business and the human development level.

3 Induced Ordered Weighted Averaging Operators

Ordered Weighed Average (OWA) operators are aggregation operators that provide a parameterized family of mean type aggregation operators that includes the minimum, the maximum, and the average [16]. In what follows, we will give some basic definitions.

Definition 1 A vector $w = (w_1, \cdots, w_n)$ is called a weighting vector if the following two conditions are verified:

$$w_d \in [0, 1], \quad 1 \leq d \leq n, \quad \text{and} \quad \sum_{d=1}^{n} w_d = 1.$$

In order to measure the similarity of other weighting vectors with the extreme weighting vectors we will introduce the concept of orness as follows:

Definition 2 The level of orness associated with the operator OWA_w is defined as

$$\alpha = \frac{1}{n-1} \sum_{d=1}^{n} (n-d) w_d \in [0, 1].$$

The level of orness measures the degree to which the aggregation behaves as the maximum operator or the minimum operator. Thus, degree 1 means that the operator is the maximum; degree 0 means that the operator is the minimum and in between we find all the other possibilities.

Definition 3 Given a weighting vector w, the OWA operator OWA_w is defined to aggregate a list of values $\{a_1, a_2, \cdots, a_n\}$ according to the following expression:

$$\text{OWA}_w(a_1, \cdots, a_n) = \sum_{d=1}^{n} w_d a_{\sigma(d)},$$

where $a_{\sigma(d)}$ is the d-th largest element in the collection $\{a_1, a_2, \cdots, a_n\}$, i.e., $a_{\sigma(1)} \geq \cdots \geq a_{\sigma(n)}$.

The ordering of the arguments can be induced by another variable called the order-inducing variable [18]. When this happens, we have a different class of operators named IOWA operators. IOWA operators have been widely used in the literature due to their suitable properties (see [10, 12, 17, 18]).

Definition 4 Given a weighting vector $w = (w_1, w_2, \cdots, w_n)$ and a vector of order-inducing variables $z = (z_1, z_2, \cdots, z_n)$, the IOWA operator $\text{IOWA}_{w,z}$ is defined to aggregate the second arguments of a list of 2-tuples $\{(z_1, a_1), \cdots, (z_n, a_n)\}$ according to the following expression:

$$\text{IOWA}_{w,z}\,(< z_1, a_1 >, \cdots, < z_n, a_n >) = \sum_{d=1}^{n} w_d a_{\eta(d)},$$

where $z_d = z(a_d)$, and the arguments (z_d, a_d) are rearranged in a way such that $z_{\eta(d)} \geq z_{\eta(d+1)}$, $1 \leq d \leq n-1$, i.e., $(z_{\eta(d)}, a_{\eta(d)})$ is the 2-tuple with $z_{\eta(d)}$ being the highest value in the set $\{z_1, z_2, \cdots, z_n\}$. If q of the $z_{\eta(d)}$ are tied, i.e. $z_{\eta(d)} = z_{\eta(d+1)} = \cdots = z_{\eta(d+q-1)}$, then

$$a_{\eta(d)} = \frac{1}{q} \sum_{k=\eta(d)}^{\eta(d+q-1)} a_k, \quad \text{and} \quad a_{\eta(d)} = a_{\eta(d+1)} = \cdots = a_{\eta(d+q-1)}.$$

In next section, we will obtain overall geographic appropriateness scores, based on a set of indicators related to the ease of doing business, using an IOWA operator. The order-inducing variable is chosen to quantify a certain property of the scores in each dimension. In our case, we are highly concerned about the variability of the scores within each indicator. Therefore, our induced variable will be the variance of the scores obtained by the geographic areas in each indicator. We will follow the procedure described in León et al. [10]. First, we will calculate an $m \times n$ matrix, M, with the scores of each geographic area, m, in each indicator, n. Then we will rearrange the columns in M according to the order-inducing variable. Third, we will determine the objective aggregation weights and finally, we will calculate the overall appropriateness for each geographic.

In order to calculate the weights we will use the method proposed by Wang and Parkan [15] in which they solve the so-called minimax disparity problem (see [10]):

$$\begin{aligned} &\text{Min. } \delta \\ &\text{s.t. } \frac{1}{n-1}\sum_{k=1}^{n-1}(n-k)w_k = \alpha \\ &w_1 + w_2 + \ldots + w_n = 1, \\ &w_k - w_{k+1} - \delta \leq 0, \qquad 1 \leq k \leq n-1, \\ &w_k - w_{k+1} + \delta \geq 0, \qquad 1 \leq k \leq n-1, \\ &w_k \geq 0, \qquad 1 \leq k \leq n-1. \end{aligned} \tag{1}$$

where $\alpha \in [0, 1]$ is the orness degree specified by the assets' manager. The overall score for each geographic area is the result of applying the IOWA operator to each element in a row with the obtained objective aggregation weights.

4 Country Ranking Based on a Fuzzy Appropriateness Measure

Let us apply the procedure previously described to the attainment of a fuzzy overall indicator of country appropriateness in terms of impact investing. Our sample is composed of 189 geographic areas (in this case, countries) evaluated in the 14 indicators for impact investing, 10 related to the ease to do business in the country and 4 related to the country level of human development (see Tables 1 and 4).

Several ways exist to determine the risk aversion of investors with regards to the accomplishment of selected criteria [7]. Risk aversion will depend on the type of investor. In this work we will use the degree of orness as an approach to risk aversion.

4.1 Country Ranking Based on the Ease to Do Business

We will first rank countries based on a fuzzy indicator of the ease to do business in the country (EDB_i). Let us summarize the information in a matrix $A = [a_{ij}]$ where by rows we have the countries ($1 \leq i \leq 189$) and by columns the indicators for the ease to do business in each country ($1 \leq j \leq 10$).

First step consists of the rearranging of columns according to the order-inducing variable, in our case, the variance of the scores in each indicator.

Table 5 displays the scores (from 0 to 100) of some of the countries in each indicator rearranged from higher to lower variability among countries. We can observe how

Table 5 Rearranging of ease to do business indicators based on their variance

Country	I_8	I_{10}	I_5	I_3	I_4	I_7	I_1	I_2	I_9	I_6
C_1	28.9	23.6	45	45.82	27.5	74.14	93.54	22.94	35.11	10
C_2	91.14	62.94	65	43.75	58.84	64.47	90.13	64.04	57.37	70
C_3	24.15	47.67	10	57.48	43.83	45.03	74.07	62.95	55.49	33.33
C_4	19.27	0	5	42.49	40.8	60.4	57.15	66.6	26.26	56.67
C_5	62.01	35.06	25	83.47	57.41	54.51	83.33	68.22	73.18	56.67
C_6	53	45.1	50	69.95	56.3	44.99	72.59	48.57	67.65	60
C_7	81.75	48.14	65	65.44	87.27	82.51	97.77	70.32	68.6	60
C_8	70.82	81.6	90	82.32	74.32	82.44	96.47	86.56	79.72	56.67
C_9	100	78.84	60	87.68	80.8	76.52	83.42	74.79	78.24	63.33
C_{10}	69.59	44.59	40	63	82.54	83.77	95.54	61.58	67.51	58.33

Table 6 Aggregating weights for different orness levels

Weights	$\alpha = 0$	$\alpha = 0.25$	$\alpha = 0.5$	$\alpha = 0.75$	$\alpha = 1$
w_1	0	0.050	0.100	0.229	1.00
w_2	0	0.050	0.100	0.199	0.00
w_3	0	0.050	0.100	0.200	0.00
w_4	0	0.050	0.100	0.140	0.00
w_5	0	0.050	0.100	0.110	0.00
w_6	0	0.050	0.100	0.080	0.00
w_7	0	0.050	0.100	0.051	0.00
w_8	0	0.050	0.100	0.021	0.00
w_9	0	0.050	0.100	0.000	0.00
w_{10}	1	0.550	0.100	0.000	0.00

the indicator related to trading across borders has more than three times variability than protecting minority investors.

Once dimensions have been rearranged according to the order-inducing variable, the variance, we obtain the aggregation weights for the ten indicators and for different orness levels, $\alpha \in [0, 1]$ (see Table 6).

The overall ease to do business scores, EDB_i, are obtained aggregating for each country C_i the weighted scores of the ease to do business indicators.

In this work, for the construction of EDB_i we apply the IOWA operator to each element in a row (obtained scores of the country in each topic or indicator) with the aggregation weights obtained in the previous step (see Table 7 which shows data for the ten first countries).

Table 7 IOWA overall ease to do business scores

Country	$\alpha = 0$	$\alpha = 0.25$	$\alpha = 0.5$	$\alpha = 0.75$	$\alpha = 1$
C_1	1.00	2.533	4.066	4.091	2.890
C_2	7.00	6.838	6.677	7.007	9.114
C_3	3.33	3.937	4.540	3.857	2.415
C_4	5.67	4.707	3.746	2.498	1.927
C_5	5.67	5.828	5.989	5.420	6.201
C_6	6.00	5.841	5.682	5.540	5.300
C_7	6.00	6.634	7.268	7.310	8.175
C_8	5.67	6.838	8.010	8.346	7.082
C_9	6.33	7.085	7.836	8.369	10.00
C_{10}	5.83	6.249	6.665	6.356	6.959

Table 8 Ranks of the countries based on IOWA for different orness levels

Rank	$\alpha = 0$	$\alpha = 0.25$	$\alpha = 0.5$	$\alpha = 0.75$	$\alpha = 1$
1	C_{72}	C_{149}	C_{149}	C_{180}	C_9
2	C_{121}	C_{121}	C_{121}	C_{45}	C_{16}
3	C_{149}	C_{72}	C_{45}	C_{121}	C_{42}
4	C_{103}	C_{179}	C_{88}	C_{57}	C_{44}
5	C_{179}	C_{103}	C_{72}	C_{88}	C_{45}
6	C_{31}	C_{88}	C_{179}	C_{149}	C_{58}
7	C_{151}	C_{31}	C_{180}	C_{62}	C_{73}
8	C_{75}	C_{125}	C_{125}	C_{179}	C_{81}
9	C_{80}	C_{163}	C_{163}	C_{163}	C_{99}
10	C_{88}	C_{45}	C_{57}	C_{31}	C_{120}

The resulting rankings for the ten first positions and for different orness levels are displayed in Table 8.

Notice that for an orness level of 0.5 equal importance is given to all the indicators and therefore, the obtained ranking using IOWA is the same than the ranking provided by the World Bank. For other orness levels different importance is given to the different indicators depending on their variability, which can be interpreted as a measure of risk.

Table 9 Rearranging of human development normalized indicators based on their variance

Country	I_{13}	I_{11}	I_{12}
C_1	0.444	0.622	0.365
C_2	0.695	0.889	0.637
C_3	0.736	0.843	0.642
C_4	0.638	0.497	0.474
C_5	0.801	0.863	0.695

Table 10 Aggregating weights for different orness levels

Weights	$\alpha = 0$	$\alpha = 0.25$	$\alpha = 0.5$	$\alpha = 0.75$	$\alpha = 1$
w_{11}	0	0.167	0.33	0.5	1
w_{12}	0	0.167	0.33	0.5	0
w_{13}	0	0.667	0.33	0	0

4.2 Country Ranking Based on Their Human Development Level

We will now rank countries based on a fuzzy indicator of their human development level, HDL_i. Let us summarize the information in a matrix $B = [b_{ik}]$ where by rows we have the countries ($1 \leq i \leq 189$) and by columns the indicators for the level of human development in each country ($1 \leq k \leq 3$).

As for the ease to do business case, the first step consists of the rearranging of columns according to the order-inducing variable, in our case, the variance of the scores in each indicator (see Table 9 for an example).

Table 9 shows indicators rearranged from higher to lower variability among countries. As we can observe, the indicator with the highest variability (risk) is the indicator for the income dimension, measuring the standard of living in the countries. Then, we have the health dimension measured by years of life expectancy at birth and finally the education index which takes into account expected and mean years of schooling. Once dimensions have been rearranged according to the order-inducing variable we obtain the aggregation weights for different orness levels, $\alpha \in [0, 1]$, (see Table 10).

The overall human development scores, HDL_i, are obtained aggregating for each country C_i the weighted scores of the human development indicators.

In this work, we calculate the indicator HDL_i applying the IOWA operator to each element in a row using the aggregation weights obtained in the previous step (see Table 11 for an example). The resulting rankings for the ten first positions and for different orness levels are displayed in Table 12.

Table 11 IOWA overall human development scores

Country	$\alpha = 0$	$\alpha = 0.25$	$\alpha = 0.5$	$\alpha = 0.75$	$\alpha = 1$
C_1	0.12	0.14	0.16	0.18	0.15
C_2	0.21	0.23	0.25	0.26	0.23
C_3	0.21	0.23	0.25	0.26	0.25
C_4	0.16	0.17	0.18	0.19	0.21
C_5	0.23	0.25	0.26	0.28	0.27

Table 12 Ranking of the countries based on human development level using IOWA

Rank	$\alpha = 0$	$\alpha = 0.25$	$\alpha = 0.5$	$\alpha = 0.75$	$\alpha = 1$
1	C_8	C_8	C_8	C_{149}	C_{137}
2	C_{121}	C_{121}	C_{125}	C_{137}	C_{90}
3	C_{45}	C_{45}	C_{164}	C_{72}	C_{149}
4	C_{79}	C_{125}	C_{45}	C_{164}	C_{24}
5	C_{125}	C_{79}	C_{121}	C_{125}	C_{125}
6	C_{120}	C_{120}	C_{120}	C_{99}	C_{177}
7	C_{62}	C_{62}	C_{79}	C_{24}	C_{99}
8	C_{179}	C_{179}	C_{149}	C_{163}	C_{164}
9	C_{178}	C_{164}	C_{62}	C_8	C_{72}
10	C_{74}	C_{178}	C_{179}	C_{120}	C_{179}

4.3 Country Ranking Based on Their Appropriateness for Impact Investing

Once overall scores have been obtained for the ease to do business and for the human development level next step consists of the ranking of the countries based on their appropriateness in terms of impact investing.

For each country C_i, we define the **degree of impact investing appropriateness**, AD_i, as an aggregation of the appropriateness degree in terms of ease to do business EDB_i and the appropriateness degree in terms of human development HDL_i.

In this work, for a given level of orness $\alpha \in [0, 1]$, the $AD_i(\alpha)$ is calculated as an aggregation of the appropriateness degree in terms of ease to do business $EDB_i(\alpha)$ and the appropriateness degree in terms of human development $HDL_i(\alpha)$.

As it is well known, aggregation of both appropriateness degrees can be done through several methods. In this work we will use the TOPSIS, Technique for Order Preference by Similarity to Ideal Solution, [8]. The standard TOPSIS method attempts to choose alternatives that simultaneously have the shortest distance from the positive ideal solution and the farther distance from the negative ideal solution. The positive ideal solution maximizes the benefit criteria and minimizes the cost criteria, whereas the negative ideal solution maximizes the cost criteria and minimizes the benefit criteria. TOPSIS makes full use of the attribute information, provides a cardinal ranking of alternatives, and does not require the attribute preferences to be independent [5, 19]. To apply this technique attribute values must be numeric, monotonically increasing or decreasing, and have commensurable units [4]. The stepwise procedure proposed by Hwang and Yoon [8] is:

Step 1. Determine the decision matrix $D = [D_{ij}]_{n \times m}$, where the number of criteria is n and the number of alternatives is m.
Step 2. Construct the normalized decision matrix. Criteria are expressed in different scaling and therefore a normalizing procedure is necessary in order to facilitate comparison (see Sect. 3 and [8]).
Step 3. Determine the weighted normalized decision matrix. It is well known that the weights of the criteria in decision making problems do not have the same mean and not all of them have the same importance. The weighted normalized value v_{ij} is calculated as:

$$v_{ij} = w_j \, r_{ij}, \quad 1 \leq i \leq m, \quad 1 \leq j \leq n.$$

where w_j is the weight associated to each criterion.
Step 4. Determine the positive ideal (PIS) and the negative ideal (NIS) solutions. The positive ideal value set A^+ and the negative ideal value set A^- are determined as follows:

$$\begin{aligned} A^+ &= \{v_1^+, \cdots, v_n^+\} = \left\{\max_i v_{ij}, \; j \in J \quad (\text{or } \min_i v_{ij}, \; j \in J')\right\} \\ A^- &= \{v_1^-, \cdots, v_n^-\} = \left\{\min_i v_{ij}, \; j \in J \quad (\text{or } \max_i v_{ij}, \; j \in J')\right\} \end{aligned}$$

where J is associated with the criteria that indicate profits or benefits and J' is associated with the criteria that indicate costs or losses.
Step 5. Calculate the separation measures. Calculation of the separation of each alternative with respect to the *PIS* and *NIS*, respectively:

$$d_i^+ = \left\{ \sum_{j=1}^{n} \left(v_{ij} - v_j^+ \right)^2 \right\}^{1/2}, \quad d_i^- = \left\{ \sum_{j=1}^{n} \left(v_{ij} - v_j^- \right)^2 \right\}^{1/2}, \quad 1 \leq i \leq m.$$

Step 6. Calculate the relative proximity to the ideal solution. Calculation of the relative proximity of each alternative to the PIS and NIS using the proximity index.

$$R_i = \frac{d_i^-}{d_i^+ + d_i^-}, \quad 1 \leq i \leq m.$$

Table 13 Ranking of the countries based on country appropriateness degree

Orness level	Ranking
$\alpha = 0$	$C_{186} \succ C_{142} \succ C_{136} \succ C_{189} \succ C_{107} \succ C_{89} \succ C_{53} \succ C_{123} \succ C_{26} \succ C_{33} \cdots$
$\alpha = 0.25$	$C_{186} \succ C_{189} \succ C_{107} \succ C_{142} \succ C_{136} \succ C_{89} \succ C_{185} \succ C_{188} \succ C_{123} \succ C_{148} \cdots$
$\alpha = 0.5$	$C_{189} \succ C_{186} \succ C_{142} \succ C_{89} \succ C_{185} \succ C_{136} \succ C_{107} \succ C_{188} \succ C_{187} \succ C_{175} \cdots$
$\alpha = 0.75$	$C_{189} \succ C_{186} \succ C_{185} \succ C_{188} \succ C_{142} \succ C_{89} \succ C_{187} \succ C_{136} \succ C_{107} \succ C_{175} \cdots$
$\alpha = 1$	$C_{188} \succ C_{189} \succ C_{185} \succ C_{136} \succ C_{186} \succ C_{89} \succ C_{142} \succ C_{32} \succ C_{102} \succ C_{123} \cdots$

The value R_i lies between 0 and 1. If $R_i = 1$, then $A_i = A_i^+$, and if $R_i = 0$, then $A_i = A_i^-$. The closer the R_i value is to 1 the higher the priority of the i-th alternative is.

Step 7. Rank the preference order. Rank the best alternatives according to R_i in descending order.

In our case the alternatives are the countries, $j = 1, \cdots, 189$, and two criteria, i, are the appropriateness degree in terms of ease to do business and the appropriateness of the country in terms of human development level. The objectives for an impact investor are to maximize the ease to do business and to minimize the human development level (as less developed countries are target geographical areas for social impact investing). The positive ideal country in terms of appropriateness for impact investing is defined as the one with the maximum degree of appropriateness in terms of ease to do business and the one with minimum human development level. Both optima are obtained from the country scores in both dimensions. The negative ideal country is defined as the country with minimum appropriateness for easy to do business and maximum human development level. For the sake of simplicity, we will consider equal weights for both objectives and we will obtain different rankings for the different orness levels using data from Tables 7 and 11. The obtained country rankings for the ten first positions are displayed in Table 13.

It is interesting to observe how the ranking of the countries completely changes when both, the ease to do business and the human level development are taken into account. For any investor, the more the ease to do business the more attractive the country is to invest in. However, for an impact investor, the more appropriate geographical areas are those presenting a low human development level. Thus, countries with bad positions in the ranking based on human development indicators rank in good positions when both aspects, ease to do business and human development are considered.

5 Conclusions

Impact investors made investment decisions in a two–stages process. Impact investors firstly target geography, impact theme, asset class and return category. This implies the establishment of several filters to the universe of investments in order to obtain the decision impact investment set. In a second step, efficient investment portfolios are obtained taking into account financial criteria (maximizing expected return, minimizing risk) and trying to maximize the social impact.

With the objective of assisting investors in the first stage of their investment decision making problem, we have defined the degree of the appropriateness of a country in terms of impact investment. This degree of country appropriateness for impact investing is based on the appropriateness degree of the countries with respect to the ease to do business and to the level of human development. The construction of the country appropriateness degree is based on Soft Computing techniques which are an attractive tool given the imprecise, ambiguous and uncertain nature of data related to social impact investment. Aggregation of the two considered dimensions, ease to do business and human development level, has been done through TOPSIS. Once the ideal and anti-ideal countries have been defined with regards to their appropriateness for impact investing we are able to obtain a ranking of the countries which shows their suitability in terms of thirteen indicators related to the ease to do business and human development.

References

1. Allman K., Escobar de Nogales X.: Impact Investment: A Practical Guide to Investment Process and Social Impact Analysis. Wiley (2015)
2. Ballestero, E., Bravo, M., Pérez-Gladish, B., Arenas-Parra, M., Plá-Santamaria, D.: Socially responsible investment: a multicriteria approach to portfolio selection combining ethical and financial objectives. Eur. J. Oper. Res. **216**(2), 487–494 (2012)
3. Ballestero E, Pérez-Gladish, B., García-Bernabeu A. (eds.): Socially Responsible Investment. A Multi-criteria Decision Making Approach. International Series in Operations Research & Management Science, vol. 219, Springer, Berlin (2015)
4. Behzadian, M., Otaghsara, S.K., Yazdani, M., Ignatius, J.: A state-of the-art survey of TOPSIS applications. Expert Syst. Appl. **39**, 13051–13069 (2012)
5. Chen, S.J., Hwang, C.L.: Fuzzy multiple attribute decision making: Methods and applications. Springer-Verlag, Berlin (1992)
6. Dorfleitner, G., Utz, S.: Safety first portfolio choice based on financial and sustainability returns. Eur. J. Oper. Res. **221**(1), 155–164 (2012)
7. GIIN, J.P. Morgan,: Eyes on the horizon. The Impact Investor Survey (2015). Accessed http://www.thegiin.org/assets/documents/pub/2015
8. Hwang, C. L., Yoon, K.: Multiple Attributes Decision Making Methods. Springer, Berlin (1981)
9. Kahneman, D., Deaton, A.: High income improves evaluation of life but not emotional well-being. Psychological and cognitive Sciences. Proc. Natl. Acad. Sci. **107**(38), 1648916493 (2010)
10. León, T., Ramón, N., Ruiz, J.L., Sirvent, I.: Using Induced Ordered Weighted Averaging (IOWA) operators for aggregation in cross-efficiency evaluations. Int. J. Intell. Syst. **29**, 1100–1126 (2014)
11. Maddison, A.: Historical Statistics of the World Economy, 12030 AD. Organisation for Economic Co-operation and Development, Paris (2010)
12. Merigó, J.M., Gil Lafuente, A.M., Zhou, L.G., Chen, H.Y.: Induced and linguistic generalized aggregation operators and their application in linguistic group decision making. Sringer Verlag, Holland, Group Decision and Negotiation (2012)
13. Oeppen, J., Vaupel, J.W.: Broken limits to life expectancy. Science **296**, 10291031 (2002)

14. Riley, J.C.: Poverty and Life Expectancy. Cambridge University Press, Cambridge, UK (2005)
15. Wang, Y.M., Parkan, C.: A minimax disparity approach for obtaining OWA operator weights. Inf. Sci. **175**(1–2), 20–29 (2005)
16. Yager, R.R.: On ordered weighted averaging aggregation operators in multi-criteria decision making. IEEE Trans. Syst. Man Cybern. **18**(1), 183–190 (1988)
17. Yager, R.R.: Induced aggregation operators. Fuzzy Sets Syst. **137**(1), 59–69 (2003)
18. Yager, R.R., Filev, D.P.: Induced ordered weighted averaging operators. IEEE Trans. Syst. Man Cybern. Part B Cybern. **29**, 141–150 (1999)
19. Yoon, K.P., Hwang, C.L.: Multiple attribute decision making. Sage Publication, Thousand Oaks, CA (1995)

A New Approach for Information Dissemination in VANETs Based on Covering Location and Metaheuristics

Antonio D. Masegosa, Idoia de la Iglesia, Unai Hernandez-Jayo, Luis Enrique Diez, Alfonso Bahillo and Enrique Onieva

Abstract Vehicular Ad-Hoc Networks (VANETs) have attracted a high interest in recent years due to the huge number of innovative applications that they can enable. Some of these applications can have a high impact on reducing Greenhouse Gas emissions produced by vehicles, especially those related to traffic management and driver assistance. Many of these services require disseminating information from a central server to a set of vehicles located in a particular region. This task presents important challenges in VANETs, especially when it is made at big scale. In this work, we present a new approach for information dissemination in VANETs where the structure of the communications is configured using a model based on Covering Location Problems that it is optimized by means of a Genetic Algorithm. The results obtained over a realistic scenario show that the new approach can provide good solutions for very demanding response times and that obtains competitive results with respect to reference algorithms proposed in literature.

A.D. Masegosa (✉) · I. de la Iglesia · U. Hernandez-Jayo · L.E. Diez · A. Bahillo · E. Onieva
DeustoTech-Fundacion Deusto, Deusto Foundation, Avda Universidades, 24,
48007 Bilbao, Spain
e-mail: ad.masegosa@deusto.es

I. de la Iglesia
e-mail: idoia.delaiglesia@deusto.es

U. Hernandez-Jayo
e-mail: unai.hernandez@deusto.es

L.E. Diez
e-mail: luis.enrique.diez@deusto.es

A. Bahillo
e-mail: alfonso.bahillo@deusto.es

E. Onieva
e-mail: enrique.onieva@deusto.es

A.D. Masegosa · I. de la Iglesia · U. Hernandez-Jayo · L.E. Diez · A. Bahillo · E. Onieva
Faculty of Engineering, University of Deusto, Avda Universidades, 24,
48007 Bilbao, Spain

A.D. Masegosa
IKERBASQUE, Basque Foundation for Science,
Maria Diaz de Haro, 3, 48013 Bilbao, Spain

C. Cruz Corona (ed.), *Soft Computing for Sustainability Science*,
Studies in Fuzziness and Soft Computing 358, DOI 10.1007/978-3-319-62359-7_9

1 Introduction

Vehicular Ad-Hoc Networks (VANETs) are communication networks in which the nodes are vehicles [12]. They have aroused the interest of the scientific community, automobile industry and institutions due to the huge number of innovative applications that they can enable [19]. Some of the most important application areas are: security (warnings about emergency break, collision at intersection, line shift, etc.); leisure and entertainment (multimedia content download, nearby points of interest etc.); traffic management (virtual traffic lights, limited access zones, electronic toll, etc.); and driver assistance (remote diagnosis; efficient and eco-driving; etc.). The two last application areas can have a big impact on the reduction of Greenhouse Gas emissions produced by vehicles as shown by some recent works [7, 11, 17]. The communications that take place within VANETs can be classified in Infrastructure-to-Infrastructure (I2I), Infrastructure-to-Vehicle (I2V), Vehicle-to-Infrastructure (V2I) and Vehicle-to-Vehicle (V2V) depending on which agent is the transmitter and the receptor. Many of the services previously mentioned require a central server that gathers all the information, generating a necessity of collecting and disseminating data from/to vehicles, respectively [8].

In this work, we focus on information dissemination from a central server to vehicles [18]. This task presents important challenges in VANETs, especially when it is made at big scale [8]. The main reasons are the diverse network topologies, the high speed of vehicles, the unequal density of them, the lack of infrastructure in roads, as well as the availability of communication technologies and their different penetration degrees. One of the approaches proposed to address the information dissemination in VANETs is the use of Virtual Infrastructures (VIs) [8]. The node of the VIs can be static (as Road Side Units (RSUs) [12]) or dynamic (vehicles). They receive information from a central server through I2I or I2V communications, to then disseminate it to nearby vehicles by using I2V and V2V communications, respectively.

The proper selection of the vehicles that act as VI plays a pivotal role in this type of approaches. It can avoid the necessity of fix infrastructure (as RSUs), reduce the network overload, or influence in the quality of the communications [8]. This selection depends on the specific service to be implemented and the requirements established by the corresponding standard [19]: messaging type, latency, message period, area range, priority or available communication technologies. Although some of the services have as target a reduced set of vehicles (e.g. emergency break or collision alert), many of them require disseminating information to the maximum number of vehicles that are located in a specific geographic area. In this last case, the objective of the VI selection method must be to identify the minimum number of nodes that are able to disseminate the information to the maximum number of vehicles while fulfilling the standard requirements [8].

In this book chapter, we propose to address this selection process as a Covering Location Problem (CLP) [9]. Roughly, the objective of these optimization problems consists on, given a set of demand nodes and a set of potential location for facilities, to determine the location and minimum number of facilities needed to cover all demand. In the case that concern us here, the demand nodes correspond to vehicles that must receive the information, whereas the potential facility locations, to those vehicles equipped with V2I communications that can, therefore, act as VI.

Basing on the communication architecture for VANETs proposed in [8] and called NAVI (Neighbor-Aware Virtual Infrastructure), we present here a new approach for information dissemination in VANETs where the process for selecting the VI is modeled as a CLP and where the underlying optimization problem is solved using a generational Genetic Algorithm (GA) with elitism [15]. To assess the validity of the proposal, we use a real scenario consisting of 45 vehicles moving in the downtown area of the city of Malaga in Spain. On one hand, the experimentation done aim to evaluate the performance of the proposal for different maximum response times which corresponds to the maximum time permitted to find a VI configuration. This factor is quite relevant due to the high demanding latency time (time elapsed between sending and receiving the message) imposed by some of the services. On the other hand, the objective of the experimentation is to test the competitiveness of our proposal by comparing it with a reference algorithm.

The next part of the chapter is structured as follows. Section 2 give some background information about information dissemination in VANETs and CLPs. The presented approach is described in Sect. 3 where we first introduce NAVI's communication architecture to then describe proposed model for VI selection based on CLPs and the GA designed to solve the underlying optimization problem. Section 4 is devoted to detail the experimental framework used to test our proposal. Concretely, it presents the testing scenario, the reference algorithm used to assess the competitiveness of our proposal, the performance measures and the implementation details. After that, in Sect. 5 we study the results obtained in both the analysis of the maximum response time and the comparison with the reference approach. Finally, we discuss the main conclusion of the work as well as the future research lines in the last section.

2 Related Work

2.1 Information Dissemination in VANETs

Due to the interest of deploying cooperative services and applications for vehicles, information dissemination in VANETs has been extensively researched [18]. The first aim of a VANET is to be an infrastructure-less self-organizing traffic information system. Therefore, the vast majority of proposed methods for information dissemination are based on a decentralized architecture where the organization of the network is managed by vehicles creating dynamic clusters enabled by short-range communication technologies. Each cluster is formed by a group of vehicles, also called cluster members (CM), and a Cluster Head (CH) which controls and executes the dissemination process, that is, it distributes information received from a central server to CMs, and performs inter-cluster communication. Vehicles create and maintain clusters based on different metrics (e.g. position, direction, speed, link quality) by periodically exchanging of status information [13]. However, when the density of the network is high, vehicles have to exchange a high number of messages to organize the network, and therefore they create additional traffic data and consume a high quantity of resources only for network management.

Other solutions try to reduce this overhead deploying infrastructure nodes at preferential locations, commonly intersections. The main problems of these approaches are the lack of flexibility and the predefined dissemination coverage [18]. Few works in the literature propose to use vehicles as mobile infrastructure. Camara et al. [3] present the virtual RSU (vRSU) concept where nodes receive and cache messages from other vRSU or access points which are located in areas with no coverage from conventional RSUs.

The majority of previously proposed methods use single-technology for information dissemination inside the VANET, namely short range communication technologies as IEEE 802.11p, similar to common WiFi networks. Recently, as an alternative to the IEEE 802.11p based VANET, the usage of cellular technologies has been investigated. The standardization of broadcast/multicast services by the Third Generation Partnership Project (3GPP) incited the research of using cellular technologies for message dissemination in VANETs [1]. However, the high cost of Long-Term Evolution (LTE) communication (commercially known as 4G network) between vehicles and base stations, the overload of base station by broadcast messages when there is high density of vehicles and the high number of hand-offs in the base station because of the high mobility of vehicles make not feasible a pure LTE based architecture for VANETs.

Recently, heterogeneous architectures have been proposed to exploit both the wide range low latency communications of cellular technologies and the low cost of IEEE 802.11p. Some works use the heterogeneous architecture to improve the efficiency of clustering [8, 20]. In [20], Remy et al. use a heterogeneous centralized architecture to reduce the clustering overhead. An interesting proposal can be found in [8] where D'Orey et al. present a method, named Neighbor-Aware Virtual Infrastructure (NAVI), for information dissemination using mobile infrastructure selected from a central entity called GeoServer. Concretely, the dissemination process is supported by a Virtual Infrastructure (VI) that represents a subset of vehicles that, through the use of LTE and IEEE 802.11p technologies, transmit the information from the central entity to those vehicles that must receive the data. In this way, NAVI allows ensuring an appropriate uplink performance, alleviating the use of fixed infrastructure, exploiting the advantages of individual technologies and being flexible enough to consider the different penetration rates of these technologies. This is the architecture we used in this study and we will describe it more in detail in Sect. 3.1.

As we explained in the introduction, the selection of the vehicles used as VI has a very high influence in the performance of this VANET architecture and it can be modeled as a CLP [9]. Concretely, the objective of this model would consists on maximizing the covering (vehicles that receive the message) while minimizing the number of vehicles used as VI. In the next section we define CLPs formally and review some literature about them.

2.2 *Covering Locations Problems*

CLPs aims at locating one or more facilities, considering that the distance or travel time from any demand point to its closest facility is reasonable [9]. A demand node is considered covered if at least one facility is located at a distance or travel time lower than or equal to a predefined threshold. This predefined threshold is known as the coverage radius, and it established how close a facility must be from a node to provide it a good quality service. For example, in some applications as emergency services, the coverage radius usually corresponds with the maximum time allowed for ambulances, fire trucks, etc. to travel for the base to the place of the incident [6, 23]; whereas in telecommunications, it corresponds to the maximum distance from a terminal to an antenna at which the quality of the signal received is good enough [14].

In the next part of this section, we describe and formulate the two most important covering location problems: the Set Covering Location Problem and the Maximum Covering Location Problem.

2.2.1 The Set Covering Location Problem (SCLP)

The objective of the SCLP consists on minimizing the number of facilities required to give full coverage to all demand nodes [21]. In this way, each demand node must have assigned a facility at a distance or time lower than the coverage radius, and all demand nodes have the same importance. The SCLP is formulated as follows:

J - set of potential locations for the facilities.

I - set of demand nodes.

S - the covering radius.

N_i - $\{j \in J | d_{ij} \leqslant S\}$, the set of potential facility locations that can cover the node i, that is, those located at a distance or time lower or equal to S, where d_{ij} is the distance between the node i and the facility potential location j.

x_j - boolean variable whose value is 1 if a facility is located in the node j, 0 otherwise.

$$Minimize : Z = \sum_{j \in J} x_j \tag{1}$$

Subject to:

$$\sum_{j \in N_i} x_j \geqslant 1 \qquad \forall i \in I \tag{2}$$

$$x_j = \{0, 1\} \qquad \forall j \in J \tag{3}$$

The objective function given in Eq. (1) minimizes the number of facilities required to cover all demand. Equation (2) ensures that each demand node i is covered by at least one facility, and finally, Eq. (3) enforces the binary restriction on the decision variable.

2.2.2 The Maximal Covering Location Problem (MCLP)

Unlike SCLP, in the MCLP the number of facilities to be located is established previously at a value p and each node can have a different importance that is established by its weight [4]. These weights are usually known as the amount of demand located at the demand node (e.g. the number of mobile phones or incidents in a certain area). The objective of the MCLP consists on finding the location of the p facilities that maximizes the amount of demand covered. The mathematical formulation of the MCLP is as follows:

I - set of the demand nodes.

J - set of potential locations for the facilities.

S - the covering radius.

N_i - $\{j \in J | d_{ij} \leqslant S\}$, the set of potential facility locations that can cover the node i, that is, those located at a distance or time lower or equal to S, where d_{ij} is the distance between the node i and the facility potential location j.

x_j - boolean variable whose value is 1 if a facility is located in the node j, 0 otherwise.

y_i - represent the coverage of node i. Its value is 1 if node i is covered ($\exists_j \in J | x_j = 1 \wedge j \in N_i$), and 0 otherwise.

a_i - amount of demand associated to node i.

p - the number of facilities to be located.

$$Maximize \quad Z = \sum_{i \in I} a_i y_i \tag{4}$$

Subject to:

$$\sum_{j \in N_i} x_j \geqslant y_i \qquad \forall i \in I \tag{5}$$

$$\sum_{j \in J} x_j = p \tag{6}$$

$$x_j = \{0, 1\} \qquad \forall j \in J \tag{7}$$

$$y_i = \{0, 1\} \qquad \forall i \in I \tag{8}$$

Equation (4) corresponds to the objective function which maximizes the demand covered by the facilities. Equation (5) establishes that a demand node is covered only when at least one facility is located at a distance or time lower or equal to the coverage radius S from it. Equation (6) ensures that the number of facilities that are located is p. Finally, Eqs. (7) and (8) restrict variables x_i and y_i to binary values.

3 Description of the New Approach for Information Dissemination in VANETs

This section is devoted to describe the new approach for information dissemination that we present in this chapter. As mentioned above, this new approach is based on the network architecture proposed for NAVI and it models the selection of the VI as a CLP [8]. To facilitate the understanding of the proposal, we first introduce NAVI's network architecture; then, we give the formal description of the CLP-based model for selecting the vehicles used as VI; and finally, the GA developed to solve the optimization problem is shown.

3.1 NAVI's Network Architecture

The architecture of NAVI was designed for an efficient collection and dissemination of information in vehicular networks through the selection of mobile infrastructure nodes (vehicles) in a scenario with multiple technologies [8]. Figure 1 shows a scheme of the general architecture of the system. The system comprises of a heterogeneous

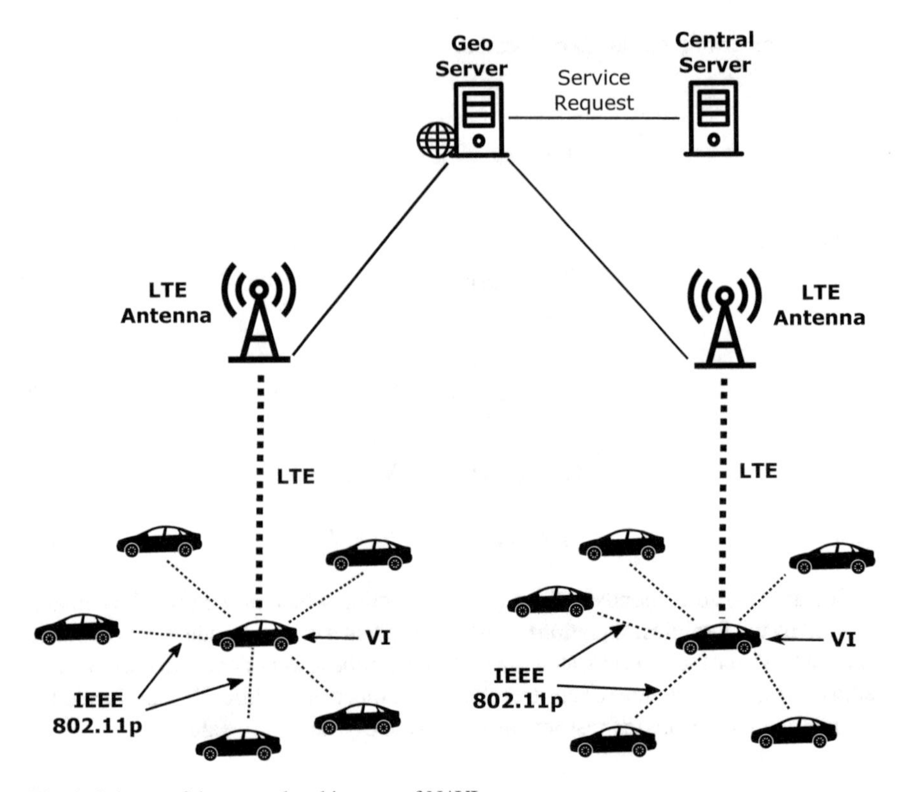

Fig. 1 Scheme of the general architecture of NAVI

network architecture consisting on short-range communication networks, as IEEE 802.11p, and long-range communication networks, as LTE. Vehicles can be categorized into three main classes in terms of network equipment: (i) short-range communication only, (ii) long-range communication only and (iii) short and long-range communication. Vehicles with short-range communication systems exchange periodic, single-hop broadcast Cooperative Awareness Messages (CAMs) containing static and dynamic vehicle information (e.g. position, speed or heading). The model considers ubiquitous availability of a long-range communication technology (e.g. UMTS, LTE) and that the resources in these networks are scarce, although a simple extension could accommodate partial deployment scenarios.

The working of the system can be divided in three main stages: (a) data collection, (b) virtual infrastructure selection and (c) data dissemination strategy execution. In this section we give a short overview of these three phases. The interested reader is referred to [8] for more details.

3.1.1 Data Collection

The periodic broadcast single-hop CAMs are used to establish when two vehicles can communicate between them. Concretely, if a vehicle A receives a CAM message from another vehicle B, it means that they can transfer information among them. In this case, we establish that B is a neighbour of A. Each vehicle has it own Neighbourhood Table (NT) with an entry for each neighbour, that is, for each vehicle from which it receives a CAM message. The entries in the NT are deleted when a certain timeout period is reached. The NTs are aggregated and transmitted periodically to the Geoserver by a subset of nodes in the Region of Interest (ROI) that can be established centrally by the GeoServer or distributed. The aggregated NTs are then used for the next phase, the VI selection.

3.1.2 Virtual Infrastructure Selection

Every certain time interval, whose length t is previously defined, the VI is updated from the information provided by the NTs. The selection of the VI is made centrally by Geoserver. This can use other data sources, as coverage information from network operators, apart from the NTs. The Geoserver launches this process each time it receives a dissemination request from the service provider.

3.1.3 Data Dissemination Strategy Execution

The vehicles selected in the previous step are in charge of performing the corresponding action: broadcast, relaying, and store-and-forward. These vehicles can also request to other nodes the propagation of this information to other vehicles. The VI nodes can also adapt the instructions of the Geoserver in case of an alteration of local conditions, and they can also be used for transferring information back to the server.

3.2 Virtual Infrastructure Selection Modeling

In this section, we describe the model proposed for the VI selection process. The objective of this process is the efficient transfer of information from a central entity to the vehicles by means of the proper configuration of the VI. Concretely, it aims to balance the amount of data transferred through long-range communication (e.g. standard LTE) and short-range communication networks (e.g. standard IEEE 802.11p) in such a way that minimizes the use of long-range networks while maximizing the coverage area. As explained in [8], although data offloading to short-rage communication networks increases the overhead in short-range networks, it results in gains in the use of cellular networks which usually have a pay-per-use system, as the common data plans offered by mobile operators.

The formulation used to model this optimization process is based on CLPs. The model can be divided into three main components: vehicles, scenario and network systems. Vehicles have access to network infrastructure resources and move in a given scenario I that it is partitioned in m zones $i \in I$. The properties of the scenario impact vehicle mobility and the communication reliability between vehicles, but we assume that during the period between the creation of the NTs and the dissemination of information they remain unaltered. Taking into account that the length of this period is usually lower than 100 ms, this assumption can be considered as realistic. The vehicles located in the scenario at the instant in which a dissemination request is received by the GeoServer it is represented by the set J. During the observation period, additional vehicles may join or leave the scenario depending on demand or routes, but as mentioned before, given the short length of the period, these changes are irrelevant. The objective of the underlying optimization process is to find the minimum subset $V \subseteq J$ that maximizes the number of zones covered in the scenario I. More formally, the model is defined as follows:

I - set of zones to cover in the region. Each zone $i \in I$ has at least one vehicle located in it.

J - set of vehicles that must be covered.

J^{VI} - subset of vehicles from J ($J^{VI} \subseteq J$) that can be potentially used as VI, that is, those equipped with short and long range communication capabilities.

CAM_{jk} - boolean variable whose value is 1 if vehicle $j \in J$ has received a CAM message from vehicle $k \in J$.

NV_j - $\{k \in J^{VI} | CAM_{jk} = 1\}$, the set of neighbours of vehicle $j \in J$.

ZV_j - zone $i \in I$ in which vehicle $j \in J$ is located.

V_i - $\{j \in J | ZV_j = i\}$, the set of vehicles located in zone i.

N_i - $\{j \in J^{VI} | (j \in V_i) \vee (\exists k \in V_i \ s.t. \ j \in NV_k)\}$, the set of vehicles that, potentially, can cover zone i, that is, those vehicle from J^{VI} that can communicate with at least one vehicle located in i.

x_j - boolean variable whose value is 1 if vehicle $j \in J^{VI}$ is selected as VI, 0 otherwise.

y_i - represent the coverage of node i. Its value is 1 if node i is covered and 0 otherwise.

p - maximum number of vehicles that can be used as VI.

$$Maximize \quad Z = (1 - F_1) \cdot F_2 \tag{9}$$

where:

$$F_1 = \frac{1}{|J^{VI}|} \sum_{j \in J^{VI}} x_j \tag{10}$$

$$F_2 = \frac{1}{|I|} \sum_{i \in I} y_i + \frac{1}{|I||J|}(-1 + \sum_{i \in I} y_i |V_i|) \tag{11}$$

Subject to:

$$\sum_{j \in N_i} x_j \geqslant y_i \qquad \forall i \in I \tag{12}$$

$$\sum_{j \in J^{VI}} x_j \leq p \tag{13}$$

$$x_j = \{0, 1\} \qquad \forall j \in J \tag{14}$$

$$y_i = \{0, 1\} \qquad \forall i \in I \tag{15}$$

Equation (9) corresponds to the objective function which combines the two objectives, maximizing the zones covered while minimizing the number of vehicles used as VI. Equation (10) defines the proportion of vehicles considered as VI. In Eq. (11), the first term of the expression corresponds to the ratio of zones covered, whereas the second one to the proportion of vehicles covered divided by $|I|$. Note that the values of this second term ranges in the interval $[0, \frac{1}{|I|}]$. In this way, a solution it is considered better if it covers more zones or at the same number of zones, if it covers more vehicles. The constraint formulated in Eq. (12) establishes that a zone i is covered only when at least one vehicle in the VI has as neighbour one of the vehicles located at i. Equation (13) ensures that the number of vehicles used as virtual infrastructure is at most p. Finally, Eqs. (14) and (15) restrict variables x_i and y_i to binary values.

To avoid dealing with feasible and unfeasible solutions arising during the search, we used the next penalization scheme:

$$Z(x) = \begin{cases} (1 - F_1(x)) \cdot F_2(x) & \text{if } p' \leq p \\ (1 - F_1(x)) \cdot F_2(x) - c(p' - p) & \text{otherwise} \end{cases} \tag{16}$$

where $p' = \sum_{j \in J^{VI}} x_j$ and c is a parameter that establish the magnitude of the penalization.

```
1  Initialize P = {ind_1, ..., ind_{P_size}} randomly;
2  ind_best ← best individual in P;
3  while Not Stopping Condition do
4      P_new ← {ind_best};
5      for (i=1 to (P_size − 1)) do
6          /* Selection step                                  */
7          ind_1 ← Tournament_Selection(P);
8          ind_2 ← Tournament_Selection(P);
9          /* Crossover step                                  */
10         ind' = uniform_crossover(ind_1, ind_2);
11         /* Mutation step                                   */
12         ind'' ← mutate(ind');
13         P_new ← P_new ∪ {ind''};
14     end
15     P ← P_new;
16     ind_best ← best individual in P;
17 end
18 return ind_best
```

Algorithm 1: Binary generational Genetic Algorithm pseudo-code

3.3 *Description of the Solver*

In this subsection we will describe the method designed to solve the optimization problem described above. We decided to use a binary GA [16] to solve it because of its good performance in CLPs [2, 10]. To be more specific, we have used a generational GA with elitism. Each individual ind corresponds to a solution and it is encoded as: $ind = (x_1, \ldots, x_{|J^{VI}|})$. The fitness of an individual ind is given by $g(ind) = Z(x_1, \ldots, x_{|J^{VI}|})$. The pseudo-code of the method can be seen in Algorithm 1.

As we can see in the pseudo-code, the algorithm first initializes the population randomly, stores the best individual in ind_{best} and then, it enters into the main loop. Within the main loop, the method starts creating a new population with only ind_{best}, in order to keep elitism. After that, it begins with the three main steps of the evolution process: selection, crossover and mutation. The selection operator used is the tournament, where the best of s individuals randomly selected from the population is chosen. The uniform crossover is applied to the two individuals picked with the tournament selection. Given two individuals $ind_1 = (x_1^1, \ldots, x_{|J^{VI}|}^1)$ and $ind_2 = (x_1^2, \ldots, x_{|J^{VI}|}^2)$, the resulting individual $ind' = (x'_1, \ldots, x'_{|J^{VI}|})$ from the uniform crossover is calculated in the next way:

$$x'_j = \begin{cases} x_j^1 \; if \; U(0,1) \leq CR \\ x_j^2 \qquad \text{otherwise} \end{cases} \tag{17}$$

where U(0, 1) is a real random number uniformly distributed in the interval [0, 1]. Regarding the mutation step, the new individual ind'' is obtained by means of the next expression:

$$x''_j = \begin{cases} B(0,1) \text{ if } U(0,1) \leq MR \\ x'_j \qquad\qquad \text{otherwise} \end{cases} \tag{18}$$

where B(0, 1) is a random binary value uniformly distributed. The mutated solution is added to the new population, and the loop starts again. Once the new population reaches P_{size} individuals, it replaces the old population. The last step of the main loop consists on updating the best individual found so far.

4 Experimental Framework

This part of the chapter is devoted to describe the experimental framework used in this study. Concretely, in Sect. 4.1 we give the details of the realistic simulation scenario employed in experimentation. Then, Sect. 4.2 we describe the method taken as reference in order to compare it versus our proposal. After that, we list the performance measures used in the experimentation and finally, Sect. 4.4 points out the details of the implementations carried out.

4.1 VANET Simulation Scenario

The simulation scenario used in this study is the same employed in [8]. The VANET simulation was done using GPS traces publicly available in NS-2 format,[1] which was taken as input in the network simulation platform. The tool SUMO (Simulation of Urban MObility) was then used to generate the mobility traces using realistic input data, including road network, vehicles routes or traffic lights among others. The location of the simulation scenario is a rectangular area of 600 m × 700 m in the downtown of the city of Malaga, Spain. The simulation period is 180 s and the maximum vehicle velocity is 50 km/h. Table 1 contains the details of the simulation parameters used in the experimentation. For more information about the simulation scenario, the interested reader is referred to [22]. Finally, we want to highlight that we have considered three different maximum transmission powers for the short-range network (IEEE 802.11p) (16, 21 and 23 dBm) and five different maximum VI sizes (2, 4, 6, 8 and 10 nodes). The transmission power controls the communication range of the vehicles and therefore the neighbourhood awareness levels: a higher transmission power implies a higher communication range that, in turn, entails a higher number

[1] http://neo.lcc.uma.es/staff/jamal/vanet/?q=node/11.

Table 1 Main simulation parameters

Type	Parameter	Value
Neighbor information	CAM frequency	1 Hz
	Neighbor table timeout	5 s
	Server update Frequency	1 Hz
Dissemination request	Frequency	1 Hz
	Dissemination area	0.44 km^2
Scenario	Type	Urban (Malaga, Spain)
	Number of vehicles	45
	Simulation duration	180 s
	Vehicle speed	10–50 km/h
	Vehicle density	113 veh/km^2
	Maximum VI size	[2, 4, 6, 8, 10] nodes
802.11p Network	Bit rate	6 Mbps
	Bandwidth	10 Mhz
	Frequency band	5.9 GHz
	Maximum Tx power	[16, 21, 23] dBm
LTE Network	eNodeB Tx power	30 dBm
	UE Tx power	10 dBm
	Propagation model	Friis Tx Eq

of neighbours per vehicle. In the CLP models, the transmission power can be seen as the coverage radius of the facilities. Regarding the maximum VI size, it limits the resource consumption in terms of LTE connections and it is represented by the parameter p in the model described in Sect. 3.2.

4.2 *Reference Algorithm*

The method used to compared our approach for VI selection is the one proposed in the original paper of NAVI architecture [8]. In that paper, they applied a Min-Max formulation to model the optimization problem and an ad-hoc greedy algorithm as solver. As in our proposal, the objective consists on maximizing the number of zones covered while minimizing the number of vehicles selected as VI, using at most p vehicles. In this section we will give a brief description of the approach for the sake of the simplicity. The interested reader is referred to the original work for further details.

The pseudocode of the greedy algorithm is given in Algorithm 2. We use the same notation employed in Sect. 3.2. The method starts computing the zones covered by each vehicle $j \in J^{VI}$ and it store it in $SC = \{SC_1, \cdots, SC_{|J^{VI}|}\}$, and it selects as VI the vehicle that covers most zones. As mentioned before, the zones covered by a

1 /* SC_j contains the zones covered by j, that is, its zone and those where it is neighbour of at least one vehicle placed on it */
2 **for** *j in J* **do**
3 | $SC_j \leftarrow \{ZV_j\} \cup (\bigcup_{k|j \in NV_k} ZV_k)$;
4 **end**
5 $L^{VI} \leftarrow J^{VI}$;
6 $j \leftarrow$ vehicle with highest $|SC|$ value in L^{VI};
7 $J' \leftarrow L^{VI} - \{j\}$;
8 /* ZC stores the zones covered so far */
9 $ZC \leftarrow SC_j$;
10 /* VI keeps the vehicles selected as VI */
11 $VI \leftarrow \{j\}$;
12 **while** $|VI| < p$ *or* $ZC = I$ **do**
13 | **for** *k in* L^{VI} **do**
14 | | /* D_k indicates the number of zones covered by vehicle k that are not already covered */
15 | | $D_k \leftarrow |ZC \cup SC_k| - |ZC|$;
16 | **end**
17 | $j \leftarrow$ vehicle with highest $|D|$ value in L^{VI} ;
18 | $L^{VI} \leftarrow L^{VI} - \{j\}$;
19 | $ZC \leftarrow ZC \cup SC_j$;
20 | $VI \leftarrow VI \cup \{j\}$;
21 **end**
22 **return** VI

Algorithm 2: NAVI's algorithm for VI selection

vehicle correspond to its zone and those where it is neighbour of at least one vehicle placed on it. After this, the algorithm enters into the main loop. Within the main loop, the method computes a dissimilarity index for each vehicle that has not been chosen as VI so far. This index basically measures the number of zones covered by the vehicle that are not already covered by the vehicles selected as VI. Once the index is calculated, the algorithm adds the vehicle with the highest dissimilarity value as VI and it starts the process again. The main loop finishes when the size of the VI is lower or equal to p (the maximum size allowed for the VI) or when the VI selected cover all zones in the region.

4.3 *Performance Measures*

The metrics used to compare the performance of the algorithms are given below:

- *Coverered Area:* percentage of zones from covered, which corresponds with those regions that has at least one vehicle that would receive the dissemination message.
- *VI size:* number of nodes selected as VI. In this way, we measure the resource consumption of the solution.

4.4 Implementation Details

To finish with the description of the experimental framework, we give here the details of the implementation done and the parameter settings.

Regarding the parameter setting for the GA used as solver, the tournament size was set to 5; the crossover rate CR was fixed to 0.5 in order to combine approximately the same amount of genes from each solution; and the mutation rate MR was established to 0.015. In the objective function, the penalization coefficient c was set to 0.2 after some preliminary experimentations.

As for the simulation scenario, the region of interest was divided into 100 rectangular zones, all of them with the same width and height. The number of vehicles to be covered was 45, and we suppose that all of them are equipped with both short and long range communication capabilities, that is, all of them can be used as VI. The positions of the vehicles were sampled every second. Given that the simulation time for the data was 180 s, in this way, we have a total of 180 instances of the problem. Taking into account that we considered three different transmission powers and five different maximum VI size, this experimentation counts with a total of $180 \times 3 \times 5 = 2700$ instance configurations. For each of these instance configurations, our proposal was run 10 times and the mean covered area and VI size were registered.

The implementation of the proposal was done in Java 8, and the experiments were run on a computer with Windows 10 Enterprise Operative System, 32GB RAM and CPU Intel Xeon E5-2650 v3 2.30GHz. The results of the NAVI method were provided by its authors.

5 Result Analysis

The experimentation done in this work has the following objectives:

- *Analyze the performance of our proposal under different Maximum Response Times.* One of the most important issues in the service supported by VANETs is the maximum time permitted to provide a VI configuration, which we will call the Maximum Response Time (MRT). The MRT depends mainly on the environment being addressed. For example, in a city, where the velocity of the vehicles is low, the MRT time is less demanding since a VI configuration will be valid for a longer time. However, in a highway where the position of the vehicles changes quickly, it is important to have a shorter MRT. This factor represents a big challenge for the method the MRTs are usually given in terms of some dozens of milliseconds.

- *Compare the performance of our proposal versus a reference algorithm.* In order to assess the competitiveness of our proposal, we will compare it against the original method proposed for the NAVI architecture that we described in Sect. 4.2.

The next part of this section is devoted to analyze the results obtained in the experimentation done for each of these objectives.

5.1 Analysis of the Performance Under Different Maximum Response Times

As mentioned before, the first objective of the experimentation was to analyse the performance of our proposal for different MRTs. The MRT can be seen as the time given to the GA described in Sect. 3.3 to find a solution for the configuration of the VI. To this end, we considered three different MRT values: 30, 60 and 90 ms. Figures 2 and 3 show boxplots with information about the distribution of the results obtained over the 180 instances with each MRT. The Y axis represents the covered area and VI size, respectively, the X axis the five Maximum VI Sizes (MVISs) and the three panels the Maximum Transmission Powers (MTxPs) considered. Each series corresponds to a MRT value. In the boxes, the central horizontal line indicates the median, the hinges of the boxes the first and third quartiles, and the whiskers the value $1.5 \cdot IQR$, where IQR is the interquartile range. The dots refer to outlier values. The graphics were generated using R programming language and the ggplot2 package.[2]

Starting with covered area, we can see in Fig. 2 that, as expected, for a fixed MTxP, a higher MVIS implies a better covering, and viceversa, for a fixed MVIS, a higher MTxP also entails a better covering. A similar rule can be established for the complexity of the problem which is negatively correlated with MTxP and MVIS. This last fact is observed in both the average and dispersion of the performance distribution. For a determined MRT, the median and the interquartile range increases and decreases, respectively, for higher values of MTxP and MVIS. Comparing the MRTs among them, we can see that 30, 60 and 90 ms provide a similar performance with small differences among them, being 10 ms the one that obtains the worst results, as expected. In any case, these differences are small, reaching only 4% points in the worst case (MTxP = 16 dBM, MVIS = 6). This shows that the performance of our proposal is robust under different MRTs in terms of covered area.

However, when we consider the VI size (see Fig. 3), we do observe important differences. These differences are bigger for lower MTxPs and higher MVISs. Again, a MRT value of 10 ms obtains the worst results, whereas values of 30, 60 and 90 ms provide similar performances, especially for 21 and 23 dBm. The difference between the median performance of this two groups of MRTs are high for 16 dBm, reaching even two vehicles, but they are progressively reduced as the power increases. However, although for higher transmission powers the average performance is similar, if we look at the worst cases, represented by the outliers, strong difference appear between the 10 ms and the rest. This indicates that the algorithm has a lower robustness in terms of VI size when the MRT is very low, but for medium and high MRTs its robustness is also good.

[2]http://ggplot2.org/.

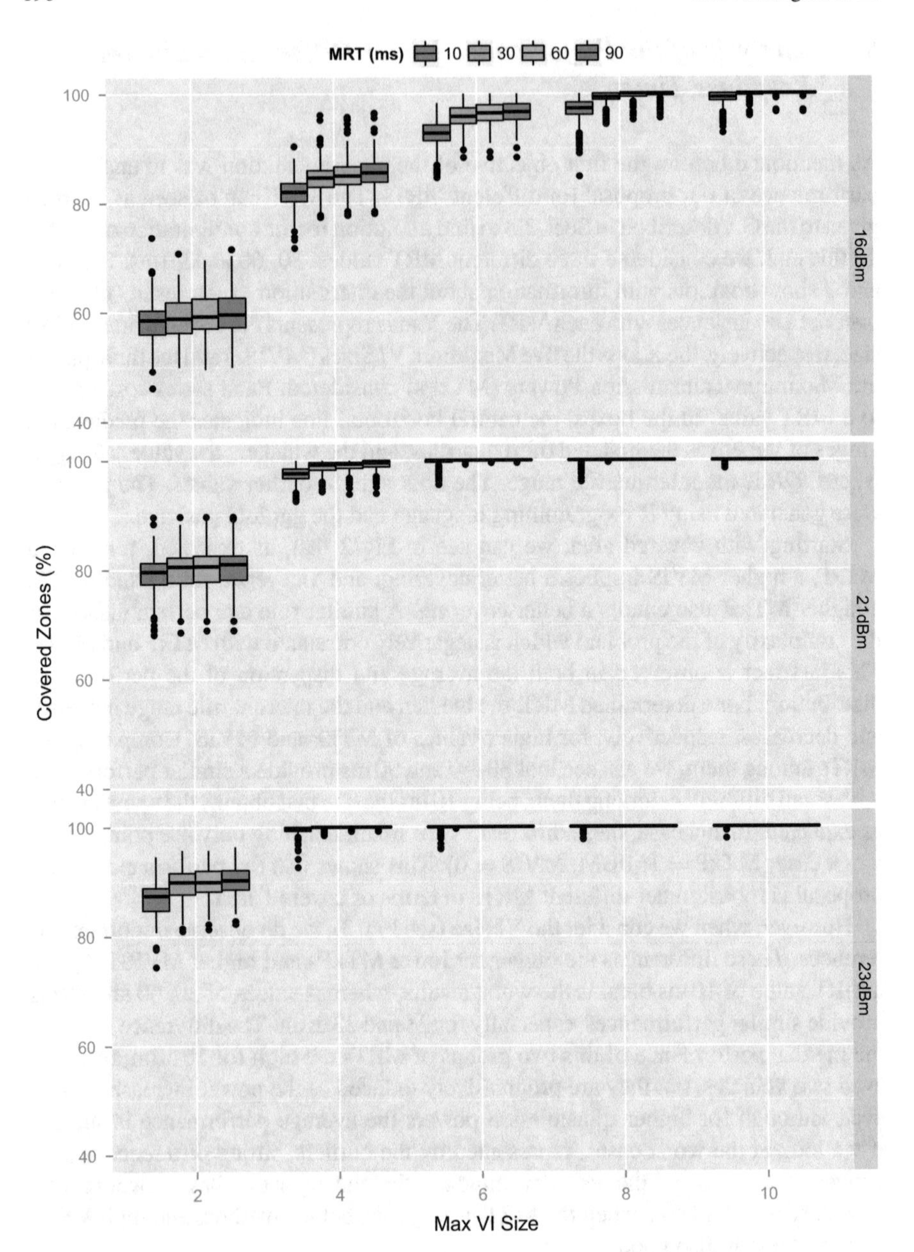

Fig. 2 Boxplot with information about the distribution of the results obtained over the 180 instances with each MRT. The Y axis represents the covered area, the X axis the five maximum VI sizes and the three panels the transmission powers considered. Each series corresponds to a MRT value. In the *boxes*, the central horizontal line indicates the median, the hinges of the boxes the first and third quartiles, and the whiskers the value $1.5 \cdot IQR$, where *IQR* is the interquartile range. The *dots* refer to outlier values

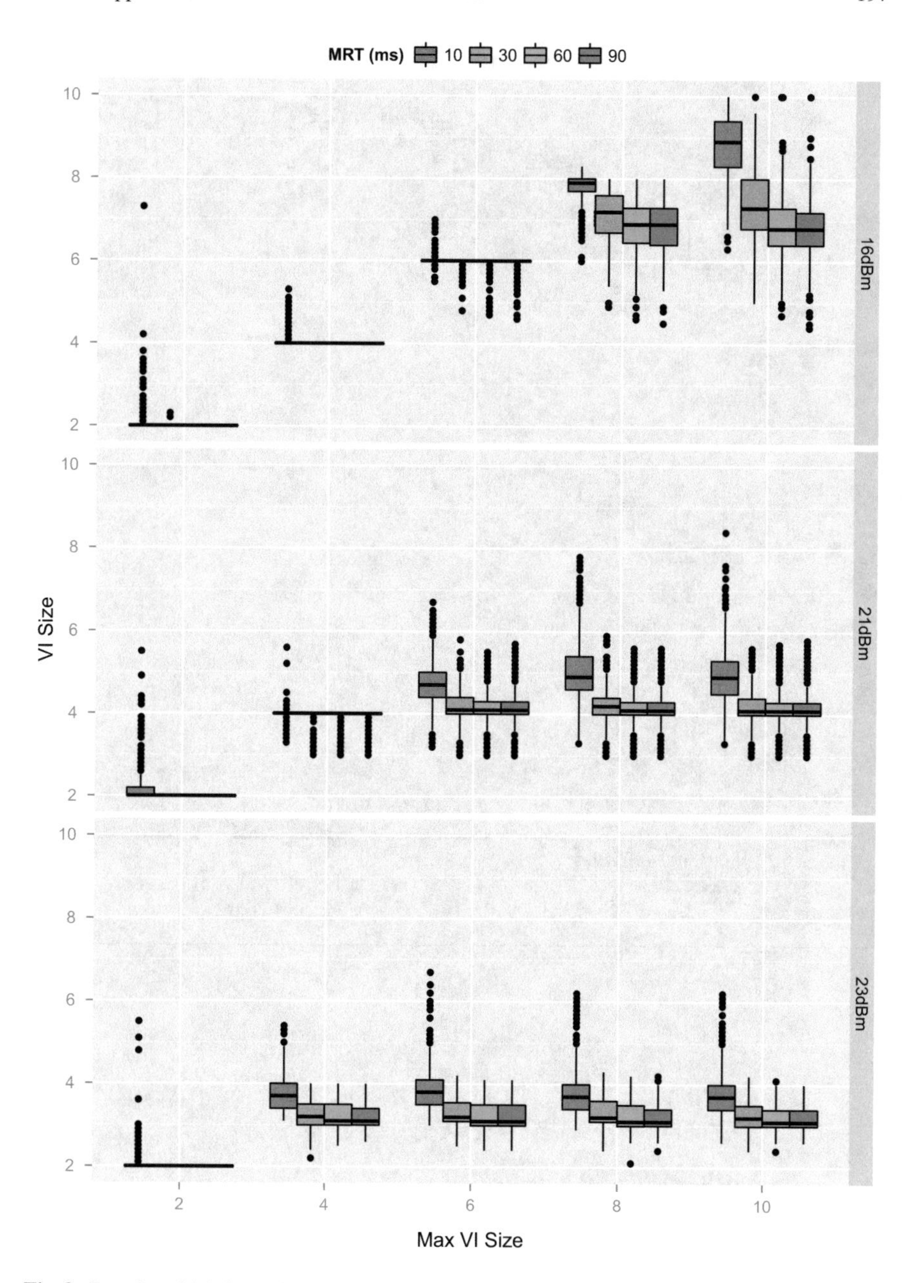

Fig. 3 Boxplot with information about the distribution of the results obtained over the 180 instances with each MRT. The Y axis represents the VI size, the X axis the five maximum VI sizes and the three panels the transmission powers considered. Each series corresponds to a MRT value. In the *boxes*, the central horizontal line indicates the median, the hinges of the boxes the first and third quartiles, and the whiskers the value $1.5 \cdot IQR$, where *IQR* is the interquartile range. The *dots* refer to outlier values

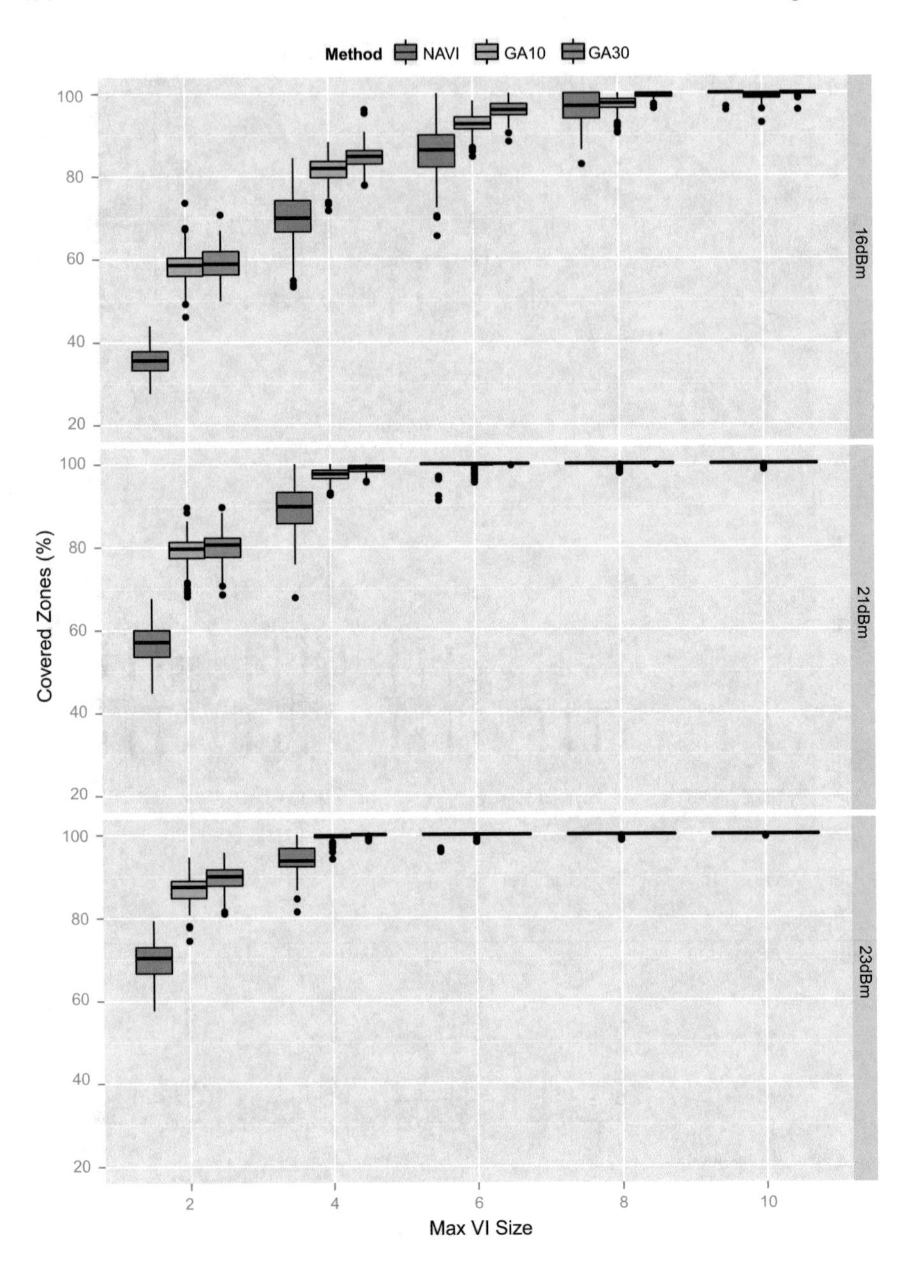

Fig. 4 Boxplot with information about the distribution of the results obtained over the 180 instances with each MRT. The Y axis represents the covered area, the X axis the five maximum VI sizes and the three panels the transmission powers considered. The series corresponds to the three methods compared: NAVI (*red*), GA10 (*green*) and GA30 (*blue*). In the *boxes*, the central horizontal line indicates the median, the hinges of the boxes the first and third quartiles, and the whiskers the value $1.5 \cdot IQR$, where *IQR* is the interquartile range. The *dots* refer to outlier values

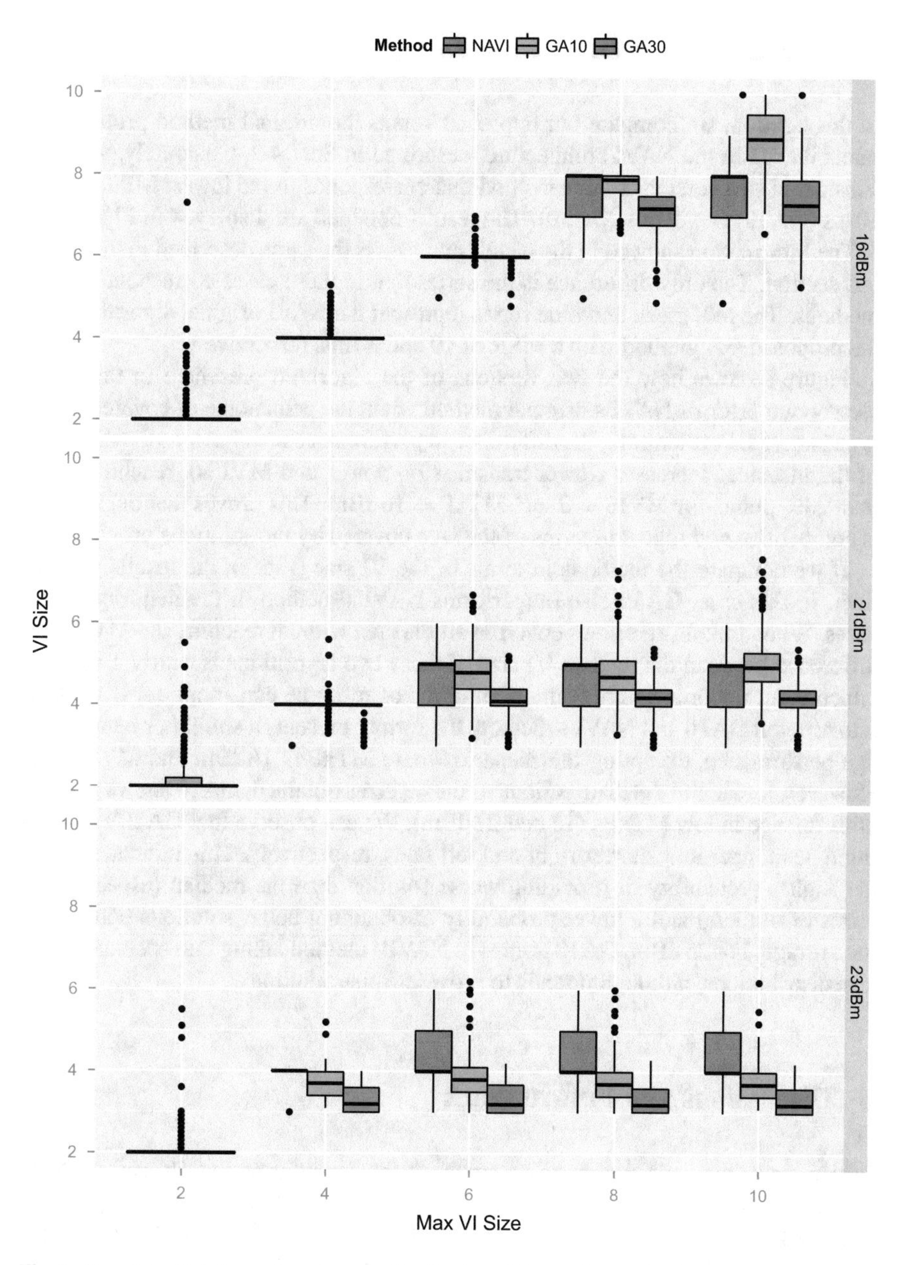

Fig. 5 Boxplot with information about the distribution of the results obtained over the 180 instances with each MRT. The Y axis represents the VI size, the X axis the five maximum VI sizes and the three panels the transmission powers considered. The series corresponds to the three methods compared: NAVI (*red*), GA10 (*green*) and GA30 (*blue*). In the *boxes*, the central horizontal line indicates the median, the hinges of the boxes the first and third quartiles, and the whiskers the value $1.5 \cdot IQR$, where *IQR* is the interquartile range. The *dots* refer to outlier values

5.2 Comparison Versus the Reference Algorithm

In this Section, we compare our approach versus the original method proposed to select the VI in the NAVI architecture, described in Sect. 4.2. Concretely, we have considered two versions of our method that corresponds to the lowest MRT values, 10 ms (GA10) and 30 ms (GA30). The results obtained are displayed in Figs. 4 and 5. The information showed in the panels and axis is the same described in the previous section. The only difference is the series that in these cases corresponds to the methods. The red, green and blue series represent the NAVI original algorithm, and the proposed GA method with a MSR of 10 and 30 ms, respectively.

Figure 4 shows how the two versions of the algorithm presented in this work clearly outperforms NAVI's original method when the percentage of covered zones is considered. Furthermore, the difference in performance are higher as the difficulty of the instances increases (lower transmission power and MVISs), reaching more than 20% points for MVIS = 2 and MTxP = 16 dBm. This proves that our method is competitive and robust in terms of the area covered by the solutions provided.

If we compare the methods in terms of the VI size (Fig. 5), the results are distinct. In this case, GA30 also outperforms NAVI's method in the majority of the cases. When the MTxP allows covering all the area without reaching the MVIS, the difference between the median VI size of these two algorithms is nearly 1 vehicle, which is an important improvement in terms of resource consumption. If we take into account GA10 and NAVI's method, the former presents a similar or better average performance, excepting the scenario where MTxP is 16 dBm and MVIS is 2. However, looking at the distributions of the VI sizes obtained, GA10 shows a worse performance in low MTxPs (16 and 21 dBm). We can observe that GA10's distribution has longer and shorter right and left tales, respectively. This indicates that it has higher probability of providing worse solution than the median (especially in terms of outliers) and a lower probability of obtaining better solutions better than the average. For 23 dBm, GA10 improves NAVI's method taking into account the VI size distributions, although it tends to provide worse solutions.

6 Discussions and Future Work

In this work, we have presented a new approach for information dissemination in VANETs based on the hybrid architecture proposed by D'Orey et al. in [8]. The proposed approach focused on the selection of the VI, modelling it as a CLP that aims at maximizing the covering while minimizing the number of facilities, represented in this case by the area covered and the number of vehicles used as VI, respectively. To solve the underlying optimization problem, we designed a generational Genetic Algorithm with elitism.

The method presented here was tested on a real scenario consisting of 45 vehicles moving on a rectangular area of 600 m $\times$ 700 m in the downtown of the city of Malaga,

Spain. Furthermore, we considered three different maximum transmission powers for the 802.11p network and five different maximum VI sizes. The performance of the solutions was measured in terms of covered area and number of vehicles used as VI (VI size). The objective of the experimentation done over this real scenario was two-fold: on one hand, analyze the performance of the proposal under different MRTs, and on the other hand, assess the competitiveness of our method comparing it with the NAVI's original algorithm for selecting the VI.

The results obtained in the experimentation showed that:

- The difficulty of the problem increases as the transmission power decreases. However, the maximum VI size has a different influence in the hardness of the instance depending on the performance measure considered. Increasing the maximum VI size leads to harder and easier instances, when we consider the covered area and the VI size, respectively.
- The MRT has a stronger influence in performance in terms of VI size than in terms of covered area, especially when the worst case is considered. In our opinion, this is due to the initialization method used. Since we initialize the solutions using a uniform random binary distribution, the individuals of the first generation have in average half of the vehicles set as VI, which would probably provide a good covered area but a very bad VI size. A short MRT sometimes may not be enough to find individuals that keep a good covering with a much lower VI size.
- Our proposal with MRT = 10 ms outperforms the covered area obtained by NAVI's algorithm but it provided worse results in terms of VI size for low transmission powers.
- When the MSR is set to 30 ms, the presented approach outperforms NAVI's original one in the two measures considered here.

From this study, several lines of research arise. In first place, we plan to work on the GA in order to improve its performance by incorporating more information about the problem. Concretely, we want to design ad-hoc initialization, crossover and mutation operators that take into account the specific features of the instance being solved as the transmission power or the maximum VI size. We also intend to address this problem using a proper multi-objective approach that optimizes simultaneously the covered area and the VI size instead aggregating both. In this way, we will obtain a pareto-front instead of a single solution. Finally, the approach presented here solves each instance of the problem from scratch. However, given that the positions of the vehicles changes in a smooth way, using information from previous VI configurations could be useful to find good solutions for the current instance faster. For this reason, we think that it will be interesting to use models and methods from the field of Dynamic Optimization Problems [5].

Acknowledgements This work has been supported by the research projects TEC2013-45585-C2-2-R and TIN2014-56042-JIN from the Spanish Ministry of Economy and Competitiveness.

References

1. Araniti, G., Campolo, C., Condoluci, M., Iera, A., Molinaro, A.: LTE for vehicular networking: a survey. IEEE Commun. Mag. **51**(5), 148–157 (2013)
2. Bouaziz, S.G., Mellouli, R., Dammak, A., Al-Hassan, M.: New variants of the covering location problem: modeling and a two-stage genetic algorithm. In: 2015 2nd World Symposium on Web Applications and Networking (WSWAN), pp. 1–6 (2015)
3. Camara, D.D., Bonnet, C., Nikaein, N., Wetterwald, M.: Multicast and virtual road side units for multi technology alert messages dissemination. In: 2011 IEEE 8th International Conference on Mobile Adhoc and Sensor Systems (MASS), pp. 947–952 (2011)
4. Church, R., Revelle, C.: The maximal covering location problem. Pap. Reg. Sci. Assoc. **32**(1), 101–118 (1974)
5. Cruz, C., González, J.R., Pelta, D.A.: Optimization in dynamic environments: a survey on problems, methods and measures. Soft. Comput. **15**(7), 1427–1448 (2011)
6. Curtin, K.M., Hayslett-McCall, K., Qiu, F.: Determining optimal police patrol areas with maximal covering and backup covering location models. Netw. Spat. Econ. **10**(1), 125–145 (2007)
7. Doolan, R., Muntean, G.M.: VANET-enabled eco-friendly road characteristics-aware routing for vehicular traffic. In: 2013 IEEE 77th Vehicular Technology Conference (VTC Spring), pp. 1–5 (2013)
8. D'Orey, P.M., Maslekar, N., de la Iglesia, I., Zahariev, N.K.: NAVI: neighbor-aware virtual infrastructure for information collection and dissemination in vehicular networks. In: 2015 IEEE Vehicular Technology Conference (VTC Spring), pp. 1–6 (2015)
9. Fisher, M.L., Kedia, P.: Optimal solution of set covering/partitioning problems using dual heuristics. Manage. Sci. **36**(6), 674–688 (1990)
10. García, S., Marín, A.: Covering location problems. Location Science, pp. 93–114. Springer International Publishing, Cham (2015)
11. Guerrero-Ibáñez, J.A., Flores-Cortés, C., Zeadally, S.: Vehicular ad-hoc networks (VANETs): architecture, protocols and applications. Next-Generation Wireless Technologies: 4G and Beyond, pp. 49–70. Springer London, London (2013)
12. Hartenstein, H., Laberteaux, K.P.: A tutorial survey on vehicular ad hoc networks. IEEE Commun. Mag. **46**(6), 164–171 (2008)
13. Hernandez-Jayo, U., Mammu, A.S.K., De-la Iglesia, I.: Reliable communication in cooperative ad hoc networks. Contemporary Issues in Wireless Communications. InTech, 2014
14. Lee, G., Murray, A.T.: Maximal covering with network survivability requirements in wireless mesh networks. Comput. Environ. Urban Syst. **34**(1), 49–57 (2010)
15. Man, K.-F., Tang, K.S., Kwong, S.: Genetic Algorithms: Concepts and Designs. Springer Science & Business Media, Berlin (2012)
16. Mitchell, M.: An Introduction to Genetic Algorithms. MIT press, Cambridge (1998)
17. Nafi, N.S., Khan, J.Y.: A VANET based intelligent road traffic signalling system. In: 2012 Australasian Telecommunication Networks and Applications Conference (ATNAC), pp. 1–6 (2012)
18. Panichpapiboon, S., Pattara-Atikom, W.: A review of information dissemination protocols for vehicular ad hoc networks. IEEE Commun. Surv. Tutor. **14**(3), 784–798 (2012)
19. Papadimitratos, P., La Fortelle, A., Evenssen, K., Brignolo, R., Cosenza, S.: Vehicular communication systems: enabling technologies, applications, and future outlook on intelligent transportation. IEEE Commun. Mag. **47**(11), 84–95 (2009)
20. Rémy, G., Senouci, S.-M., Jan, F., Gourhant, Y.: LTE4V2X: LTE for a centralized vanet organization. In: 2011 IEEE Global Telecommunications Conference (GLOBECOM), pp. 1–6 (2011)
21. Toregas, C., Swain, R., Revelle, C., Bergman, L.: The location of emergency service facilities. Oper. Res. **6**, 1363–1373 (1971)
22. Toutouh, J., Garcia-Nieto, J., Alba, E.: Intelligent OLSR routing protocol optimization for VANETs. IEEE Trans. Veh. Technol. **61**(4), 1884–1894 (2012)
23. Yang, L., Jones, B.F., Yang, S.-H.: A fuzzy multi-objective programming for optimization of fire station locations through genetic algorithms. Eur. J. Oper. Res. **181**(2), 903–915 (2007)

Product Matching to Determine the Energy Efficiency of Used Cars Available at Internet Marketplaces

Mario Rivas-Sánchez, Maria P. Guerrero-Lebrero, Elisa Guerrero, Guillermo Bárcena-Gonzalez, Jaime Martel and Pedro L. Galindo

Abstract The growth of the Internet has fuelled the availability of e-commerce marketplaces and search engines must face with a huge amount of ambiguity and inconsistencies in the data. Product matching aims at disambiguating descriptions of products belonging to different websites in order to be able to recognize identical products and to merge the content from those identical items. In this work first we evaluate some similarity measures for string matching and then, we apply a complete product matching methodology to the retail market of used cars. We use a reference or master list of items and information about a wide variety of used cars offers. The resulting linkage allows energy efficiency assignment of the model identified.

1 Introduction

Trade of used cars conducted electronically has grown extraordinarily, and an important concern of these e-commerce marketplaces is access to a unified catalogue where prices and energy efficiency of similar appliance models can be easily compared. In Spain Coches.net [25] is one of the most popular websites for selling and buying new and used cars. This marketplace accounts each month with more than 5.4 million unique visitors, 11.6 million visits and 190 million page views. While it provides information regarding each car's condition, energy efficiency and supposed history, the "official" information about each car model is held by the Ministry of Industry, Energy and Tourism Agency in the IDAE (Instituto para la Diversificación y Ahorro de la Energía) dataset [24], where users and dealers can access to.

M. Rivas-Sánchez (✉) · J. Martel
Itelligent Information Technologies. Parque Tecnológico CEEI, El Puerto de Sta Maria (Cádiz), Cádiz, Spain
e-mail: mrivas@intelligent.es

M.P. Guerrero-Lebrero · E. Guerrero (✉) · G. Bárcena-Gonzalez · P.L. Galindo
Department of Computer Science and Engineering, University of Cádiz, Cádiz, Spain
e-mail: elisa.guerrero@uca.es

C. Cruz Corona (ed.), *Soft Computing for Sustainability Science*,
Studies in Fuzziness and Soft Computing 358, DOI 10.1007/978-3-319-62359-7_10

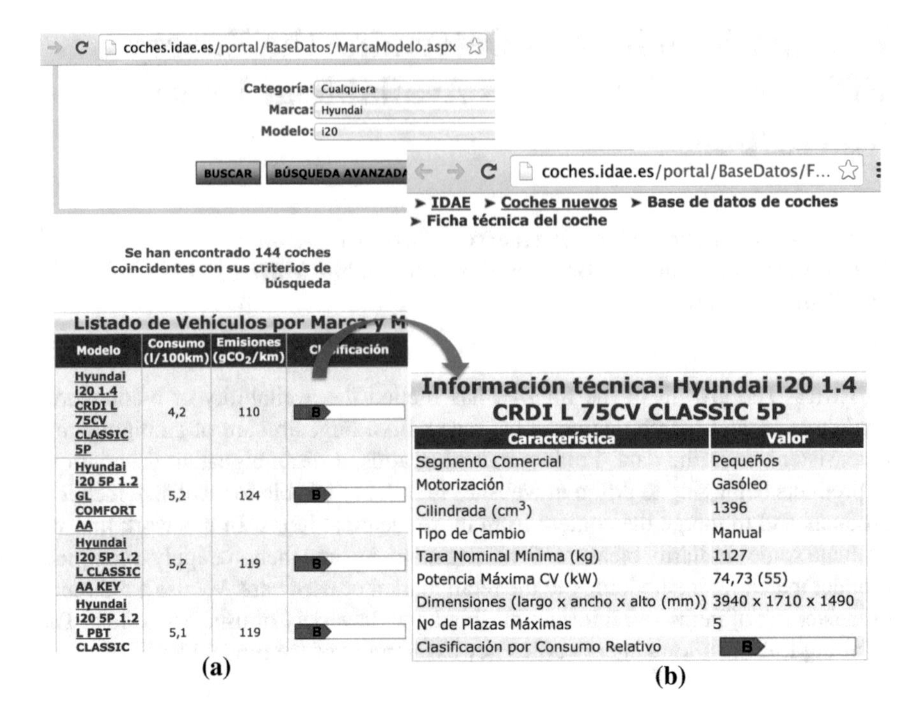

Fig. 1 Two screenshots from IDAE dataset, **a** searching for a car specifying brand and model and **b** detailed information about the first entry encountered in the dataset that matches the search

Multiple sources of products means that listed values may differ. Each pricing or energy efficiency guide receives data from different sources and makes different judgments about that data. Figures 1 and 2 show this ambiguous situation, IDAE web is in charge of maintaining the "objective" information about every car model, while marketplaces show a list of cars for sale, doing their best to highlight the main features of their products.

The identification and matching of different items as the same product is essential to obtain accurate and reliable results. This is the task of Product Matching, identifying, matching and merging records that correspond to the same product from several data sources [1, 4, 22, 23], allowing the development of tools for product monitoring, product comparison and pricing analysis.

In this work we evaluate some similarity measures for string matching and describe the complete procedure to obtain a product linkage between the offers in the retail market of used cars and the IDAE dataset in order to determine the real efficiency index of these cars.

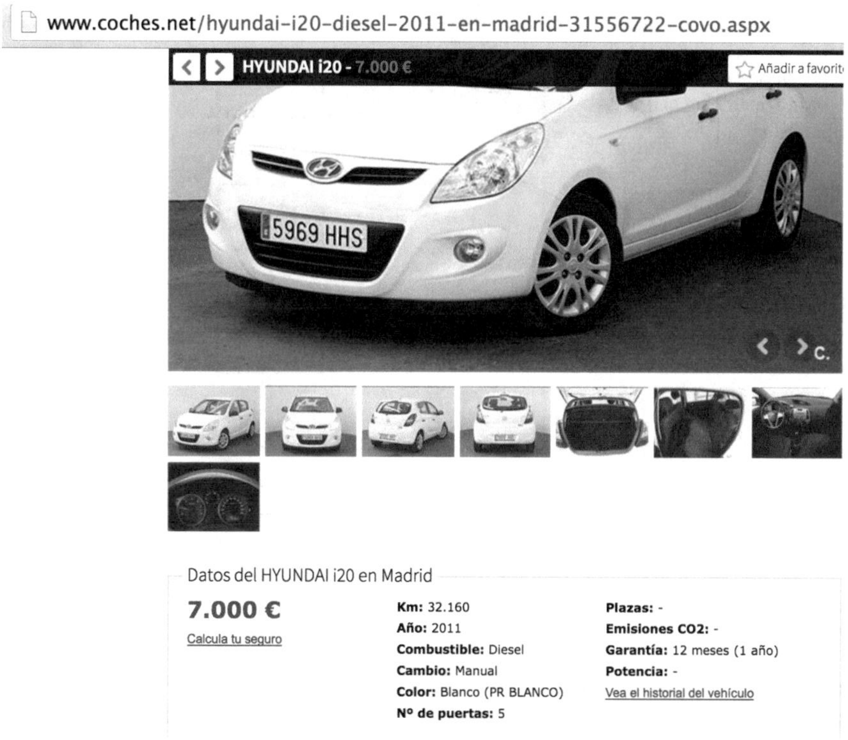

Fig. 2 Screenshot from coches.net dataset, information about a specific car for sale

2 Product Matching Process

The matching process is comprised of four main stages: pre-processing, indexing, matching and supervision. Figure 3 shows the flowchart corresponding to this process. Preprocessing allows obtaining consistent data to indexing [7], that groups data into blocks in order to reduce the number of comparisons made in the matching stage, where the similarity between two records is evaluated, applying a linkage procedure [14]. If the matching is good the product is automatically linked, otherwise the matching process requires manual supervision, the last step of the complete process. Next subsections describe these stages in more detail.

2.1 Preprocessing

The preprocessing stage is essential to ensuring cleansing and normalization of data generated from different sources, allowing the unification of the descriptions and the

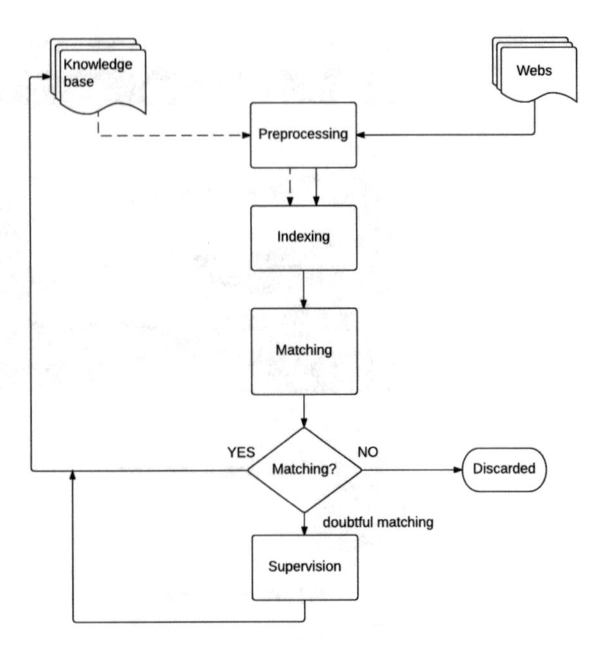

Fig. 3 Matching process flowchart with the four main steps: preprocessing, indexing, matching and supervision

generation of structured information [12]. There are a number of text features in this context (punctuation, lower and uppercase, special characters, etc.) that can make difficult to process automatically [3, 7], thus several tasks must be carried out:

1. Convert text in lowercase
2. Remove all accented characters
3. Remove all punctuation marks
4. Replace multiple whitespace with a single space

2.2 *Indexing*

Product matching is based on the comparison between records. This second stage allows reducing the potential number of comparisons between records in the dataset and creates a set of candidate records.

In the event that only two different sources are used, potentially each record from one dataset needs to be compared with all records in the other dataset, then the number of comparisons required are $m\,x\,n$, where m and n are the number of products that belong to first and second sources, respectively.

Indexation aims at decreasing the number of comparisons [15]. The general approach is to create one or several blocks or clusters, according to the so-called blocking keys that results in similar records being successfully grouped into the same blocks. Comparisons are made among the members of the same block, therefore the smaller size the better. In this way search space reduction and matching efficiency increase can be possible.

The reduction ratio provides information about how many candidates record pairs were generated by an indexing technique compared to all possible record pairs. A high reduction ratio means that an indexing technique has removed a significant number of record pairs from the full comparison space. Equation 1 describes this ratio, where, n_M is the total number of matched records, n_N is the total number of non-matched records, with $n_M + n_N = m x n = T$, where T is the number of total records [4].

S_M is the total number of matched records in the block, S_N is the total number of non-matched records in the block, and N is the number of records in block ($N = S_M + S_N$).

$$rr = 1 - \frac{(S_M + S_N)}{(n_M + n_N)} \tag{1}$$

2.3 Matching

The matching is based on the application of some approximate string similarity measure that maps a pair of strings to a real number. From character-based approaches, through token-based or the use of hybrid methods, there exists a wide variety of metrics that can be used to obtain some idea of how similar or different two attribute values are [20, 21].

These metrics are normalised, ranging from 0 meaning that both strings are completely different, to 1 indicating exact coincidence, in-between values indicating intermediate similarity or dissimilarity. The problem to solve is situated in a really specific context therefore, disambiguating descriptions of products are not relevant in this work because databases only contain information on cars.

In this work different approaches have been considered, although the most popular character-level measure is based on edit-distance functions11 that count the number of edit operations to convert one string into another, other alternatives based on token-based calculations (TFIDF) and some hybrid approaches (Monge–Elkan) combining the advantages of character-based and token-based distance functions have been broadly used for product matching. Among of all them we have selected Monge–Elkan, TFIDF and 5-shingles to be compared, using as secondary similarity function Jaro metric.

In [5] a comparison of string distance metrics was presented, they investigated different approaches, including edit-distance metrics, fast heuristic string comparators, token-based distance metrics, and hybrid methods. They combined some of these metrics to improve their performance. Based on this work, here we consider the combination of some of these metrics, in particular the combined schemes TFIDF and Monge–Elkan and TFIDF, Monge–Elkan and 5-shingles together.

2.3.1 Jaro Metric

Jaro metric [9, 10] is based on the number c of the common characters between two strings, A and B, and the number of transpositions, Eq. (2) describes this metric:

$$sim_{Jaro} = \begin{cases} ifc = 0 & 0 \\ otherwise & \frac{1}{3}\left(\frac{c}{|A|} + \frac{c}{|B|} + \frac{c-t}{c}\right) \end{cases} \tag{2}$$

where t is half the number of transpositions, |A| and |B| denotes number of characters in strings A and B respectively, and if $c = 0$ then $sim_{Jaro} = 0$ by definition.

If a_i is a character belonging to string A, it is said that is *common with* B when there is a character b_j in B, such as $b_j = a_i$ when considering a sliding window of size max(|A|, |B|)/2. Jaro metric is particularly suitable for short strings where typographical errors can be expected.

2.3.2 Monge–Elkan

The Monge–Elkan method [16] constitutes a simple but effective method to measure the similarity between two text strings, *A* and *B*, containing several tokens (|A| and |B| respectively), and using an internal similarity function, *sim'*. Its calculation is given by Eq. (3):

$$sim(A, B) = \frac{1}{|A|}\sum_{i=1}^{|A|} \max(sim'(A_i, B_j))_{j=1}^{|B|} \tag{3}$$

The Monge–Elkan method preserves the properties of the internal character-based measure (e.g. ability to deal with misspellings, typos, OCR errors) and deals successfully with missing or disordered tokens [11].

2.3.3 Q-Shingles

Instead of single tokens, this metric uses subsequences of tokens, so-called shingles of length q, that are obtained by sliding a window of length q [4, 5]. Then the similarity between two shingles is calculated as the number of items contained in the intersection (common q-shingles) of the two sets divided by the number of items contained in the union of the two sets (Eq. 4):

$$sim(A, B) = \frac{c_{common}}{c_A + c_B - c_{common}} \tag{4}$$

The choice of q affects the resemblance calculated, if q is very small permutations of the same token can be more frequent, whereas a too large value of q could be oversensitive to small alterations.

2.3.4 TFIDF

Term frequency–inverse document frequency (TFIDF) is one of the most commonly used term weighting and term selection measurements in today's information retrieval systems [8, 17–19]. The number of times a term occurs in a document is called term frequency (TF), while IDF stands for the inverse of the number of documents that contain (or are indexed by) the term in question [2]. TFIDF is defined as the product of both statistics (Eq. 5):

$$TFIDF = TF \cdot \frac{1}{DF} \tag{5}$$

As TF gives higher weights to terms that occur more frequently in a document and DF to terms that occur less frequently, the TFIDF weight increases proportionally to the number of times a word appears in the document, but it is offset by the frequency of the word in the corpus, which helps to adjust for the fact that some words appear more frequently in general.

The cosine similarity [3] between the vector of TFIDF tokens values, $w_{i,t}$, is given by Eq. (6)

$$sim_{cos}(S_i, S_j) = \frac{1}{W_i \cdot W_j} \sum_{t=1}^{n} w_{i,t} \cdot w_{j,t} \tag{6}$$

where S_i and S_j are two strings decomposed by token vectors T, $w_{i,t}$ denotes the TFIDF value of each token t and w_i can be calculated as follows (Eq. 7):

$$W_i = \sqrt{\sum_{t=1}^{n} w_{i,t}^2} \tag{7}$$

This metric can be seen as a comparison between strings on a normalized space because not only the magnitude of each word count (TFIDF) of each string is taken into account, but the angle between the strings.

2.3.5 Classification

$k-$NN classifies a case to the class most common amongst its k nearest neighbours calculated by the cosine similarity measurement.

The fuzzy k-NN classifier [13] generalizes the crisp k-NN classifier. Rather than assigning a class label to an input pattern, the fuzzy k-NN algorithm assigns a membership value as a function of the pattern's distance from its k-nearest neighbours. If these values are not high enough a supervision of the classification will be needed. Equation 8 is used in order to calculate the class membership of an input pattern x:

$$\mu_{\omega i}(x) = \frac{\sum_{j=1}^{k} \mu_{\omega i}(x_j) d^{-2/(m-1)}}{\sum_{j=1}^{k} d_j^{-2/(m-1)}} \quad (8)$$

where $x_1, x_2, \ldots, x_k$ denote the k-nearest neighbour labelled reference patterns of x, and $d_j = ||x - x_j||$ is the distance measurement between x and its jth nearest neighbour x_j. The pattern is assigned to the class given by Eq. 9:

$$\omega(x) = argmax_{i=1}^{M} (\mu_{wi}(x)) \quad (9)$$

The fuzzification parameter m determines how heavily the distance is weighted when calculating the class membership. As m increases toward infinity, the term $d_j^{-2/(m-1)}$ approaches 1 regardless of the distance. Consequently, the neighbours x_j are more evenly weighted. As m decreases toward 1, however, the closer neighbours are weighted far more heavily than those further away. This has the effect of effectively reducing the number of neighbours that contribute to the class membership value of the input data point.

3 Results and Discussion

In this section we describe the results obtained in each stage of the process to obtain a product linkage between the offers in the retail market of used cars and the IDAE dataset. We evaluate the different string similarity metrics and select the most suitable to be applied in this particular case.

3.1 Preprocessing

Table 1 shows some examples of this stage, first column shows the original input and second column the output obtained after applying the preprocessing tasks to cleanse the text and remove accented characters, punctuation marks and duplicate whitespaces [6].

Table 1 Example of the application of the preprocesing tasks

Preprocessing input	Preprocessing output
Alfa Romeo GIULIETTA 1.4 TB M-Air 170CV TCT SPRINT	Alfa Romeo giulietta 1.4 tb mair 170cv tct sprint
Honda CR-V 1.6 i-DTEC 120 4 × 2: Comfort, Elegance 15YM	Honda crv 1.6 idtec 120 4 × 2 comfort elegance 15ym
Citroën C-Elysée PureTech 82	Citroen celysee puretech 82
Ford Focus 2014 Sportbreak 1.0 Ecoboost Auto-Start-Stop (125cv)"	Ford focus 2014 sportbreak 1.0 ecoboost autostartstop 125cv

3.2 *Indexing*

We work with 20011 different products from IDAE dataset and with 7591 records from coches.net, then a total amount of 151,903,501 comparisons would be needed to match each product of coches.net to IDAE dataset.

From IDAE dataset we have the following attributes: car brand, car model, car name, consumption, emissions, energy classification, segment car, fuel type, engine size, transmission, tare weight, horsepower, size and seats. While from coches.net: car brand, car model, car name, picture, price, city, fuel type, transmission, colour, seats and horsepower.

Then the attributes considered to form the blocking keys are: car brand, car model, fuel type, transmission and seats. These attributes have been selected according to the following requirements:

1. They must reduce the number of candidates for the matching step.
2. They must be available in most of the products considered; otherwise the block does not reduce the number of candidates.
3. They must be of high quality. In many situations if the quality of the block values is low and data contain errors, the block assigned to the object could not contain the real element to link.

This selection has led to get the products of the IDAE dataset clustered in 2804 blocks. Figure 4 shows the size of the 50 biggest blocks. The first block, the biggest one, contains 218 products.

The histogram in Fig. 5 shows the number of blocks (y axis) of every size (x axis), for instance, there exists 609 blocks of size 1, 529 blocks containing 2 products, 295 with 3 elements, etc.

Figure 6 shows groups of blocks in intervals of 5, meaning that the most common blocks are those of size from 1 to 5 elements. In fact the average size is 7.13 elements per block.

Therefore, in the worst case any element of the biggest block (218 elements) will be matched, then the minimum Reduction Ratio value is given by using Eq. 1 and substituting N by 218 and n_M and n_N by 20011, the total number of elements in the dataset:

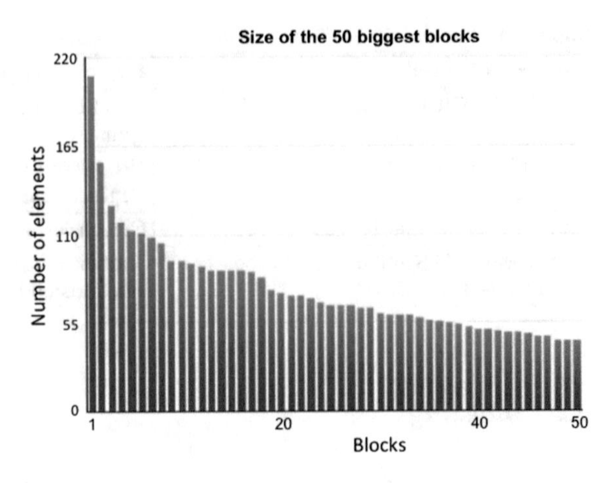

Fig. 4 Size (*y axis*) of the 50 biggest blocks (*x axis*). The first block, the biggest one, contains 218 products

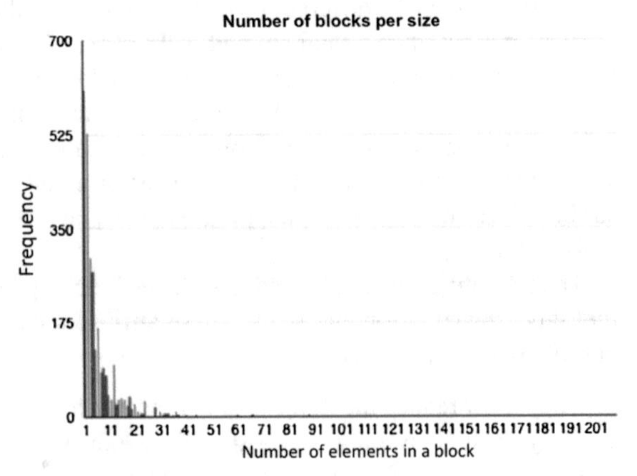

Fig. 5 Histogram of the number of blocks per size. The horizontal axis is labeled *Number of elements in a block*, taking values from 1 to 218 items. The vertical axis represents the number of blocks of each size, and it is numbered from 0 to 700

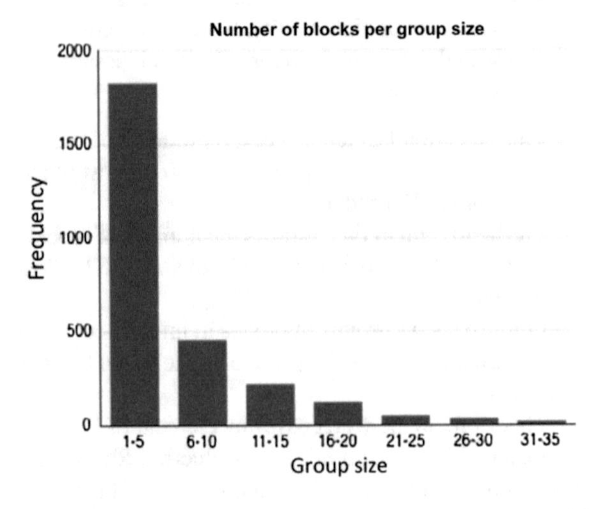

Fig. 6 Histogram of the number of blocks per size grouped in intervals of 5. In this case block size are clustered in groups of 5, then first vertical bar represents the number of blocks (frequency) which contain from 1 to 5 elements, second bar the number of blocks which contain from 6 to 10 elements, and so on

$$rr_{min} = 1 - \frac{218}{20011} = 0.9896 \quad (10)$$

While the best case occurs when the block just contains one element, giving the maximum Reduction Ratio when $N = 1$:

$$rr_{max} = 1 - \frac{1}{20011} = 0.9996 \quad (11)$$

Finally the average Reduction Ratio is computed taking into account the averaged number of elements in a block ($N = 7.13$ elements in this problem):

$$\overline{rr} = 1 - \frac{7.13}{20011} = 0.9996 \quad (12)$$

3.3 Metrics Assessment and Classification

After blocking IDAE dataset, each model description from coches.net must be analysed in order to find the corresponding entry within the IDAE partitions.

k-NN decision rule could be used in order to link every car model from coches.net to the corresponding entry in IDAE dataset. Since IDAE dataset entries are unique, initially the size of a block corresponds to the number of classes per block, and then k should be set up to 1 (initially there are no more than one sample per class). A feedback process has been implemented in order to incorporate all the classified models from coches.net into the dataset, in this way the number of examples per class is successively increased for future experiments.

In order to test this preliminary fuzzy classifier with $k = 1$, 930 samples from coches.net were manually labelled and used for validation purposes. Table 2 shows the error and success classification rates when using different string similarity metrics, in particular Monge–Elkan using Jaro as secondary metric, 5-shingles and TFIDF, and the combined schemes TFIDF+Monge–Elkan and TFIDF+Monge–Elkan + 5-shingles together.

As shown in Table 2, if the metrics are considered alone, Monge–Elkan is the worse, while TFIDF outperforms the others, providing a success rate of 88.19%. The error rate of Monge–Elkan is the highest, 29%, however in conjunction with TFIDF the classification results improve significantly providing a success rate of 88.71%, although slightly better than TFIDF alone. The incorporation of 5-shingles to this combination does not provide better classification results; therefore it is not worth taking into consideration 5-shingles for using in combined metrics.

Table 2 Classification results obtained by the selected string similarity metrics

Metric	Error rate (%)
Monge–Elkan	29.00
5-shingles	14.5
TFIDF	11.81
TFIDF + Monge–Elkan	11.29
TFIDF + Monge–Elkan+ 5-shingles	11.29

4 Conclusions and Future Work

We have presented a comparison of different string similarity measures as well as a complete procedure to product matching applied to determine the efficiency energy index of used cars for sale. Preprocessing tasks allows obtaining consistent data that can be tokenized and indexed in different blocks, in this way the number of comparisons is drastically reduced and simplifies the next step, the matching of products. The combined approach of Monge–Elkan plus TFIDF provides the best results for classification purposes. In contrast to crisp k-NN algorithm, which yields only the best-matching class, the class membership values returned by the fuzzy k-NN can provide much richer information about the ambiguity of the similarities identified between the new data and the historical examples. The preliminary results with $k = 1$ have yielded to a success rate of almost 89% even in this case the fuzzy k-NN can be considered more nuanced than the crisp version. It is expected to obtain higher success rates as the number of examples per class increases, and using higher k values.

The geospatial information in coches.net and the energy efficiency indexes from IDAE database can be used after the product matching process as a starting point for further analysis about the energy efficiency in the Spanish used cars market. For example, a study to identify Spanish provinces with the most polluting, or the most energy efficient car fleet, can be used to optimize a renew car fleet strategy.

References

1. Agrawal, R., Ieong, S.: Aggregating web offers to determine product prices. In: Proceedings of the 18th ACM SIGKDD International Conference on Knowledge Discovery and Data Mining, pp. 435–443, ACM (2012)
2. Aizawa, A.: An information-theoretic perspective of tf-idf measures. Inf. Process. Manag. **39**(1), 45–65 (2003)
3. Bilenko, M., Basil, S., Sahami, M.: Adaptive product normalization: using online learning for record linkage in comparison shopping. In: Fifth IEEE International Conference on Data Mining, pp. 8-pp. IEEE (2005)
4. Christen, P.: Data Matching: Concepts and Techniques for Record Linkage, Entity Resolution, and Duplicate Detection. Springer Science & Business Media, Berlin (2012)

5. Cohen, W.W., Ravikumar, P.D., Fienberg, S.E.: A comparison of string distance metrics for name-matching tasks. In: IIWeb, Vol. 2003, pp. 73–78 (2003)
6. Eisenstein, J.: What to do about bad language on the internet. In: HLT-NAACL, pp. 359–369 (2013)
7. Han, J., Kamber, M., Pei, J.: Data Mining: Concepts and Techniques. Elsevier, Amsterdam (2011)
8. Hong, T.P., Lin, C.W., Yang, K.T., Wang, S.L.: Using TF-IDF to hide sensitive itemsets. Appl. Intell. **38**(4), 502–510 (2013)
9. Jaro, M.A.: Advances in record linkage methodology as applied to the 1985 census of Tampa Florida. J. Am. Stat. Assoc. **84**(406), 414–420 (1989). doi:10.1080/01621459.1989.10478785
10. Jaro, M.A.: Probabilistic linkage of large public health data files. Stat. Med. **14**(5:7), 491–498 (1995)
11. Jimenez, S., Becerra, C., Gelbukh, A., Gonzalez, F.: Generalized Mongue-Elkan method for approximate text string comparison. In: International Conference on Intelligent Text Processing and Computational Linguistics, pp. 559–570. Springer, Berlin, Heidelberg (2009)
12. Kannan, A., Givoni, I.E., Agrawal, R., Fuxman, A.: Matching unstructured product offers to structured product specifications. In: Proceedings of the 17th ACM SIGKDD International Conference on Knowledge Discovery and Data Mining, pp. 404–412. ACM (2011)
13. Keller, J.M., Gray, M.R., Givens, J.A.: A fuzzy k-nearest neighbor algorithm. IEEE Trans. Syst. Man Cybern. **4**, 580–585 (1985)
14. Köpcke, H., Thor, A., Thomas, S., Rahm, E.: Tailoring entity resolution for matching product offers. In: Proceedings of the 15th International Conference on Extending Database Technology, pp. 545–550. ACM (2012)
15. Leskovec, J., Rajaraman, A., Ullman, J.D.: Mining of Massive Datasets. Cambridge University Press, Cambridge (2014)
16. Monge, A., Elkan, C.: The field matching problem: Algorithms and applications. In: Proceedings of The Second International Conference on Knowledge Discovery and Data Mining, (KDD) (1996)
17. Paik, J.H.: A novel TF-IDF weighting scheme for effective ranking. In: Proceedings of the 36th International ACM SIGIR Conference on Research and Development in Information Retrieval, pp. 343–352. ACM (2013)
18. Ren, F., Sohrab, M.G.: Class-indexing-based term weighting for automatic text classification. Inf. Sci. **236**, 109–125 (2013)
19. Salton, G., Buckley, C.: Term-weighting approaches in automatic text retrieval. Inf. Process. Manag. **24**(5), 513–523 (1988)
20. Singhal, A.: Modern information retrieval: a brief overview. IEEE Data Eng. Bull. **24**(4), 35–43 (2001)
21. Thor, A. (2010). Toward an adaptive string similarity measure for matching product offers. In: GI Jahrestagung (1), pp. 702–710
22. Winkler, W.E.: String Comparator metrics and enhanced decision rules in the Fellegi–Sunter model of record linkage. In: Proceedings of the Section on Survey Research Methods, pp. 354–359. American Statistical Association (1990)
23. Winkler, W.E.: Overview of Record Linkage and Current Research Directions. Research Report Series, RRS (2006)
24. http://www.idae.es Instituto para la diversificación y ahorro de la energía
25. http://www.coches.net Web para la compra y venta de coches usados

Fault Diagnosis in a Steam Generator Applying Fuzzy Clustering Techniques

Adrián Rodríguez Ramos, Rayner Domínguez García, José Luis Verdegay Galdeano and Orestes Llanes Santiago

Abstract In this chapter the design of a fault diagnosis system using fuzzy clustering techniques for a BKZ-340-140 29M steam generator in a thermoelectric power station is presented. The application aims to study the advantages of these techniques in the development of a fault diagnosis method with the characteristic to be robust to external disturbances and sensitive to small magnitude faults. The wavelet transform (WT) is used for isolating noise present in measurements. The fault diagnosis system was designed for the water-steam circuit of the steam generator by its great incidence in the correct operation of the generation blocks. The obtained results indicate the feasibility of the proposal.

1 Introduction

Actually, is pressing to have a sustainable develop system to guarantee the protection of the environment, the rational and efficient use of raw materials and energy and the human safety. Catastrophical accidents produced by industrial faults as the happened at Bhopal, India in December 3rd, 1984 [41]; at Chernobyl, Ukraine, in April 26th, 1986 [3, 32] or at Texas, USA, in March 23rd, 2005 [11], to only cite some examples, should be avoided.

On the other hand, in the current industries, there is a marked need to improve the efficiency in the processes in order to produce with higher quality and besides keeping the environmental and industrial safety regulations [13, 14, 42]. Non planned stops

A. Rodríguez Ramos (✉) · R. Domínguez García · O. Llanes Santiago
Departamento de Automática y Computación, Instituto Superior Politécnico
José A. Echevarría CUJAE, La Habana, Cuba
e-mail: adrian.rr@electrica.cujae.edu.cu

O. Llanes Santiago
e-mail: orestes@tesla.cujae.edu.cu

J.L. Verdegay Galdeano
Departamento de Ciencias de la Computación e Inteligencia Artificial, E.T.S.
Ingeniería Informática, Universidad de Granada, España, Spain
e-mail: verdegay@decsai.ugr.es

C. Cruz Corona (ed.), *Soft Computing for Sustainability Science*,
Studies in Fuzziness and Soft Computing 358, DOI 10.1007/978-3-319-62359-7_11

and faults in equipments can have an unfavorable impact in the availability of the systems, in the safety of operators and in the environment. In an industrial context, the safety is associated to a set of specifications to be accomplish for reducing the risk of accidents. With this in mind, it is indispensable that supervisory systems and automatic control are incorporated in the industrial processes such that allow holding a satisfactory operation, compensating the effects of disturbances and the changes which happen in them. Furthermore, the faults need to be detected and isolated, being these tasks associated to the fault diagnosis systems [15]. Within the methods of fault diagnosis there are the ones based on models [8, 27, 35, 42, 43, 45] and the ones based in historical data of the process [1, 2, 6, 7, 10, 21, 26, 28, 36].

The first approach uses mathematical tools to obtain models that describe the operation of the processes. Their basic principle is determined by the generation of residues based on discrepancies between the measurable signals of the real process and the values obtained from these models. This entails an elevated knowledge about the characteristics of the process, their parameters and the zone of operation. The above can be occasionally very difficult due to the complexity of some industrial processes. On the other hand, the approach based in historic data is divided into two groups. In one group are the methods, that use the Artificial Intelligence (AI) tools and the other group that makes an statistical analysis of these data. Generally, methods that use this approach do not need a mathematical model and they do not require much prior knowledge of the process parameters [44]. These characteristics constitute an advantage in complex systems, where relationships among variables are nonlinear or unknown and where it is very difficult determining an analytical model that describes efficiently the dynamics of the process.

The development of fault diagnosis methods, robust to external disturbances and sensitive to incipient faults is considered as a current problem in the industries where a fault would provoke magnus economic and human losses [23].

Making an analysis of the different techniques developed in the last years for control and fault diagnosis tasks, it is significative the increment in the use of the fuzzy logic techniques, specially the fuzzy clustering methods [4, 5, 16, 24, 30, 31, 34, 40]. Evaluating the characteristics of this technique, it was considered to be advantageous for developing a robust fault diagnosis system regarding to external disturbances, sensible to faults of small dimensions and easy to implement, for complex process as is the case of the steam generators or boilers.

The objective of the present chapter is the design of a fault diagnosis system for a boiler BKZ 340-140-29M, present in some Cuban industries, using the fuzzy clustering techniques.

The structure of this chapter is the following: in Sect. 2, corresponding to Materials and methods a general description of the boiler BKZ 340 140-29M and the design of the fault diagnosis system is presented. In Sect. 3 are presented the design of the experiments and the analysis of results. Finally, the conclusions are presented.

2 Materials and Methods

2.1 *General Description of the Steam Generator BKZ 340 140-29M*

A steam generator or boiler is an equipment very used in some industries as for example the chemical industry and the electricity generation industry. Among their different types, are the watertube steam generators that allow high pressures in their outlet and have a great generation capacity. A steam generator is an equipment that transform the energy obtained by the combustion of a fuel in thermic energy and transfers it to water to produce steam that is used in other equipment or systems. It is a multivariable dynamic system with a big quantity of interactions among its variables and parameters, that are affected by multiple disturbances. The steam generation is a continuous process which variables can take different values in a certain range so they need to be controlled for the safety and efficiency of the operation of the system. Nowadays, there are a great variety of boilers, which can be classified according to the type of fuel, shaft, supporting systems, heat transfer and disposal of fluids. The steam generators type BKZ-340-140-29M show a single steam dome and they are designed to work with a turbine type K-100-130-3600-2T3 to produce electric energy, having a 340 ton/hour overheated-vapor capacity and each one consists of an oven, a steam dome and four overheating stages. At the outlet of the last overheating stage of the steam generator, the steam goes to the turbine and returns to the second stages of the reheater. For more information see [39].

2.2 *Statistical Analyses of Faults*

For the steam generator BKZ-340-140-29M, three fundamental circuits are defined that include all the necessary equipment for production, treatment and distribution of steam, named: water-steam, air-gases and combustion.

After a statistical study of all faults that took place in the last three years in a generator of this type used in a thermoelectric plant [39], it was determined the sub-system that most affected the operation of the steam generator. It was the water-steam sub-system, with the 67% of the total of interruptions. For this reason, it was decided the implementation of a fault diagnosis system for this circuit. The economizer and superheater were the most affected technological equipment. After a review of the steam production process and taking into account the opinion of experts, it was determined the faults to be diagnosed:

- **F1**: Pore in the reheater.
- **F2**: Pore in the economizer.
- **F3**: Fault in the opening of the reducing device.
- **F4**: Pore in the boiler.

- **F5**: Secondary combustion.
- **F6**: Infiltrations in the oven.

2.3 Variable Symptoms

A study of which variables should be used to detect and isolate the faults was done. For this, individual interviews to the process engineers of the plant were performed. As final result of this analysis the variables selected were: variation of the pressure in the oven (v_1, mm H_2O), variation of generation power (v_2, MW), and variation in the temperature of gases in the outlet (v_3, C).

The variable v_3 requires a more detailed explanation. As it is typical in the BKZ 340-140-29M generators, on the outlet of the economizer two regenerative air heaters (RAH) are located. The behavior of the temperature of the gases on the outlet of these mechanisms is related to the state of the economizer. The appearing of pores in it provokes a sudden variation in the behavior of the temperature of the gases in the outlet. Incorporating the analysis of the temperature for both RAH in the design of the fault diagnosis system will result in an increase of its complexity. However, in order to diagnose the state of the economizer, it is only necessary to watch the maximum value of the variation of the temperature in both RAH. For this reason a pre-processing mechanism of the both values was made to use only the biggest temperature variation as value of v_3. In addition to the symptom variables (v_1, v_2, v_3), a fourth variable (v_4) is monitored. This variable corresponds to the variation of the oil flow used as a breaker of the diagnosis system. The diagnosis system was designed to work when the generation system is in steady state. In the case that the operation point of the generation system is changed, this variable (v_4) will be not in the adequate range and it provokes the disconnection of the diagnosis system. The variation of this variable should be between 0.2 kg/h, according to the opinion of the experts.

2.4 Proposal of Design of the Diagnosis System

A trained classifier with techniques of fuzzy clustering supports the classification sketch shown in Fig. 1. In the data pre-processing stage the Wavelet Transform (WT) is applied to isolate the noise in the measurements, allowing the enhancement of the robustness of the diagnosis system. For the decomposition of the signal, the Discrete Wavelet Transform (DWT) is applied using Multiresolution Analysis (MRA) with five levels and the function Daubechies as wavelet mother. The parameters used for the decomposition of the signal are used frequently in applications of de-noise in fault diagnosis, achieving excellent results [18, 33, 46]. The classifier presents an off-line learning stage and an online recognition stage. In the training stage the historical data of the process are used to train (modeling the functional stages through the clusters) a

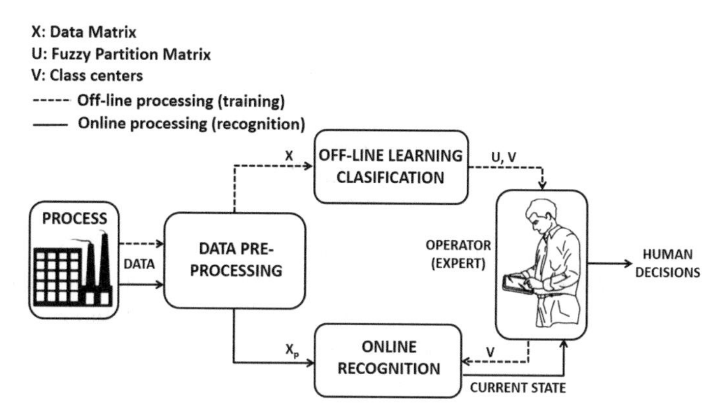

Fig. 1 Diagnosis scheme based in fuzzy classifiers

fuzzy classifier. After the training, the classifier is used online (recognition) in order to process every new sample taken from the process. The result is to offer information in real-time to the operator of the system's state.

The clustering methods group the data in different classes based on a measure of similitude. In the processes, data correspond to the samples taken by means of a SCADA system (Supervisory Control and Data Acquisition), and the classes can be associated to operational states. In the case of statistical classifiers, each sample is compared to the center of the class by means of a similitude measure to determine if it belongs or not to a class and, in the case of the diffuse classifiers, the comparison is made to determine the membership degree to each class.

The class of n data vectors $X = [x_1, x_2, \ldots, x_N]$ where N corresponds to the number of samples, is divided into c groups or classes (using a method of clustering). The fuzzy clustering allows getting the membership degrees matrix $U = [\mu_{ik}]_{c\times N}$ where μ_{ik} represents the degree of fuzzy membership of the sample k to the ith class and should fulfill the conditions shown in (1). From the analysis of the degree of fuzzy membership for each sample k, it can be determined the class in which the sample is located. In general, the higher membership degree determines the class that the sample is assigned, as it is observed in (2).

$$M_{fc} = \left\{ U \in \mathbb{R}^{cxN} \middle| \mu_{ik} \in [0, 1], \forall i, k; \sum_{i=1}^{c} \mu_{ik} = 1, \forall k; 0 < \sum_{k=1}^{N} \mu_{ik} < N, \forall i \right\} \tag{1}$$

$$C_i = \{i : max\{\mu_{ik}\}, \forall i, k\} \tag{2}$$

For the fuzzy clustering, different methods have been proposed. Among them, the most known are the ones based in distance. Within these methods, there are the Fuzzy-C means (FCM) and Gustafson-Kessel means (GKM) algorithms that use optimization criteria, as it is appreciated in (3). This makes possible to group the data according to the similitude among themselves.

$$J\left(X;U,v\right)=\sum_{i=1}^{c}\sum_{k=1}^{N}\left(\mu_{ik}\right)^{m}\left(d_{ik}\right)^{2} \tag{3}$$

In these algorithms, the similitude is evaluated by means of the distance function d_{ik}, according to the Eq. (4). This distance function is measured between the data and the centers of classes $v = v_1, v_2, ..., v_c$, being $A \in \Re^{n\times n}$ the norm induction matrix, where n is the quantity of measured variables.

$$d_{ik}^{2}=\left(x_{i}-v_{k}\right)^{T}A\left(x_{i}-v_{k}\right) \tag{4}$$

The exponent $m > 1$ in (3), is an important factor that regulates the fuzziness of the resulting partition. The measure of dissimilarity J is the square distance between each data point and the clustering center v_k. This distance is pondered by the potency of the membership degree $(\mu_{ik})^m$. The value of the cost function J is a measure of the pondered total quadratic error in the inside of the group generated by the representation of c clustering defined by the clustering center v_k. Statistically, J can be seen as a measure of the total variance of x_i regarding v_k.

The minimization of the C-means functional J represents a nonlinear optimization problem that can be solved using a varied collection of available methods. The most popular method is Picard's simple iteration through first-rate conditions for stationary J points, known as the fuzzy c means algorithm (FCM). The optimization problem is then stated in the following way:

$$min\left\{J\left(X;U,v\right)\right\}=\sum_{i=1}^{c}\sum_{k=1}^{N}\left(\mu_{ik}\right)^{m}\left(d_{ik}\right)^{2} \tag{5}$$

with:

$$\sum_{i=1}^{c}\mu_{ik}=1,\ \mu_{ik}\in\left[0,1\right],\forall i,k \tag{6}$$

The stationary points of the objective function can be found by the addition of limits to J by means of Lagrange multipliers, as is showed in (7).

$$J\left(X;U,v,\lambda\right)=\sum_{i=1}^{c}\sum_{k=1}^{N}\left(\mu_{ik}\right))^{m}\left(d_{ik}\right)^{2}+\sum_{k=1}^{N}\lambda_{k}\left(\sum_{i=1}^{c}\mu_{ik}-1\right) \tag{7}$$

and by setting the gradients of J with respect to U, v and λ to zero. If $d_{ik}^2 > 0, \forall i, k$ and $m > 1$, then $(U, v) \in M_{fc}$ can minimize to J only if:

$$\mu_{ik} = \frac{1}{\sum_{j=1}^{c} \left(d_{ik,A}/d_{jk,A}\right)^{2/(m-1)}} \tag{8}$$

$$v_i = \frac{\sum_{k=1}^{N} (\mu_{ik})^m x_k}{\sum_{k=1}^{N} (\mu_{ik})^m} \tag{9}$$

In (9) should be noted that v_i is the pondered media of the data elements that belongs to a clustering, i.e., it is the center of the clustering c.

2.5 *Off-Line Training*

In this stage a historical data set that represent the different operation states of the process (classes) for training the fuzzy classifier is used, where the center of each one of classes $v = v_1, v_2, ..., v_N$ is determined. These clustering algorithms are iterative procedures where N data are grouped in c classes. The user chooses the number of c classes. The centers of the classes are initialized in a random form and they are modified during this iterative process. In a similar way the membership degrees matrix U is modified until it will be stabilized, that's to say $\|U_t - U_{t-1}\| < \varepsilon$, where ε is a tolerance limit. This training process is described in the Algorithm 1, where the FCM and GKM algorithms are used. The only difference is that in the case of FCM, the Euclidean distance is used allowing the data grouping in hyper-spheres, while the GKM algorithm uses the Mahalanobis distance, allowing the data grouping in hyper-ellipses.

Algorithm 1 Training (FCM and GKM)

Fixing: $1 < c < N, \varepsilon > 0, m > 1$, number of iterations
Initialize: $U^{(0)} \in M_{fc}$
for $l = 1$ to $l = Itr_{max}$ **do**
  Calculating the centers of the clusterings with (9)
  Calculating the distances with (4)
  Updating the partition matrix (8)
  Verifying stopping criterion
end for
Optimal solution found: U^{best}, v_i^{best}

2.6 *Online Recognition*

In this stage, the distance between the on-line observations and the centers of every class determined in the training stage according to Eq. (4) is calculated in every instant of time. Next, the fuzzy membership degree of the sample k raised to the i^{-th} class is obtained. Starting from the analysis of the fuzzy membership degree for each sample k, it can be determined the class in which the sample is located, where the major degree of membership determines the class that the sample is assigned. This procedure is described in the Algorithm 2.

Algorithm 2 Recognition with FCM and GKM

for $k = 1$ to $k = N$ **do**
  Calculating the distances the observation k to class centers with (4)
  Calculating the degree of membership of the k to the i^{-th} with (8)
  Determining to which class belongs the observation k with (2)
end for

2.7 *Parameters of the Algorithm FCM and GKM*

- **Stopping criterion**: The algorithms FCM and GKM stop the iterations when the norm of difference of U in two successive iterations is minor than the parameter ε. The usual election is $\varepsilon = 0.001$, although $\varepsilon = 0.01$ works well in the majority of the cases, and this reduce the computation time. In the case that the algorithm does not stop for this condition, it finishes for the number of iterations initially established.
- **Parameter m**: The weighted exponent m regulates the fuzziness of the resulting partition. It can be considered that for $m \to \infty$, all patterns will have the same membership degree μ_{ik} to each *cluster* and for $m \to 1$ one pattern will belong to an only cluster.

Analysis of the parameter *m*
In most of the literature concerning to the fuzzy clustering methods based on distance, values between 1, 2 and 2 are used as m pondering exponent, achieving excellent results. However, a lot of authors conclude that this parameter depends on important way of the set of work data [4, 5, 30]. Considering this, a modification of the Algorithm (1) is proposed in the training stage without necessity to fix the parameter m.

In this stage, an optimization algorithm is applicable to estimate the parameter m through optimizing a validity index. This will allow obtaining an improved U partition matrix and therefore a better position of the centers of each one of the classes that characterize the different operation states of the system, determined during the off-line learning. Later, the estimated m value will be used during the online recognition contributing to a better classification of the operation states.

The validity measures are indexes that permit to evaluate quantitatively the result of a clustering method and comparing its behavior when its parameters vary. Some indexes evaluate the resulting U matrix, while others focus on the geometric resulting structure.

In this case, it is used the partition coefficient (PC) [19, 25, 47] as it is shown in the Eq. (10), which measures the degree of fuzziness of the partition U. It is considered that if the partition U is less fuzzy, the clustering is better. Seen in a different way, they allow measuring the degree of overlapping among the classes.

$$PC = \frac{1}{N}\sum_{i=1}^{c}\sum_{k=1}^{N}(\mu_{ik})^2 \tag{10}$$

In this case, the optimum comes up when PC is maximized. In this case, each pattern belong to an only group. Minimum comes up when each pattern likewise belongs to each group.

2.8 *Optimization Algorithm*

Inside the fault diagnosis field, metaheuristic algorithms have been widely used, obtaining excellent results [9, 20, 22]. They can locate efficiently the neighborhood of the global optimum in the most of the occasions and with an acceptable computational time. There is a large number of metaheuristic algorithms, in their original and improved versions. Some examples are Genetic Algorithm, Differential Evolution, Particle Swarm Optimization and Ant Colony Optimization.

In this chapter, Differential Evolution is used due to its advantages for its simple structure, higher speed and robustness [9]. These characteristics prove to be suitable when a fault detection in real time is required.

Differential evolution (DE) was proposed in 1995 to solve optimization problems [37, 38]. On its beginnings this algorithm was developed to improve Goldberg's Genetic Algorithms [12] and Simulated Annealing [17]. DE is an evolutionary algorithm based in populations, that uses methods derived from Biology like inheritance, mutation, natural selection and crossover.

The idea behind DE is generating a population of new feasible solutions based on perturbed solutions belonging to the population of solutions obtained until that time. This generation scheme is based on three operators: Mutation, Crossover and Selection. The configuration of DE can be summarized using the following notation:

$$DE/\mathbb{X}^j/\gamma/\lambda \tag{11}$$

where $\mathbb{X}^j$ denotes the solution to be perturbed in the iteration *j-sima*, γ the number of solution pairs of the population that will be used in the disturbance of $\mathbb{X}^j$ and λ indicates the distribution function that will be used during the crossover. In this case has been considered the configuration $DE/\mathbb{X}^{j(best)}/1/Bin$, where $\mathbb{X}^{j(best)}$ indicates the best individual of the Z population and Bin the Binomial Distribution function. This mutation operator is expressed in the following way:

$$\mathbb{X}^{j+1} = \mathbb{X}^{j(best)} + F_S(\mathbb{X}^{j(\alpha)} - \mathbb{X}^{j(\beta)}) \tag{12}$$

where $\mathbb{X}^{j+1}, \mathbb{X}^{j(best)}, \mathbb{X}^{j(\alpha)}, \mathbb{X}^{j(\beta)} \in \mathbb{R}^n$, $\mathbb{X}^{j(\alpha)}$ and $\mathbb{X}^{j(\beta)}$ are elements of the Z population and F_S is the escalation factor. For complementing the mutation operator, the crossover operator is defined for each component $\mathbb{X}_n$ of the solution vector:

$$\mathbb{X}_n^{j+1} = \begin{cases} \mathbb{X}_n^{j+1}, & \text{if } R < C_R \\ \mathbb{X}_n^{j(best)}, & \text{in other case} \end{cases} \tag{13}$$

where $0 \leq C_R \leq 1$, is the crossover constant that is another control parameter in DE. R is a random number that is generated by the distribution λ that in this case it is the binomial distribution.

Finally, the selection operator results as follows:

$$\mathbb{X}^{j+1} = \left\{ \mathbb{X}^{j+1}, \text{ if } F\left(\mathbb{X}^{j+1}\right) \leq F\left(\mathbb{X}^{j(best)}\right) \mathbb{X}^{j(best)}, \text{ in other case} \right. \tag{14}$$

The control parameters in DE are the size of the population Z, the crossover constant C_R and the escalation factor F_S. In [37] some simple rules are given for the selection of the parameters in dependence of the type of application. In this case the objective function F is the validity measure shown in (10), where the individuals of the population Z are represented by the parameter m. The general diagram of these new algorithms is the same and it is shown as follows:

Algorithm 3 Training (FCM-DE and GKM-DE)

Fixing: Z, F_S, C_R, number of iterations
Generating initial population of Z solutions.
Selecting better solution $\mathbb{X}^{best}$.
for $l = 1$ to $l = Itr_{max}$ **do**
  Applying Mutation according to (12)
  Applying Crossover according to (13)
  Applying Selection according to (14)
  Updating $\mathbb{X}^{best}$
  Verifying Stopping criterion
end for
Found optimal solution: $\mathbb{X}^{best}$

3 Design of Experiments and Analysis of Results

A very important step in the design of diagnosis systems based in supervised learning and classification tools, is verifying the quality of the performed task. The most used criteria for this analysis are: Confusion Matrix (CM), the global classification error and the Receiver Operating Characteristic (ROC).

The confusion matrix, also known as matrix of uniform cost, is an indicator that allows visualizing the individual and relative errors of the classifier. In this chapter, the confusion matrix shows how the diagnosis system confuses different operation states, allowing to evaluate the probabilistic functions of every stage of fault. In the rows of the confusion matrix the predicted states for the classifier are represented and the real states in the columns. Each C_{ij} element of a matrix indicates the number of times that the classifier confuses the state i with the j in a set of N experiments.

The results obtained from the application of the proposed methodology to fault diagnosis in the water-steam sub-system of a steam generator BKZ-340-140-29M in a thermoelectric power station is presented. In order to simplify the analysis of the results, two sets of data belonging to the normal operation of the process and to the occurrence of a pore in the overheater (fault 1) were used. In the training process of the classifier 400 observations were used (200 for each class) and for the online recognition other 800 samples were used (400 for each class). The data sets that are shown in Figs. 2 and 3, were taken when the generation system was in steady state, in other words, the variable v_4 was in the adequate range.

The algorithms 1 and 2 were run 9 times for m = [1.2, 1.3, 1.4, 1.5, 1.6, 1.7, 1.8, 1.9 and 2] with $\varepsilon = 0.001$ and 100 iterations. The results shown in Figs. 4 and 5 where the WT was not applied in the data preprocessing, reflect that for algorithms

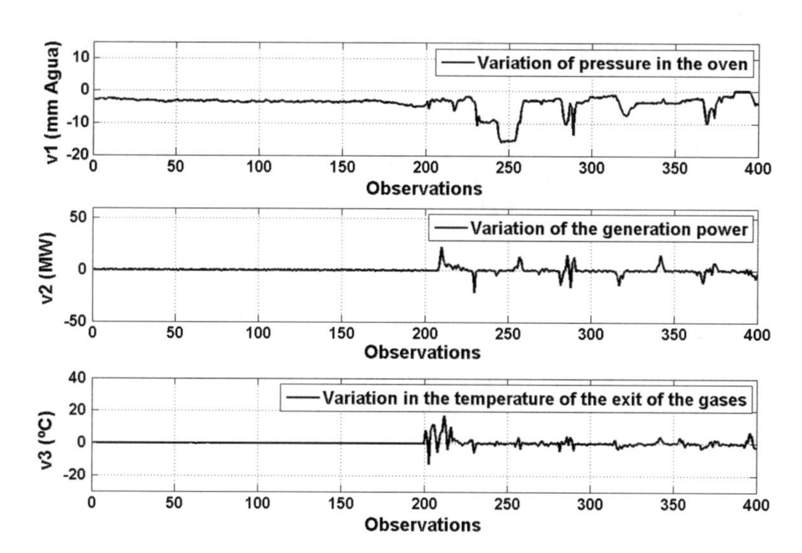

Fig. 2 Training data

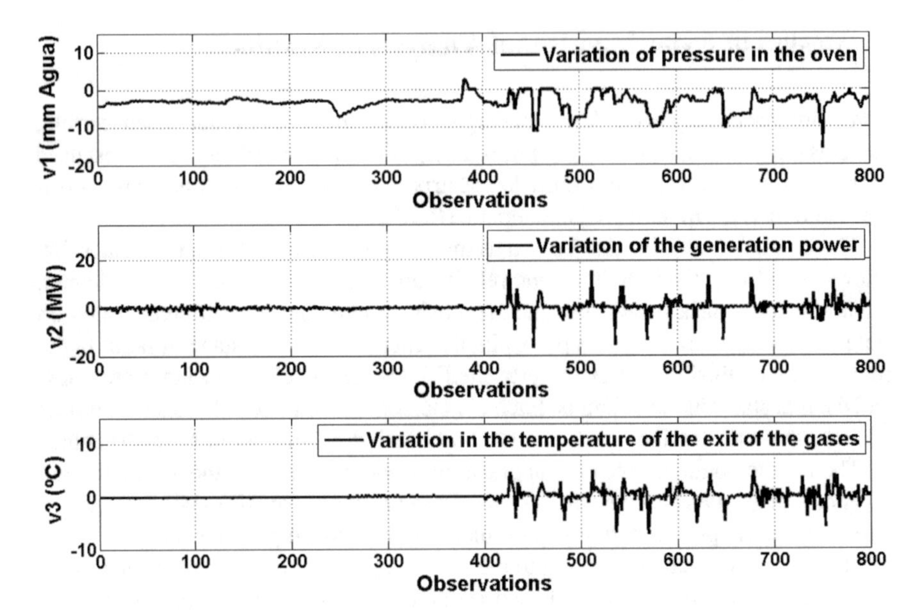

Fig. 3 Recognition data

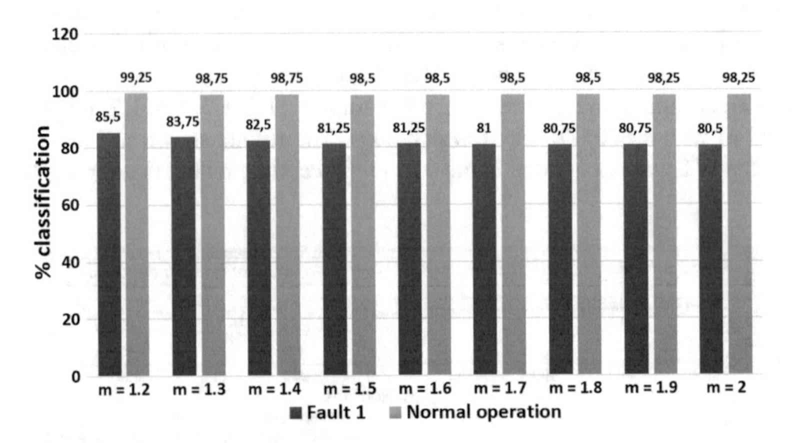

Fig. 4 Classification with FCM without applying TW

FCM and GKM the best classification of the operation states are obtained for the smaller value of m. The Figs. 6 and 7 show the results when the WT is applied.

For the implementation of the algorithms FCM-DE and GKM-DE, three strategies of the differential evolution algorithm were considered, one with more inclination toward diversification (DE_1), the second with more tendency to intensification (DE_2) and the other with similar capacity of diversification and intensification (DE_3). The parameters of bigger influence in the type of search (diversification and intensification) are Z and C_R [29, 37]. In DE_1 the value of the parameter $C_R = 0.9$, for

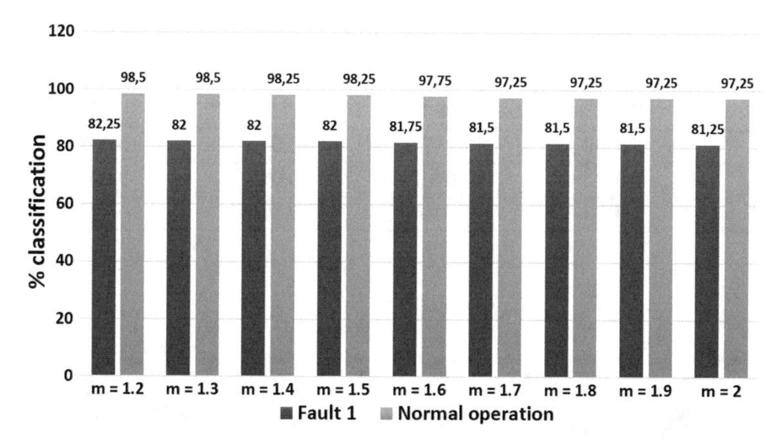

Fig. 5 Classification with GKM without applying TW

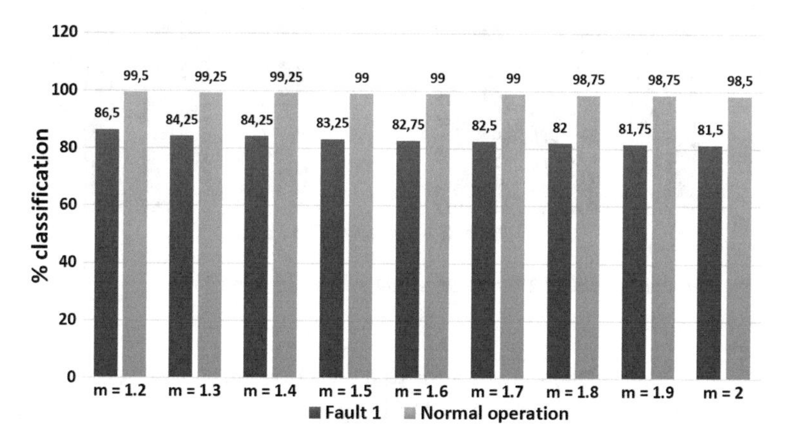

Fig. 6 Classification with FCM applying TW

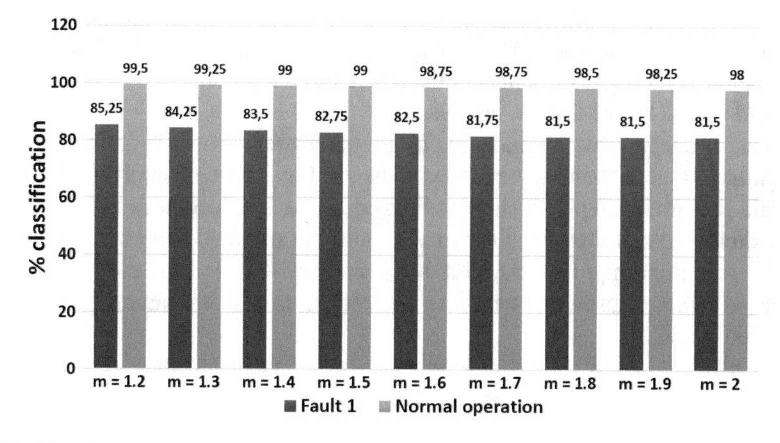

Fig. 7 Classification with GKM applying TW

Table 1 Result of FCM-DE and GKM-DE

Strategy	Mean $\widehat{m}$		Variance $\widehat{m}$	
	FCM	GKM	FCM	GKM
DE_1	1.0738	1.1172	0.0030	0.0078
DE_2	1.0736	1.0907	0.0050	0.0012
DE_3	1.0675	1.0464	0.0034	0.0011

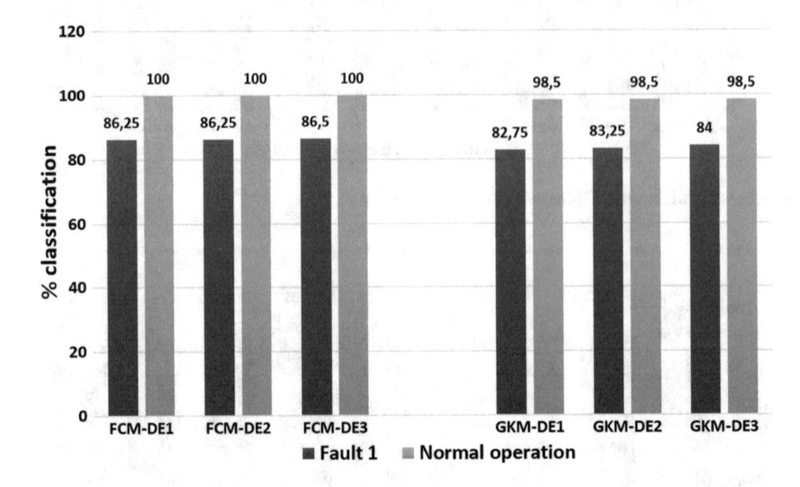

Fig. 8 Classification with FCM-ED and GKM-ED without applying TW

DE_2, $C_R = 0.1$ and finally $C_R = 0.5$ for strategy DE_3. All strategies have the same parameter $F_S = 0.1$, also the same size of the population: $Z = 10$ is considered and the space of search: $m = [1;2]$.

During the training stage, each algorithm was run 10 times and the arithmetical mean and the variance of the estimates were calculated. Results are shown in Table 1. In order to determine what DE's strategy was better, the statistical Friedman's test was applied, and the result indicate that there are no significant differences in the obtained results by the three strategies.

Later, the mean values of the estimates for parameter m are used during the online recognition stage, obtaining the results shown in Fig. 8 without applying the WT in the data pre-processing. The results obtained by the application of the Friedman's test indicate no significative difference among the results. However it is possible to observe a light improvement with DE_3, since $m \rightarrow 1$ because each pattern will belong to an only cluster and therefore less confusion will exist among the different operation states.

When using the WT to isolate the present noise in the measurements, light improvements are obtained as observed in Fig. 9.

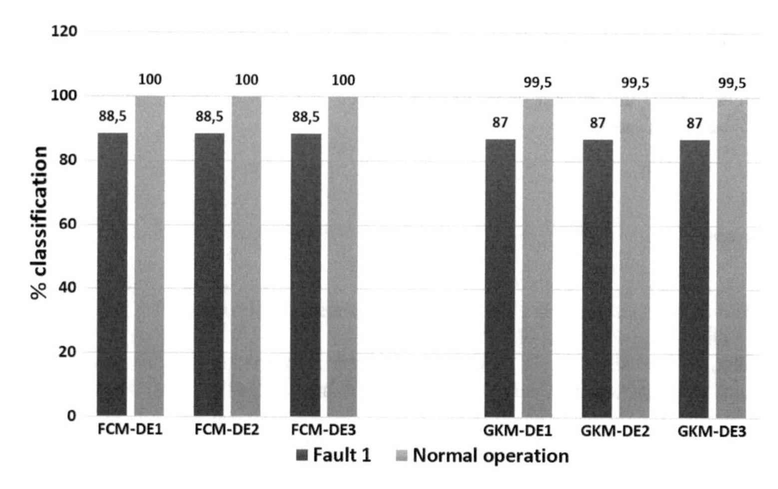

Fig. 9 Classification with FCM-ED and GKM-ED applying TW

3.1 *Analyses of Robustness and Sensibility*

As the obtained results demonstrate, the diagnosis diagram based in fuzzy clustering methods allowed, in this real process, classifying the two operation states with high percents where the data are corrupted with noise. This demonstrate the robustness of the proposed diagnostic system.

As fault 1 is one of the fault generated by pores, in this case in the overheater, it is possible to analyze the sensitivity of the diagnostic system taking into account the expert criteria that this type of fault only is detectable by their, when the pore already has a considerable size. Results indicate that in spite of the fact that this fault becomes of small magnitude, the proposed diagnosis system allows classifying correctly with over 80% of certainty this fault in the process.

4 Conclusions

The main result of this chapter is the feasibility to design a fault diagnosis system based in techniques of fuzzy clustering for achieving robustness in the presence of external disturbances and noise without losing sensibility to detect incipient or small-magnitude faults.

It was demonstrated the influence of the parameter m in obtaining the U partition matrix and therefore in the position of the centers of each one of the classes that characterize the different operation states of the system. A modification of the algorithms during the training stage was performed, estimating with the DE algorithm the best value of m to be used in online recognition. Results evidence the feasibility of the proposal. At last, WT demonstrated its capacity of isolating the noise from the measurements for improving the robustness in fault diagnosis.

Acknowledgements In what concerns to authors from the University of Granada, Spain, this research was supported in part under TIN2014-55024-P from the Spanish Ministry of Economy and Competitiveness and in part under project P11-TIC-8001 from the Andalusian Government (both with FEDER funds).

References

1. Alghazzawi, A., Lennox, B.: Monitoring a complex refining process using multivariate statistics. Control Eng. Pract. **16**, 294–307 (2008)
2. Azadeh, A., Ebrahimip, V.: A fuzzy inference system for pump failure diagnosis to improve maintenance process: the case of a petrochemical industry. Expert Syst. Appl. **37**, 627–639 (2010)
3. Baverstock, K., Williams, D.: The Chernobyl accident 20 years on: an assessment of the health consequences and the internatioanl response. Environ. Health **114**(9), 1312–1317 (2006). doi:10.1289/ehp.9113
4. Bedoya, C., Uribe, C., Isaza, C.: Unsupervised feature selection based on fuzzy clustering for fault detection of the tennessee eastman process. Advances in Artificial Intelligence. Lecture Notes in Artificial Intelligence, pp. 350–360. Springer, Berlin (2012)
5. Bota, J., Isaza, C., Kempowsky, T., LeLann, M.V., Aguilar-Martn, J.: Automation based on fuzzy clustering methods for monitoring industrial processes. Eng. Appl. Artif. Intell **26**, 1211–1220 (2013)
6. Camacho, O., Padilla, D., Gouveia, J.L.: Fault diagnosis based on multivariate statistical techniques. Rev. Tec. Ing. Univ. Zulia **30**, 253–262 (2009)
7. Chen, J., Roberts, C., Weston, P.: Fault detection and diagnosis for railway track circuits using neuro-fuzzy systems. Control Eng. Pract. **16**, 585–596 (2010)
8. Ding, S.X.: Model-Based Fault Diagnosis Techniques. Springer, Berlin (2008)
9. Echevarra, L.C., Llanes Santiago, O., Silva Neto, A.J.: An approach for fault diagnosis based on bio-inspired strategies. Studies in Computational Intelligence, vol. 284, pp. 53–63. Springer, Berlin (2010)
10. Fan, J., Wang, Y.: Fault detection and diagnosis of non-linear non-gaussian dynamic processes using kernel dynamic independent component analysis. Inf. Sci. **259**, 369–379 (2014)
11. Fatal accident investigation report: isomerisation unit explosion, final report. Technical Report (2005). http://www.bp.com/liveassets/bp_internet/us/bp_us_english/STAGING/local_assets/downloads/t/final_report.pdf
12. Goldberg, D.E.: Genetic Algorithms in Search, Optimization and Machine Learning. Addison-Wesley, Reading, Massachusetts (1989)
13. Heng, A., Zhang, S., Tan, A.: Rotating machinery prognostics: state of the art, challenges and opportunities. Mech. Syst. Signal Process. **23**, 724–739 (2009)
14. Hwang, I., Kim, S., Kim, Y., Seah, C.: A survey of fault detection, isolation and reconfiguration methods. IEEE Trans. Control Syst. Technol. **18**, 636–656 (2010)
15. Isermann, R.: Fault-Diagnosis Applications: Model-Based Condition Monitoring: Actuators, Drives, Machinery, Plants, Sensors and Fault-Tolerant Systems. Springer (2011)
16. Jyoti, K., Singh, S.: Data clustering approach to industrial process monitoring, fault detection and isolation. Int. J. Comput. Appl. **17**, 41–45 (2011)
17. Kirkpatrick, S., Gelatt, C., Vecchi, M.P.: Optimization by simulated annealing. Science **220**, 671–680 (1983)
18. Kunpeng, Z., San, W., Soon, H.: Wavelet analysis of sensor signals for tool condition monitoring: a review and some new results. Int. J. Mach. Tools Manuf. **49**, 537–553 (2009)
19. Li, C., Zhou, J., Kou, P., Xiao, J.: A novel chaotic particle swarm optimization based fuzzy clustering algorithm. Neurocomputing **83**, 98–109 (2012)

20. Liu, Q., Lv, W.: The study of fault diagnosis based on particle swarm optimization algorithm. Comput. Inf. Sci. **2**, 87–91 (2009)
21. Liu, J., Zhao, F., Liu, Y.: Learning kernel parameters for kernel fisher discriminant analysis. Pattern Recognit. Lett. **34**(9), 1026–1031 (2013)
22. Lobato, F., Steffen Jr., F., Silva Neto, A.: Solution of inverse radiative transfer problems in two-layer participating media with differential evolution. Inverse Probl. Sci. Eng. **18**, 183–195 (2009)
23. Miguel, L.J.D., Blázquez, L.F.: Fuzzy logic-based decision-making for fault diagnosis in a dc motor. Eng. Appl. Artif. Intell. **18**, 423–450 (2005)
24. Omran, M., Engelbrecht, A., Salman, A.: An overview of clustering methods. Intell. Data Anal **11**, 583–605 (2007)
25. Pakhira, M., Bandyopadhyay, S., Maulik, S.: Validity index for crisp and fuzzy clusters. Pattern Recognit. **37**, 487–501 (2004)
26. Pang, Y.Y., Zhu, H.P., Liu, F.M.: Fault diagnosis method based on KPCA and selective neural network ensemble. Adv. Mater. Res. **915**, 1272–1276 (2014)
27. Patan, K.: Artificial Neural Networks for the Modelling and Fault Diagnosis of Technical Processes. Springer, Berlin (2008)
28. Patan, K., Witczak, M., Korbicz, J.: Towards robustness in neural network based fault diagnosis. Int. J. Appl. Math. Comput. Sci. **18**(4), 443–454 (2008)
29. Price, K., Storn, R., Lampinen, J.: Differential Evolution: A Practical Approach to Global Optimization. Springer, Berlin (2005)
30. Qin, S.J.: Survey on data-driven industrial process monitoring and diagnosis. Annu. Rev. Control **36**, 220–234 (2012)
31. Qin, S.J.: Survey on data-driven industrial process monitoring and diagnosis. Annu. Rev. Control **36**(2), 220–234 (2012)
32. Raloff, J.: Chernobyl: intangible fallout from reactor. Nature **321**, 5067 (1986). doi:10.1038/321186a0
33. Rengaswamy, R., Dash, S., Maurya, M., Venkatasubramanian, V.: A novel interval-halving framework for automated identification of process trends. AIChE J. **50**, 149–162 (2004)
34. Silva, S., Dias, M., Lopez, V., Brennan, M.: Structural damage detection by fuzzy clustering. Mech. Syst. Signal Process **22**, 1636–1649 (2008)
35. Simani, S., Patton, R., Fantuzzi, C.: Model Based Fault Diagnosis in Dynamic Systems Using Identification Techniques. Springer, Berlin (2002)
36. Sina, S., Sadough, Z.N., Khorasani, K.: Dynamic neural network-based fault diagnosis of gas turbine engines. Neurocomputing **125**, 153–165 (2014)
37. Storn, R., Price, K.: Differential evolution: a simple and efficient adaptive scheme for global optimization over continuous spaces. International Computer Science Institute (1995)
38. Storn, R.M., Price, K.V.: Differential evolution: a simple and efficient adaptive scheme for global optimization over continuous spaces. J. Global Optim. **12**, 1–16 (1997)
39. Torres Vizcaya, M., Rodrguez Barrios, T., Prieto Moreno, A., Llanes Santiago, O.: Diagnstico de fallos en el generador de vapor BKZ-340-140-29m. Revista Ingeniería Electrónica, Automática y Comunicaciones (RIELAC) **32**, 31–41 (In Spanish) (2011)
40. Uribe, C., Isaza, C.: Unsupervised feature selection based on fuzzy partition optimization for industrial processes monitoring. In: IEEE International Conference on Computational Intelligence for Measurement Systems and Applications (CIMSA), Ottawa, pp. 1–5 (2011)
41. Varma, R., Varma, D.: The bhopal disaster of 1984. Bull. Sci. Tecnol. Soc. **25**(1), 37–45 (2005). doi:10.1177/0270467604273822
42. Venkatasubramanian, V., Rengaswamy, R., Kavuri, S.N.: A review of process fault detection and diagnosis, part 1: quantitative model-based methods. Comput. Chem. Eng. **27**, 293–311 (2003)
43. Venkatasubramanian, V., Rengaswamy, R., Kavuri, S.N.: A review of process fault detection and diagnosis, part 2: qualitative models and search strategies. Comput. Chem. Eng. **27**, 313–326 (2003)

44. Wang, J., Hu, H.: Vibration-based fault diagnosis of pump using fuzzy technique. Measurement **39**, 176–185 (2009)
45. Witczak, M.: Modelling and Estimation Strategies for Fault Diagnosis of Non-Linear Systems. Springer, Berlin (2007)
46. Wu, J., Hsu, C.: Fault gear identification using vibration signal with discrete wavelet transform technique and fuzzy-logic inference. Expert Syst. Appl. **36**, 3785–3794 (2009)
47. Wu, K., Yang, M.: A cluster validity index for fuzzy clustering. Pattern Recognit. **26**, 1275–1291 (2005)

An Updated Review on Watershed Algorithms

R. Romero-Zaliz and J.F. Reinoso-Gordo

Abstract Watershed identification is one of the main areas of study in the field of topography. It is critical in countless applications including sustainability and flood risk evaluation. Beyond its original conception, the watershed algorithm has proved to be a very useful and powerful tool in many different applications beside topography, such as image segmentation. Although there are a few publications reviewing the state-of-the-art of watershed algorithms, they are now outdated. In this chapter we review the most important works done on watershed algorithms, including the problem over-segmentation and parallel approaches. Open problems and future work are also investigated.

1 Introduction

One of the main topics in the field of topography are *watersheds* [90]. Knowing the right watersheds and their corresponding catchment basins is essential in countless applications in areas such as Civil Engineering, Hydrology, Environmental Science, Ecology, Limnology, Urban Planning, Agriculture and so on [2, 52, 66, 86]. For instance, determining areas where a flood risk exists can help experts to make a decision to forbid the urban construction in such areas [35]; studying the movement of water within the hydrological cycle need the use of catchment basins as units [59]; analyzing water quality inside lakes or reservoirs depends on several factors included at catchment basin scale [71]; determining territorial boundaries (e.g., such as in the case of Hudson Bay basin) require the use of watersheds [80], etc.

R. Romero-Zaliz (✉)
Department of Computer Science and Artificial Intelligence, University of Granada, Granada, Spain
e-mail: rocio@decsai.ugr.es

J.F. Reinoso-Gordo
Department of Architectonic and Engineering Graphic Expression, University of Granada, Granada, Spain
e-mail: jreinoso@ugr.es

C. Cruz Corona (ed.), *Soft Computing for Sustainability Science*,
Studies in Fuzziness and Soft Computing 358, DOI 10.1007/978-3-319-62359-7_12

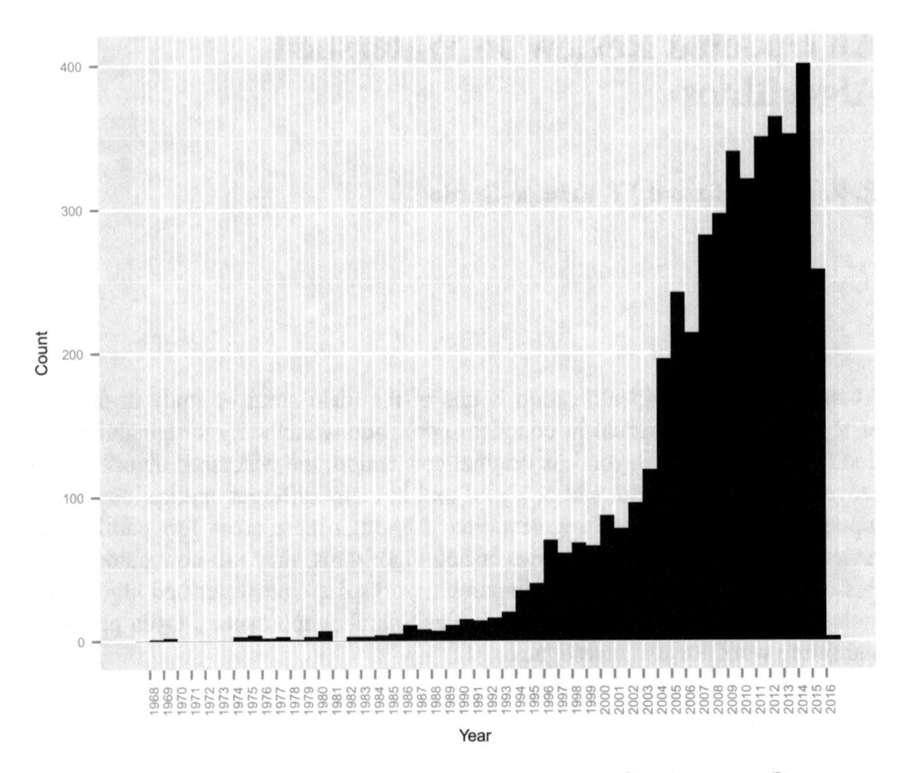

Fig. 1 Number of publications per year. Data obtained from Scopus © and CiteseerX © in October 13th, 2015

In the late '60 and '70s mathematicians and engineers started to work in the first algorithmic solutions for determining catchment basin's boundaries and watersheds [38, 43, 72]. Later on, researchers implemented a standardized version which they called the *watershed algorithm* [22]. The watershed algorithm has proved to be a very useful and powerful tool in many different application fields such as cartography [7, 32, 50, 65, 94], general image segmentation [61, 73, 87, 92, 97], video related issues [17, 19, 34, 95, 98], etc. There are several publications on biological and/or medical applications such as analysis of MRI images [3, 28, 46, 56, 77], cell images [4, 12, 14, 24, 85], ultrasound images [16, 33, 41, 45, 76], mammogram images [13, 23, 40, 79, 89], microscopy images [1, 6, 37, 42], among others.

Although there are a few publications reviewing the state-of-the-art of watershed algorithms [43, 60] and applications [18, 29, 49, 75], they are now outdated. In the last few years there has been an increasing number of publications in journals and conferences (Figs. 1 and 2).

Nowadays, most of the research in watershed algorithms are specifically devoted to image segmentation, but there are still some applications to real topographical watersheds, as shown in the world map of Fig. 3.

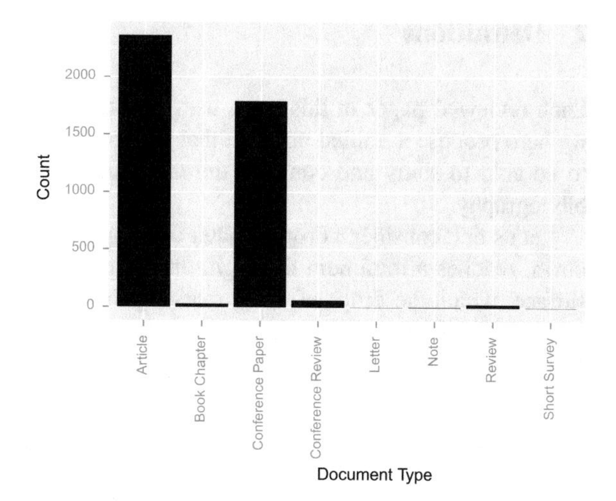

Fig. 2 Number of publications by document type

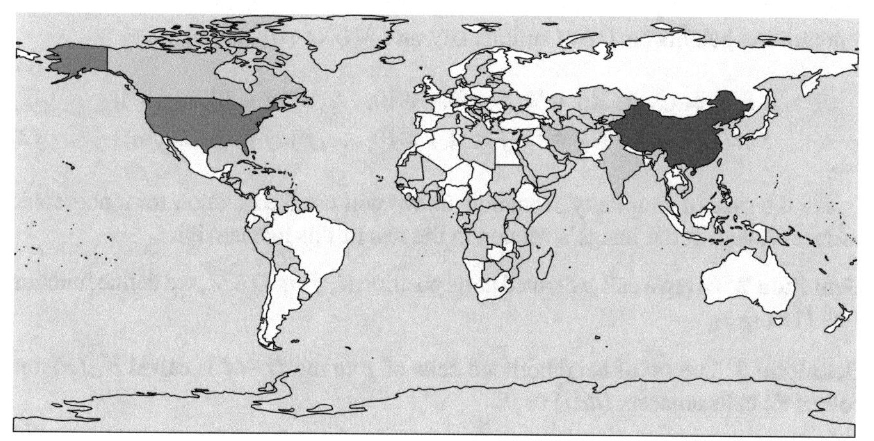

Fig. 3 Publications apply to topographic watersheds by country. The number of publications is color-coded from *white* to *green*, where *darker green* represent a higher number of publications in the given country

The rest of this survey is organized as follows. Section 2 introduces the basic common definitions used in the reviewed papers. Section 3 shows the main algorithms and strategies used for watershed determination. Section 4 is devoted to one of the main problems in watershed algorithms: over-segmentation. Section 5 reviews parallel approaches. Finally, Sect. 6 is reserved for conclusions and discussion.

2 Definitions

Each reviewed paper in this work uses different notation and terminology. Thus, we here propose a unified notation that will be used throughout this work in order to be able to study and compare the main algorithms and strategies cited in the bibliography.

Let us first consider a drop of water on a topographic surface. The water streams down, reaches a minimum of height and stops there. The set of all points of the surface, which the drops of water reaching this minimum can come from, can be associated with each minimum. Such a set of points is a **catchment basin** of the surface. The lines, which separate different catchment basins, are called **watersheds** or **watershed lines**.

Definition 1 We consider a topographic surface represented by a grid structure called *Digital Elevation Model (DEM)*, composed of cells, analogous to a digital image (IMG), composed of pixels. Every cell/pixel has a natural number as value representing heights on DEM or intensity on IMG of size $n \times m$.

$$DEM = \{x_{ij} \in \mathbb{R} | i \in \mathbb{N}, j \in \mathbb{N}, i \in (0, \ldots, n), j \in (0, \ldots, m)\} \tag{1}$$
$$IMG = \{x_{ij} \in \mathbb{N} | i \in \mathbb{N}, j \in \mathbb{N}, i \in (0, \ldots, n), j \in (0, \ldots, m)\} \tag{2}$$

For the sake of simplicity, from now on we will use the notation for topographic surface instead of the image's version in the rest of this manuscript.

Definition 2 Given a cell p defined as its position (i, j) in DEM, we define function I as $I(p) = x_{ij}$.

Definition 3 The set of neighborhood cells of p in the DEM is called $N_G(p)$ and collect all cells adjacent (adj) to p.

$$N_G(p) = \{p' \in I | adj(p, p')\} \tag{3}$$

Definition 4 A path P of length l between two cells p and p' in DEM is a $(l + 1)$-tuple of adjacent cells $(p_0, p_1, \ldots, p_{l-1}, p_l)$ such that $p_0 = p$, $p_l = p'$.

Definition 5 Cells belonging to the same connected plateau CP must satisfy the following condition

$$\forall p, p' \in CP, \exists P = (p0, p_1, \ldots, p_l) | p_0 = p \wedge p_l = p' \wedge \forall p_i \in P, I(p_i) = I(p) = I(p') \tag{4}$$

Definition 6 A minimum area M of DEM is a connected plateau of cells from which it is impossible to reach a cell of lower altitude without having to climb (Fig. 4).

$$M = \{p | \forall p' \notin CP(p) \wedge I(p') \leq I(p) \rightarrow \forall P = (p0, p_1, \ldots, p_l) | p_0 = p \wedge p_l = p', \exists i \in [1, l] | I(p_i) \geq I(p_0)\} \tag{5}$$

Fig. 4 Connected plateau

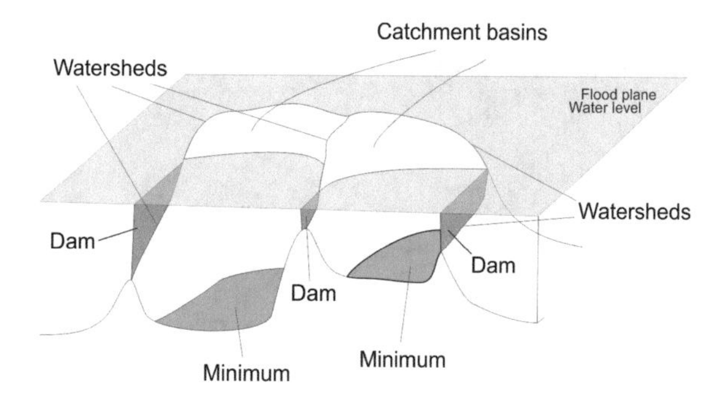

Fig. 5 Geodesic distance

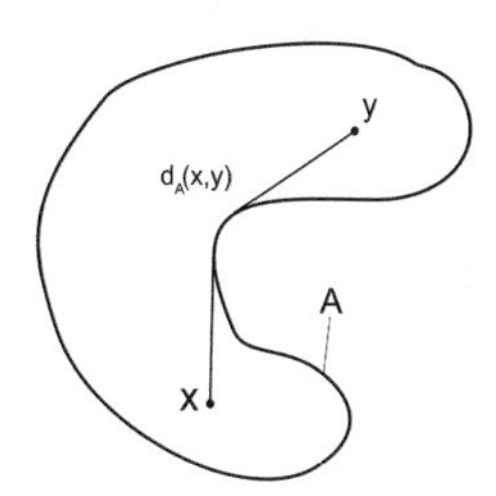

For a particular altitude h we denote the minimum area as M_h.

Definition 7 $T_h(I)$ stands for the threshold of I at level h where h is the value taken by I and $h \in [h_{min}, h_{max}]$.

$$T_h(I) = \{p | I(p) \leq h\} \tag{6}$$

Definition 8 The geodesic distance $d_A(p, p')$ between two cells p and p' in A is the minimun of the length of the paths which join p and p' and are totally included in area A (Fig. 5).

$$d_A(p, p') = min(\{length(P), P = (p, \ldots, p') \text{ which is totally included in } A\}) \tag{7}$$

Definition 9 Let A and B be two sets of cells of a given DEM (Fig. 6), where $B \subset A$. B is composed by k areas not adjacent: $B_1, B_2, \ldots, B_k$ (black areas in Fig. 6) but connected through A. The geodesic influence zone $IZ_A(B_i)$ of a component of B in A is the locus of the cells of A whose geodesic distance to B_i is smaller than their geodesic distance to any other component of B (blue color in Fig. 6 is the $IZ_A(B_1)$).

$$IZ_A(B_i) = \{p \in A, \forall j \in [1, k] \wedge j \neq i | d_A(p, B_i) < d_A(p, B_j)\} \tag{8}$$

Definition 10 The cells of A which do not belong to any geodesic influence zone constitute the skeleton by influence zones (SKIZ) of B inside A, denoted $SKIZ_A(B)$

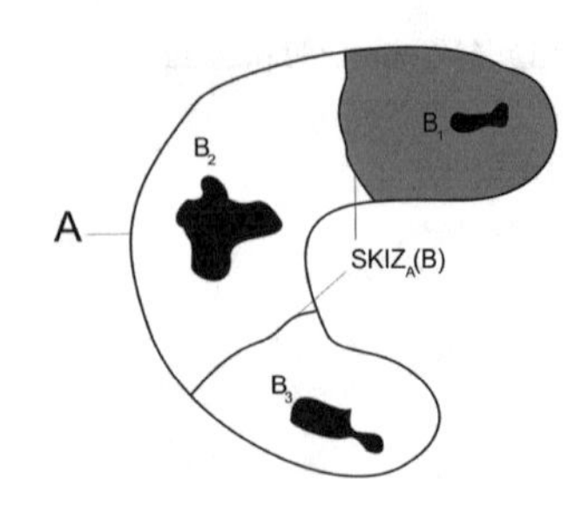

Fig. 6 Geodesic influence zone. A small example with $k = 3$

$$SKIZ_A(B) = A \setminus IZ_A(B) \text{ with } IZ_A(B) = \bigcup_{i \in [1,k]} IZ_A(B_i) \tag{9}$$

Definition 11 The set of catchment basins of DEM is equal to the set $X_{h_{max}}$ composed of not adjacent areas and obtained after the following recursion

$$\begin{cases} X_h = T_h(I) \ , \ h = h_{min} \\ X_{h+1} = M_{h+1} \cup IZ_{T_{h+1}(I)}(X_h) \ , \ h \in (h_{min}, h_{max} - 1] \end{cases} \tag{10}$$

Definition 12 The watersheds are $X^C_{h_{max}}$, i.e. the complement set of $X_{h_{max}}$.

3 Algorithms

Naturally, the first algorithms for computing watersheds are found in the field of topography. Lets recall that topographic surfaces are numerically handled through DEMs, these are arrays of numbers that represent the spatial distribution of terrain altitudes [90]. In image segmentation, its main application, the idea of the watershed construction is quite simple: a gray scale image can be considered as a topographic relief, the gray scale value of a pixel being the altitude at that particular point [69]. Using this analogy we can now review all papers regardless of its application.

There are a few reviews devoted to watershed algorithms in the bibliography. The first one appeared in the early 90s, when Beucher and Meyer publish a book chapter introducing what they have called the "watershed transformation" (basically the flooding process explained below) along with the principles of morphological segmentation and morphological tools [9]. This transformation was rapidly adopted by many other researchers for their own particular applications [30, 96].

Later on in 2003, Najman and Couprie study the behavior of watershed algorithms. Through the introduction of the concept of "pass value" they show that most classical watershed algorithms do not allow the retrieval of some important topological features of the image [60]. An important consequence of this result is that it is not possible to compute sound measures such as depth area or volume of basins using most classical watershed algorithms.

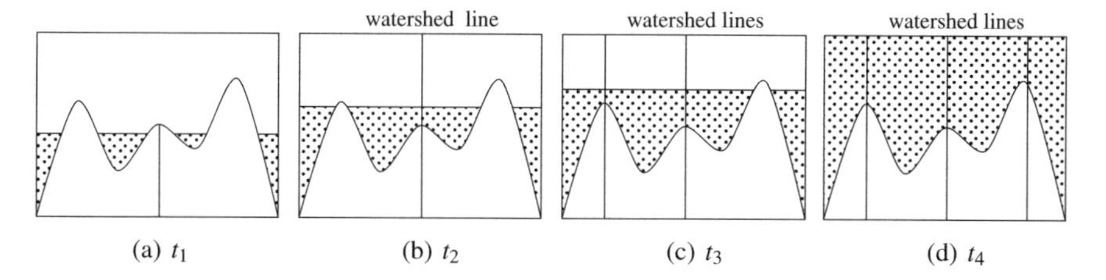

Fig. 7 Immersion-based watersheds

Finally and more than 5 years later, Körbes and Lotufo present a communication in a symposium reviewing fifteen watershed algorithms in a comprehensive way [43]. Afterwards several novel algorithms were developed and thus needed an updated review.

But let us first introduce the two main strategies for determining watersheds, we will study each of them separately in the next subsections.

3.1 By Immersion

The first strategy that we will analyze is called immersion (also called flooding). It was first developed for contour detection in images and introduced by Beucher and Lantuejoul in 1979 [8].

Later, an algorithmic-based definition for the identification of watersheds by immersion was introduced by Vincent and Soille in 1991 [90]. By analogy to the idea of immersion, we can figure that we have pierced holes in each regional minimum of our topographic surface. We then slowly immerse our surface into a lake. Starting from the minima of lowest altitude, the water will progressively fill up the different catchment basins. Now, at each position where the water coming from two different minima would merge, we build a "dam". At the end of this immersion procedure, each minimum is completely surrounded by dams, which delimit its associated catchment basin. The whole set of dams which has been built provides a division of the surface in its different catchment basins. These dams correspond to the watersheds of our surface [90] (Fig. 7). Vincent and Soille algorithm was developed for image segmentation and involves two major steps: (1) sorting of pixels in increasing order of gray value (the gray value of a pixel being the altitude at a particular point), and (2) fast computation of geodesic influence zones by breadth-first scanning of each threshold level using a first-in-first-out (FIFO) data structure (Algorithm 1).

Meijster and Roerdink propose in 1998 [51] an algorithm with two stages: (1) transform a lower complete image using a FIFO-queue algorithm, and (2) calculate the watershed using graph theory and removing the old FIFO-queue (Algorithm 4).

Lotufo and Falcao, in 2000, reviews the watershed in the graph framework of a shortest-path forest problem using a lexicographic path cost formulation. This

Algorithm 1 Vincent and Soille's immersion watershed approach [90]

Require: I_i ▷ *I_i original image of size $n \times m$*
Ensure: I_o ▷ *I_o image of the labeled watersheds*

1: MASK $\leftarrow -2$
2: WSHED $\leftarrow 0$
3: $queue \leftarrow \emptyset$
4: **for all** $p \in I_o$ **do**
5: | $I_o(p) \leftarrow -1$
6: **end for**
7: $current\_label \leftarrow 0$
8: $dist \leftarrow 0$
9: Initialize each pixel of I_d with 0 ▷ *I_d work image of distances*
10: Sort the pixels of I_i in the increasing order of their gray values
11: $h_{min} \leftarrow min(I_i)$
12: $h_{max} \leftarrow max(I_i)$

13: **for** $h \leftarrow h_{min}$ **to** h_{max} **do** ▷ *geodesic SKIZ of level $h-1$ inside level h*
14: | **for all** p such that $(I_i(p) = h)$ **do** ▷ *pixels accessed through the sorted array*
15: | | $I_o(p) \leftarrow$ MASK
16: | | **if** $(\exists p' \in N_G(p) | (I_o(p') > 0)$ **or** $(I_o(p') =$ WSHED$))$ **then**
17: | | | $I_d(p) \leftarrow 1$
18: | | | fifo_push($queue$, p)
19: | | **end if**
20: | **end for**
21: | $dist \leftarrow 1$
22: | fifo_push($queue$, p^0) ▷ *p^0 is a fictitious pixel*
23: | **repeat**
24: | | $p \leftarrow$ fifo_get($queue$)
25: | | **if** $(p = p^0)$ **then**
26: | | | **if** $queue \neq \emptyset$ **then**
27: | | | | fifo_push($queue$, p^0)
28: | | | | $dist \leftarrow dist + 1$
29: | | | | $p \leftarrow$ fifo_get($queue$)
30: | | | **end if**
31: | | **end if**

formulation reflects the behavior of the ordered queue-based watershed algorithm [47].

In 2001, Chen and Shi [15] modify the original Vincent and Soille immersion-based watershed algorithm to correct some issues: (1) incorrect labeling when a point p is at the same distance from three or more adjacent catchment basins (i.e., it will be labeled as catchment basin instead as watershed), (2) unnecessary computation of geodesic distance between points which belong to the labeled, (3) memory consumption, and (4) incapacity to obtain information about catchment basins during the processing. The modified algorithm proposed introduces a third step call "gushing

```
32:     for all p' ∈ N_G(p) do                ▷ p' belongs to an already labeled basin or to the
33:       if ((0 < I_d(p') < dist) or (I_o(p') = WSHED)) then                          watershed
34:         if (I_o(p') > 0) then
35:           if ((I_o(p) = MASK) or (I_o = WSHED)) then
36:             I_o(p) ← I_o(p')
37:           else if (I_o(p) ≠ I_o(p')) then
38:             I_o(p) ← WSHED
39:           end if
40:         else if (I_o(p) = MASK) then
41:           I_o(p) ← WSHED
42:         end if
43:       else if ((I_o(p') = MASK) and (I_d(p') = 0)) then
44:         I_d(p') ← dist + 1
45:         fifo_push(queue, p')
46:       end if
47:     end for
48:   until (queue = ∅)
49:   for all p|(I_i(p) = h) do                ▷ checks if new minima have been discovered
50:     I_d(p) ← 0                              ▷ the distance associated with p is reset to 0
51:     if (I_o(p) = MASK) then
52:       current_label ← current_label + 1
53:       fifo_push(queue, p)
54:       I_o(p) ← current_label
55:       while (queue ≠ ∅) do
56:         p' ← fifo_get(queue)
57:         for all p'' ∈ N_G(p') do
58:           if (I_o(p'') = MASK) then
59:             fifo_push(queue, p'')
60:             I_o(p'') ← current_label
61:           end if
62:         end for
63:       end while
64:     end if
65:   end for
66: end for
```

step", that together with the immersion step, produces a fast recognition of pixels of labeled and gets the flood level of the catchment basin.

Later on, Rambabu et al. first, in 2003, propose a new algorithm based on hill climbing simulation, that avoided multiple scanning of the original matrix by using different queues to store pixels [69]. Then, in 2008, Rambabu and Chakrabarti gives an updated and corrected version of this same algorithm [70].

In 2005, Shen and Chang [74] present a nearest-neighbor graph (NNG) based watershed algorithm. The main idea behind this work is to transform the image into a NNG and then partitioned by discovering the defined geographic features in the

Algorithm 2 Meijster and Roerdink's Lower_complete function

Require: I_i ▷ *I_i original image of size $n \times m$*
Ensure: I_{lc} ▷ *I_{lc} lower complete image*

1: $queue \leftarrow \emptyset$
2: **for all** $p \in I_i$ **do**
3: $I_{lc}(p) \leftarrow 0$
4: **if** $\exists p' \in N_G(p) | (I_i(p') < I_i(p))$ **then**
5: fifo_push($queue, p$)
6: $I_{lc} \leftarrow -1$
7: **end if**
8: **end for**
9: $dist \leftarrow 1$
10: fifo_push($queue, p^0$) ▷ *p^0 is a fictitious pixel*
11: **while** $(queue \neq \emptyset)$ **do**
12: $p \leftarrow$ fifo_get($queue$)
13: **if** $(p = p^0)$ **then**
14: **if** $(queue \neq \emptyset)$ **then**
15: fifo_push($queue, p^0$)
16: $dist \leftarrow dist + 1$
17: **end if**
18: **else**
19: $I_{lc} \leftarrow dist$
20: **for all** $p' \in N_G(p) | ((I_i(p') = I_i(p))$ **and** $(I_{lc}(p') = 0))$ **do**
21: fifo_push($queue, p'$)
22: $I_{lc}(p') \leftarrow -1$ ▷ *to prevent from queuing twice*
23: **end for**
24: **end if**
25: **end while**
26: **for all** $p \in I_i | (I_{lc}(p) \neq 0)$ **do**
27: $I_{lc} \leftarrow dist \times I_i(p) + I_{lc}(p) - 1$
28: **end for**

first step. The initial population result is also transformed into the NNG again and the recursively distilled by the proposed algorithm.

3.2 By Rainfall

The second strategy is called rainfall and has two main steps: (1) the weakest edges are removed by "drowning" the image, creating a number of "lakes" grouping all the pixels that lie below a certain threshold (this is useful to reduce the influence of noise, and reduces the over-segmentation), and (2) the direction of a raindrop from each pixel would flow if it would fall on the topographic activity surface. This steepest descent neighbor and the pixel under consideration are then merged, finally enabling the localization of the remaining edges and segments [21] (Fig. 8).

Algorithm 3 Meijster and Roerdink's Resolve function [51]

Require: p ▷ *p input pixel*
Ensure: ce ▷ *ce canonical element (label assigned to a minimum) of p or WSHED in case p lies on a watershed*

1: Create array sln where $sln[p, i]$ is a pointer to the i-th steepest lower neighbor of pixel p
2: WSHED $\leftarrow -1$
3: $i \leftarrow 1$
4: $ce \leftarrow 1$
5: **while** $(i \leq 4)$ **and** $(ce \neq$ WSHED$)$ **do**
6: **if** $((sln[p, i] \neq p)$ **and** $(sln[p, i] \neq$ WSHED$))$ **then**
7: $sln[p, i] \leftarrow$ Resolve$(sln[p, i])$
8: **end if**
9: **if** $(i = 1)$ **then**
10: $ce \leftarrow sln[p, 1]$
11: **else if** $(sln[p, i] \neq ce)$ **then**
12: $ce \leftarrow$ WSHED
13: **for** $j \leftarrow 1$ **to** 4 **do**
14: $sln[p, i] \leftarrow$ WSHED
15: **end for**
16: **end if**
17: $i \leftarrow i + 1$
18: **end while**

Algorithm 4 Meijster and Roerdink's immersion watershed approach [51]

Require: I_i ▷ *I_i original image of size $n \times m$*
Ensure: I_o ▷ *I_o image of the labeled watersheds*

1: WSHED $\leftarrow 0$
2: $I_{lc} \leftarrow$ Lower_complete(I_i) ▷ *transform the image (Algorithm 2)*
3: **for all** $p \in I_{lc}$ **do**
4: $I_o(p) \leftarrow$ Resolve(p) ▷ *see Algorithm 3*
5: **end for**

Mortensen and Barret in 1999 introduces an optimization variation for the rainfall-based watershed algorithm using a tobogganing technique, which makes a much more computationally efficient algorithm [57]. In this version, tobogganing over-segments an image into small regions by sliding in the derivative terrain. The basic idea is that given the gradient magnitude of an image, each pixel determines a slide direction by finding the pixel in a neighborhood with the lowest gradient magnitude. Pixels that "slide" to the same local minimum are grouped together, thus segmenting the image into a collection of small regions.

A year later, Bieniek and Moga [11] propose an efficient algorithm based on connected components that generates the same results as the original Meyer's algorithm [53] but with a simpler algorithmic construction and, hence, a lower complexity (it can label all catchment basins by only scanning the image four times).

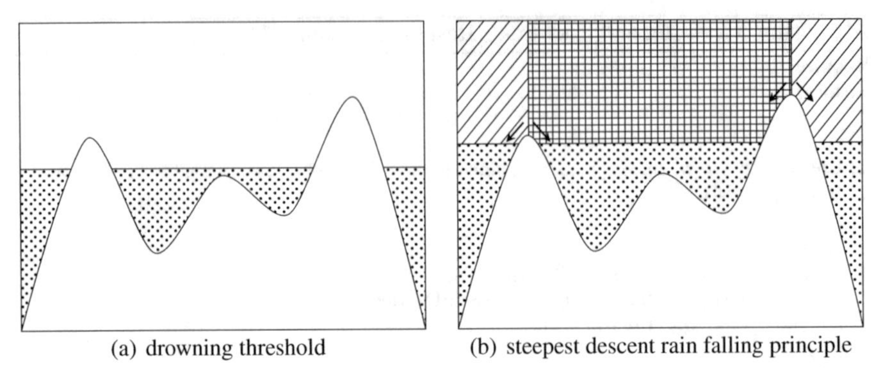

(a) drowning threshold (b) steepest descent rain falling principle

Fig. 8 Rainfall-based watersheds

Later on, in 2007, Osma-Ruiz et al. implement a more efficient algorithm by computation of shortest paths [63] (Algorithm 5). This algorithm produces the same result as the previous works using only two scans (plus another to initialize the data structures), thus decreasing the running time obtained in [11, 81].

One of the latest work in the area was developed by Świercz and Iwanowski in 2010 [82]. In their work, a mechanism called directional code is used to code descent paths, where each visited pixel receives a temporary marking in the output label array. The value of this temporary marking represents the position of pixel in the image. A path can be therefore viewed as a series of pointers to pixels stored in the output array.

3.3 Mixed Approaches

In 2005 Sun et al. modifies Bieniek and Moga's work simulating raining to generate the connected components using chain code instead of pixel address. Afterwards, they simulate flooding to label catchment basins by tracing chain codes [81]. This algorithm not only can label catchment basins by scanning the image only four times, but also is more helpful to the following image processing.

Cousty et al. in 2009 introduces the first work that mathematically prove equivalence to both immersion-based and rainfall-based watersheds [20]. For this purpose, the authors propose a new definition of watershed, called *watershed cut* and give a liner-time algorithm to compute the watershed cuts of an edge-weighted graph (pre-processed image). The proposed algorithm does not require any sorting step or the use of any sophisticated data structure such as a hierarchical queue or a representation to maintain unions of disjoint sets.

Algorithm 5 Osma-Ruiz's rainfall watershed approach [63]

```
Require: I_i                                   ▷ I_i original image of size n × m
Ensure: I_o                                    ▷ I_o image of the labeled watersheds

 1: UNVISITED ← −8                             ▷ Step 1: Initialization
 2: PENDING ← −9
 3: for all p ∈ I_i do
 4: |  I_o(p) ← UNVISITED
 5: end for
 6: qPending ← ∅
 7: qEdge ← ∅
 8: qInner ← ∅
 9: qDescending ← ∅

10: ncatch ← 1                                 ▷ Step 2: Identifying regional minima and
                                                 steepest descending paths
11: for all p ∈ I_i do
12: |  if (I_o(p) = UNVISITED) then            ▷ if the point has not been
13: |  |  for all p' ∈ N_G(p) do                  analyzed yet, study it
14: |  |  |  if (I_i(p) = I_i(p')) then         ▷ this is a plateau
15: |  |  |  |  if (qPending = ∅) then
16: |  |  |  |  |  I_o(p) ← PENDING
17: |  |  |  |  |  queue_push(qPending, p)
18: |  |  |  |  end if
19: |  |  |  |  I_o(p') ← PENDING
20: |  |  |  |  queue_push(qPending, p')
21: |  |  |  else if (I_i(p') = min(I_i(N_G(p)))) then
22: |  |  |  |  min ← p'
23: |  |  |  end if
24: |  |  end for
25: |  |  if (qPending ≠ ∅) then                ▷ if p belongs to a plateau
26: |  |  |  while (qPending ≠ ∅) do               make it lower-complete image
27: |  |  |  |  p' ← queue_pop(qPending)           if not, p is considered a minimum
28: |  |  |  |  min ← ∅                            unless there is a lower neighbor
29: |  |  |  |  if (p ≠ p') then                ▷ calculations already done for seed
30: |  |  |  |  |  for all p'' ∈ N_G(p') do      ▷ put in the queue all the points
31: |  |  |  |  |  |  if (I_i(p'') = I_i(p)) then   in the plateau
32: |  |  |  |  |  |  |  if (I_o(p'') = UNVISITED) then
33: |  |  |  |  |  |  |  |  I_o(p'') ← PENDING
34: |  |  |  |  |  |  |  |  queue_push(qPending, p'')
35: |  |  |  |  |  |  |  end if
```

```
36:                 else if (I_i(p'') = min(I_i(N_G(p')))) then min ← p''
37:                 end if
38:             end for
39:         end if
40:         if (min ≠ ∅) then                    ▷ classify p' as either an edge or inner point
41:             min ← I_o(p')
42:             queue_push(qEdge, p')
43:         else
44:             queue_push(qInner, p')
45:         end if
46:     end while
47:     if (qEdge ≠ ∅) then                       ▷ if the plateau has no edge points,
48:         if (qInner ≠ ∅) then                                             it is a minimum
49:             while (qEdge ≠ ∅) do                            else, make it lower complete
50:                 p' ← queue_pop(qEdge)
51:                 for all p'' ∈ N_G(p') do
52:                     if ((I_i(p'') = I_i(p)) and (I_o(p'') = PENDING)) then
53:                         I_o(p'') ← p'
54:                         queue_push(qEdge, p'')
55:                     end if
56:                 end for
57:             end while
58:         end if
59:     else
60:         while (qInner ≠ ∅) do
61:             p' ← queue_pop(qInner)
62:             I_o(p') ← ncatch
63:         end while
64:         ncatch ← ncatch + 1
65:     end if
66: else
67:     if (min = ∅) then
68:         I_o(p) ← ncatch
69:         ncatch ← ncatch + 1
70:     else
71:         min ← I_o(p)
72:     end if
73: end if
74: end if
75: end for
```

```
76: for all p ∈ I_i do                         ▷ Step 3: Assignment of pixels to catchment basins
77:   p' ← p
78:   while (I_o(p') ≤ 0) do                                         ▷ it is not a minimum
79:     queue_push(qDescending, p')
80:     p' ← p'_ref                                     ▷ p'_ref point pointed to by p'
81:   end while
82:   while (qDescending ≠ Ø) do
83:     p'' ← queue_pop(qDescending)
84:     I_o(p'') ← I_o(p')
85:   end while
86: end for
```

4 Over-Segmentation

Researcher noticed that it was hard to apply watershed transformations since they are very sensitive to noise. One of the main drawbacks of the classical watershed algorithm is a phenomenon known as over-segmentation. There are several approaches to reduce the impact of this issue, which can be categorized into two approaches: *pre-processing* and *post-processing*. Several pre-processing and post-processing methods were reviewed by Bieniecki in 2004 applied to color images. Between the studied pre-processing methods, the author analyze noise removal by using a median filter, color morphology and other smoothing filters. Between the post-processing methods, the author research merging basins by gradient watersheds on graphs, basin dynamics, inclusionary and exclusionary cues, image component labeling and multi-scale gradient analysis [10].

4.1 Pre-processing

The most efficient pre-processing techniques are based on markers. In a marker-based algorithm, the gradient image is first modified by a marker image, which is a binary image with the object interiors (markers) being set to 0 and the uncertainty areas being set to 255 (Fig. 9). Each marker indicates the presence of an object [25].

In Moga and Gabbouj watershed transformation is augmented to perform with the aid of a priori supplied image markers. In this method pixels are first clustered based on spatial proximity and gray level homogeneity with the watershed transformation. Boundary-based region merging is the effected to condense non-marked regions into marked catchment basins The agglomeration strategy works with a weighted neighborhood graph representation of the over-segmented image [55].

Some years later Gao et al. propose a marker-based algorithm using a disjoint set data structure with a linear complexity [25]. Later on they present an updated algorithm to extract the regional minima from the low frequency components in the gradient map. The extracted minima constitute the binary marker image. Then the

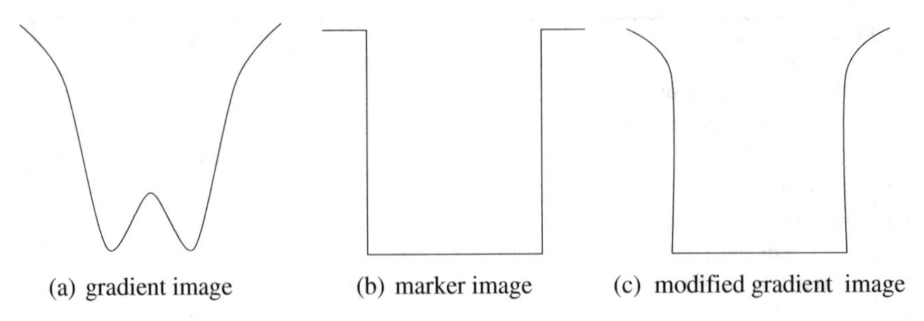

Fig. 9 Modification process of the gradient image by the marker image

original gradient map is modified by suppressing its all-intrinsic minima around these extracted markers. Thus, compared with traditional approaches, both the spurious minima are more effectively removed and meanwhile the boundaries of objects are more effectively protected [26].

In 2009, Zhu et al. apply a multi-scale alternating sequential filtering by reconstruction to simplify the input image in order to remove local minima which are caused by irregular gray disturbance and noise, and preserve important contour information [99]. After this procedure is done, there still exists some local minima problem which is reduced by a marker-extracted method that uses minima imposition to make a marked image before watershed transformation. Markers are a set of components marking flat regions of an image. The mark extraction can suppress all of is intrinsic minima. Finally, the watershed algorithm is applied to the modified gradients by the markers.

A stochastic version of the watershed algorithm is obtained by choosing randomly in the image the seeds from which the watershed regions are grown. In the 2009's work Tolosa et al. explore two seed-generation processes to avoid over-segmentation. The first is a non-uniform Poisson process, the density of which is optimized on the basis of the opening granulometry. The second process positions the seed randomly within disks centered on the maxima of a distance map [84].

Procházka et al. proposes in 2010 a smoothing procedure to reduce noise and over-segmentation. This procedure first applies wavelet image de-noising, and afterwards a smoothing phase. The main idea in this phase is to remove image elements smaller than the size of structuring element and to fill gaps between pixels and smoothes their outer edges [68].

Also in 2010, Moumoun et al. introduces a filtering step to eliminate insignificant minima that take into account not only the depth of the pixel but also the change in concavity [58]. This is done by defining a hierarchy between minima using a generic graph. In this graph each region is represented by a node. The minimum curvature of a region is associated with the corresponding node. The arcs express the adjacencies relations between regions and two regions are called adjacent if they have adjacent faces. The weight of each arc is defined by the depth measurement already mentioned.

4.2 Post-processing

Gies and Bernard propose in 2004 a merging phase of basin using statistical information. With regions of an unspecified size, the merging criterion must follow statistical rules accounting for the region size. The authors consider the regions as the outcome of stochastic processes. The merging criterion is based on the knowledge of region area and statistical regional measures, determining the statistical reliability of the merging [27].

Patino publishes in 2005 an approach that characterizes each of the segmented regions and then employs the composition of fuzzy relations to group together similar regions. A fuzzy c-means algorithm along with fuzzy relations can group together similar gray-level values, but only between adjacent regions [64].

Jianhua et al. propose a more complex strategy for basin merging in 2005. Their method pre-segments the image by watersheds and then merges it by Immune Clonal Algorithm (ICA). To implement the task, several operators are proposed such as the DC operator, the Proportional Creation of the First generation operator, and fitness function based on JND and average gray value [36].

Moumoun et al. also propose a post-processing technique based on region merging by the depth of the watershed segmentation. This depth is defined by the difference between the height of the saddle point and the minimum of adjacent regions [58].

4.3 Other Approaches

There are very few approach to reduce over-segmentation that do not use pre-processing or post-processing. Swiercz and Iwanowsky propose in 2011 a "water-ball method designed to counter over-segmentation during the actual calculation of watersheds. This proposal can be perceived as a composite method for object extraction, combining several techniques and mechanisms to produce satisfactory macro-scale results [83]. The "water-ball method uses two distinct mechanisms. The first one consists of a "rolling ball" based on the simulation of a larger object rolling down the slope (contrary to classic rainfall methods where a single drop of water is used). This ball has the ability to cross small ridges and ignore small, insignificant local minima. The second mechanism, weakening with edge enhancement, makes it possible to eliminate low, insignificant interior boundaries.

5 Parallel Approaches

When watershed algorithms are applied in large images or big DEMs, serial versions can take many hours to return the solution. therefore, researchers dedicated their effort to implement parallelized versions, but do to the recursive nature of the watershed transformation, its parallelization is not a trivial task [54] and cannot suit real-time operations in most cases [62].

5.1 Parallelization Using Distributed Memory

Most of the existing papers on parallel watershed algorithms are based on a divide-and-conquer approach [54, 55, 67, 72, 93] with regular domain decomposition into blocks of data. Each block is then processed by a different processor independently, thus using distributed memory. Afterwards all blocks are merged together. In 2010 Świercz and Iwanowski [82] present an interesting approach to reduce the merging process by pre-calculating the adjacent sections between the different blocks. Thus, the merging process reduces to a bulk memory copy of each processed block.

5.2 Parallelization Using Shared Memory

There are a few publications that uses a shared memory approach where there is a need of constant synchronization between the processors that process adjacent blocks. In Wagner et al. [91] the authors propose a chromatic ordering that allows to gain a correct segmentation without an examination of adjacent domains or a final relabeling. Later, VanNeerbos et al. [88] parallelize topological watersheds in such a way that border pixels between threads are not calculated at the same time. To do this, each thread process its tile in different stages and synchronizing between all threads is performed after each stage. Also in 2011 Mahmoudi et al. [48] introduce an adapted parallelization strategy called split, distribute and merge strategy which allows efficient parallelization of a large class of topological operators including, mainly, smoothing, skeletonization, and watershed algorithms. To achieve a good speedup they focus on task scheduling.

5.3 Parallelization Using Graphics Processor Units

There is a rising tendency in research in parallelization using Graphics Processor Units (GPUs) and watershed algorithms are not an exception. Kauffmann and Piche [39] describe a cellular automaton (CA) to perform the watershed transform in N-D images. Due to the local nature of CA algorithms they show that they are designed to run on massively parallel processors and therefore, be efficiently implemented on low cost consumer GPUs. A few years later Körbes et al. review the advances in watershed processing on GPU architecture [44] on two algorithms: one inspired by the drop of water paradigm and depth-search approaches; and one based on cellular automata to process a shortest-path forest with sum cost function. In 2012 Hucko and Srámek [31] present the first GPU watershed algorithm able to process data larger than the available memory, as the whole data has to be present in the memory of the device. In their manuscript they present two versions of a streamed multi-pass algorithm for watershed computation on a GPU. As the slice-based streaming approach is used both variants are capable of processing data exceeding the size of the available graphics accelerator memory. Another interesting approach is shown by Quedada-Barriuso et al. [5] where the watershed transform based on a cellular automaton, especially when the synchronization rules are relaxed. In particular, they compare a synchronous and an asynchronous implementation of the algorithm. One of the main applications for watersheds are medical related issues, therefore Smistad et al. recently published a comprehensive survey on medical image segmentation on GPUs [78].

6 Conclusions

Recent advances in watershed algorithms focus on the use of different data structures to improve the efficiency of the proposed algorithms. Most of the work done has been developed for image segmentation purposes, specially on medical and biological images. Also, parallel approaches seem to be an area of constant development. Curiously, many of the most interesting proposals were published in conferences instead of in journals. This tendency seems to change in the latest years of research.

Although there are many different watershed algorithmic solutions, there are still many problems to solve. Thus, there is an increasing amount of publications over the last few years. That is, watershed algorithms are still an open problem and there are more and more new fields starting to use and develop this kind of techniques from different points of view.

Another open issue is the oversegmentation watershed algortihms produce. Although there are several approches to minimize this problem, there is still no perfect solution available yet. Oversegmentation can be an important issue when determining countours on images that will be used for counting purposes (such as blood cells).

Many applications that use watershed algorithms work with a small dataset or datasets. For real-world applications in topography that includes large regions of terrain, watershed algorithms became slow and inefficient. Sometimes they are not capable to work since memory becomes a crucial issue. For this kind of big data processes the only available solution is the partition of the dataset into smaller sets and the posterior merge of each partial solution.

We believe there is still a lot of improvement to be developed in this field. For instance, there is a lack of techniques for real-time and streaming applications. Also, we could not retrieve many studies on real watersheds in many countries, thus indicating that many catchment basins worldwide have not been studied yet.

References

1. Ancin, H., Roysam, B., Dufresne, T., Chestnut, M., Ridder, G., Szarowski, D., Turner, J.: Advances in automated 3-d image analysis of cell populations imaged by confocal microscopy. Cytometry **25**(3), 221–234 (1996)
2. Andreatta, S., Wallinger, M., Posch, T., Psenner, R.: Detection of subgroups from flow cytometry measurements of heterotrophic bacterioplankton by image analysis. Cytometry **44**(3), 218–225 (2001)
3. Angel Viji, K., Jayakumari, J.: Performance evaluation of standard image segmentation methods and clustering algorithms for segmentation of mri brain tumor images. Euro. J. Sci. Res. **79**(2), 166–179 (2012)
4. Arslan, S., Ozyurek, E., Gunduz-Demir, C.: A color and shape based algorithm for segmentation of white blood cells in peripheral blood and bone marrow images. Cytometry Part A **85**(6), 480–490 (2014)
5. Barriuso, P.Q., Heras, D., Argüello, F.: Efficient gpu asynchronous implementation of a watershed algorithm based on cellular automata, pp. 79–86, (2012)
6. Becattini, G., Mattos, L., Caldwell, D.: A novel framework for automated targeting of unstained living cells in bright field microscopy. pp. 195–198 (2011)
7. Beucher, S., Bilodeau, M.: Road segmentation and obstacle detection by a fast watershed transformation. pp. 296–301 (1994)
8. Beucher, S., Lantuéjoul, C.: Use of watersheds in contour detection. workshop published (1979)
9. Beucher, S., Meyer, F.: The morphological approach to segmentation: the watershed transformation. mathematical morphology in image processing. Opt. Eng. **34**, 433–481 (1993)
10. Bieniecki, W.: Oversegmentation avoidance in watershed-based algorithms for color images. pp. 169–172 (2004)
11. Bieniek, A., Moga, A.: An efficient watershed algorithm based on connected components. Pattern Recognit. **33**(6), 907–916 (2000)
12. Bullet, J., Gaujoux, T., Borderie, V., Bloch, I., Laroche, L.: A reproducible automated segmentation algorithm for corneal epithelium cell images from in vivo laser scanning confocal microscopy. Acta Ophthalmol. **92**(4), e312–e316 (2014)
13. Camilus, K., Govindan, V., Sathidevi, P.: Pectoral muscle identification in mammograms. J. Appl. Clin. Med. Phys. **12**(3), 215–230 (2011)
14. Chen, L.-C., Nguyen, T.-H., Lin, S.-T.: Viewpoint-independent 3d object segmentation for randomly stacked objects using optical object detection. Meas. Sci. Technol. **26**(10) (2015)
15. Chen, T.: Gushing and immersion alternative watershed algorithm, pp. 246–248 (2001)
16. Cheng, J.-Z., Chen, K.-W., Chou, Y.-H., Chen, C.-M.: Cell-Based Image Partition and Edge Grouping: A Nearly Automatic Ultrasound Image Segmentation Algorithm for Breast Cancer Computer Aided Diagnosis. vol. 6915 (2008)
17. Chien, S.-Y., Chen, L.-G.: Reconfigurable morphological image processing accelerator for video object segmentation. J. Signal Process. Syst. **62**(1), 77–96 (2011)
18. Christ, M.., Parvathi, R.: Segmentation of medical image using clustering and watershed algorithms. Am. J. Appl. Sci. **8**(12), 1349–1352 (2011)
19. Chung, K.-L., Lai, Y.-S., Huang, P.-L.: An efficient predictive watershed video segmentation algorithm using motion vectors. J. Inf. Sci. Eng. **26**(2), 699–711 (2010)
20. Cousty, J., Bertrand, G., Najman, L., Couprie, M.: Watershed cuts: Minimum spanning forests and the drop of water principle. IEEE Trans. Pattern Anal. Mach. Intell. **31**(8), 1362–1374 (2009)
21. De Smet, P., Pires, R.L.V.: Implementation and analysis of an optimized rainfalling watershed algorithm. SPIE Int. Soc. Opt. Eng. **3974**, 759–766 (2000)
22. Digabel, H., Lantuejoul, C.: Iterative algorithms. pp. 85–89 (1978)
23. Dubey, R., Hanmandlu, M., Gupta, S.: A comparison of two methods for the segmentation of masses in the digital mammograms. Comput. Med. Imaging Graph. **34**(3), 185–191 (2010)

24. Elsalamony, H.: Detecting distorted and benign blood cells using the hough transform based on neural networks and decision trees. In: Deligiannidis, L., Arabnia, H. (eds.) Emerging Trends in Image Processing, Computer Vision and Pattern Recognition, pp. 457–473 (2014)
25. Gao, H., Xue, P., Lin, W.: A new marker-based watershed algorithm. vol. 2, pp. II81–II84 (2004)
26. Gao, L., Yang, S., Xia, J., Liang, J., Qin, Y.: A new marker-based watershed algorithm (2007)
27. Gies, V., Bernard, T.: Statistical solution to watershed over-segmentation. Int. Conf. Image Process. **3**, 1863–1866 (2004)
28. Guo, Z., Xin, Y., Liu, S., Lv, X., Li, S.: Comparisons of fat quantification methods based on mri segmentation, pp. 1817–1821 (2014)
29. Hagyard, D., Razaz, M., Atkin, P.: Analysis of watershed algorithms for greyscale images. Proc. 3rd IEEE Int. Conf. Image Process. **3**, 41–44 (1996)
30. Held, C., Wenzel, J., Webel, R., Marschall, M., Lang, R., Palmisano, R., Wittenberg, T.: Using multimodal information for the segmentation of fluorescent micrographs with application to virology and microbiology. In: Conference proceedings : ... Annual International Conference of the IEEE Engineering in Medicine and Biology Society. IEEE Engineering in Medicine and Biology Society. Conference, pp. 6487–6490 (2011)
31. Hucko, M., Srámek, M.: Streamed watershed transform on gpu for processing of large volume data, pp. 137–141 (2012)
32. Idbraim, S., Mammass, D., Aboutajdine, D., Ducrot, D.: An automatic system for urban road extraction from satellite and aerial images. WSEAS Trans. Signal Process. **4**(10), 563–572 (2008)
33. Ikedo, Y., Fukuoka, D., Hara, T., Fujita, H., Takada, E., Endo, T., Morita, T.: Development of a fully automatic scheme for detection of masses in whole breast ultrasound images. Med. Phys. **34**(11), 4378–4388 (2007)
34. Jabid, T., Mohammad, T., Ahsan, T., Abdullah-Al-Wadud, M., Chae, O.: An edge-texture based moving object detection for video content based application. pp. 112–116 (2011)
35. Jiafu, L., Yan, L., Wenfeng, Z., Jing, L.: Storm floods risk assessments by ga-bp: a case study of seven countries in Asia. Int. J. Adv. Comput. Technol. **3**(10), 323–329 (2011)
36. Jianhua, L., Shuang, W., Licheng, J.: Method to reduce over-segmentation of images using immune clonal algorithm. vol. 6786 (2007)
37. Jouini, M., Vega, S., Mokhtar, E.: Multiscale characterization of pore spaces using multifractals analysis of scanning electronic microscopy images of carbonates. Nonlinear Process. Geophys. **18**(6), 941–953 (2011)
38. JW, L., JA, D.: Optimal identification of lumped watershed models. Water Resour. Res. **5**(3), 583–590 (1969)
39. Kauffmann, C., Piche, N.: A cellular automaton for ultra-fast watershed transform on gpu (2008)
40. Kekre, H., Sarode, T., Gharge, S.: Vector quantization for tumor demarcation of mammograms. Commun. Comput. Inf. Sci. **70**, 157–163 (2010)
41. Kollorz, E., Angelopoulou, E., Beck, M., Schmidt, D., Kuwert, T.: Using power watersheds to segment benign thyroid nodules in ultrasound image data, pp. 124–128 (2011)
42. Kong, J., Cooper, L., Kurc, T., Brat, D., Saltz, J.: Towards building computerized image analysis framework for nucleus discrimination in microscopy images of diffuse glioma, pp. 6605–6608 (2011)
43. Körbes, A., Lotufo, R.: Analysis of the watershed algorithms based on the breadth-first and depth-first exploring methods. pp. 133–140 (2009)
44. Körbes, A., Vitor, G., De Alencar Lotufo, R., Ferreira, J.: Advances on watershed processing on gpu architecture. In: Lecture Notes in Computer Science. Lecture Notes in Artificial Intelligence. Lecture Notes in Bioinformatics, vol. 6671, pp. 260–271, LNCS (2011)
45. Linguraru, M., Howe, R.: Texture-based instrument segmentation in 3d ultrasound images, vol. 6144 II (2006)
46. Liu, J., Chen, K.: Novel method of mri medical image segmentation combining watershed algorithm and wkfcm algorithm. Appl. Mech. Mater. **121–126**, 4518–4522 (2012)

47. Lotufo, R., Falcao, A.: The ordered queue and the optimality of the watershed approaches. Math. Morphol. Appl. Image Signal Process. **18**, 341–350 (2000)
48. Mahmoudi, R., Akil, M.: Real time topological image smoothing on shared memory parallel machines, vol. 7871 (2011)
49. Malpica, N., De Solórzano, C., Vaquero, J., Santos, A., Vallcorba, I., García-Sagredo, J., Del Pozo, F.: Applying watershed algorithms to the segmentation of clustered nuclei. Cytometry **28**(4), 289–297 (1997)
50. Mei, T., Li, D., Qin, Q.: Application of knowledge based watershed transform approach to road detection, vol. 6045 II (2005)
51. Meijster, A., Roerdink, J.B.T.M.: A disjoint set algorithm for the watershed transform. In: Proceedings EUSIPCO '98, IX European Signal Processing Conference, pp. 1665–1668 (1998)
52. Mendonca, A.S., Rezende, R.A.: Application of geographical information systems and stochastic hydrology to the siting and design of water reservoirs. In: International Geoscience and Remote Sensing Symposium. IGARSS'99, vol. 2, pp. 1220–1222 (1999)
53. Meyer, F.: Topographic distance and watershed lines. Signal Process. **38**(1), 113–125 (1994)
54. Moga, A., Cramariuc, B., Gabbouj, M.: Parallel watershed transformation algorithms for image segmentation. Parallel Comput. **24**(14), 1981–2001 (1998)
55. Moga, A., Gabbouj, M.: Parallel marker-based image segmentation with watershed transformation. J. Parallel Distrib. Comput. **51**(1), 27–45 (1998)
56. Mohan, E., Sugumaran, R., Venkatachalam, K.: Automatic brain and tumor segmentation in mri using fuzzy classification with integrated bayesian model. Int. J. Appl. Eng. Res. **9**(24), 25859–25870 (2014)
57. Mortensen, E.N., Barrett, W.A.: Toboggan-based intelligent scissors with a four-parameter edge model. In: CVPR, IEEE Computer Society, pp. 2452–2458 (1999)
58. Moumoun, L., El Far, M., Chahhou, M., Gadi, T., Benslimane, R.: Solving the 3d watershed over-segmentation problem using the generic adjacency graph (2010)
59. Muzylev, E., Uspensky, A.: Modelling the Hydrological Cycle of River Basins Using High Resolution Satellite Information, pp. 241–248. IAHS-AISH Publication, Wallingford (2001)
60. Najman, L., Couprie, M.: Watershed algorithms and contrast preservation. In: Lecture Notes in Computer Science. Lecture Notes in Artificial Intelligence and Lecture Notes in Bioinformatics, vol. 2886, pp. 62–71 (2003)
61. Nithya, A., Kayalvizhi, R.: Extended fuzzy switching median filter and morphological algorithm for medical image segmentation. ARPN J. Eng. Appl. Sci. **10**(1), 80–90 (2015)
62. Noguet, D.: Massively parallel implementation of the watershed based on cellular automata, pp. 42–52 (1997)
63. Osma-Ruiz, V., Godino-Llorente, J., Sáenz-Lechón, N., Gómez-Vilda, P.: An improved watershed algorithm based on efficient computation of shortest paths. Pattern Recognit. **40**(3), 1078–1090 (2007)
64. Patino, L.: Fuzzy relations applied to minimize over segmentation in watershed algorithms. Pattern Recognit. Lett. **26**(6), 819–828 (2005)
65. Peng, B., Xu, A., Li, H., Han, Y.: Road extraction based on object-oriented from high-resolution remote sensing images (2011)
66. Perry, E., Norton, S., Kamman, N., Lorey, P., Driscoll, C.: Deconstruction of historic mercury accumulation in lake sediments, Northeastern United States. Ecotoxicology **14**(1–2), 85–99 (2005)
67. Plaza, A., Plaza, J., Valencia, D., Martinez, P.: Parallel segmentation of multi-channel images using multi-dimensional mathematical morphology (2008)
68. Procházka, A., Vysata, O., Jerhotova, E.: Wavelet use for reduction of watershed transform over-segmentation in biomedical images processing (2010)
69. Rambabu, C., Rathore, T., Chakrabarti, I.: A new watershed algorithm based on hillclimbing technique for image segmentation. In: TENCON 2003. Conference on Convergent Technologies for Asia-Pacific Region, vol. 4, pp. 1404–1408 (2003)
70. Rambabu, C., Chakrabarti, I.: An efficient hillclimbing-based watershed algorithm and its prototype hardware architecture. J. Signal Process. Syst. **52**(3), 281–295 (2008)

71. Randhir, T., Lee, J., Engel, B.: Multiple criteria dynamic spatial optimization to manage water quality on a watershed scale. Trans. Am. Soc. Agric. Eng. **43**(2), 291–299 (2000)
72. Roerdink, J., Meijster, A.: The watershed transform: definitions, algorithms and parallelization strategies. Fundamenta Informaticae **41**(1–2), 187–228 (2000)
73. Rong, J., Pan, Y.-L.: Accuracy improvement of graph-cut image segmentation by using watershed. Adv. Mater. Res. **341–342**, 546–549 (2012)
74. Shen, W.-C., Chang, R.-F.: A nearest neighbor graph based watershed algorithm, pp. 6300–6303 (2005)
75. Sheshadri, H., Kandaswamy, A.: Application of watershed algorithms to mammogram image analysis. IETE Tech. Rev. **23**(3), 173–178 (2006)
76. Shrimali, V., Anand, R., Kumar, V.: Current trends in segmentation of medical ultrasound b-mode images: a review. IETE Tech. Rev. **26**(1), 8–17 (2009)
77. Sinha, K., Sinha, G.: Efficient segmentation methods for tumor detection in mri images (2014)
78. Smistad, E., Falch, T., Bozorgi, M., Elster, A., Lindseth, F.: Medical image segmentation on gpus - a comprehensive review. Med. Image Anal. **20**(1), 1–18 (2015)
79. Sridhar, B., Reddy, K., Prasad, A.: An artificial neural network and mathematical morphology based computer aided detection system for cancer detection in mammograms. Int. J. Appl. Eng. Res. **9**(23), 21079–21097 (2014)
80. Su, H., Yang, Z.-L., Dickinson, R., Wilson, C., Niu, G.-Y.: Multisensor snow data assimilation at the continental scale: the value of gravity recovery and climate experiment terrestrial water storage information. J. Geophys. Res. Atmospheres **115**(10) (2010)
81. Sun, H., Yang, J., Ren, M.: A fast watershed algorithm based on chain code and its application in image segmentation. Pattern Recognit. Lett. **26**(9), 1266–1274 (2005)
82. Świercz, M., Iwanowski, M.: Fast, parallel watershed algorithm based on path tracing. In: Lecture Notes in Computer Science. Lecture Notes in Artificial Intelligence and Lecture Notes in Bioinformatics, vol. 6375, pp. 317–324, LNCS(PART 2) (2010)
83. Swiercz, M., Iwanowski, M.: Waterball-iterative watershed algorithm with reduced oversegmentation. Adv. Intell. Soft Comput. **95**(4), 385–394 (2011)
84. Tolosa, S., Blacher, S., Denis, A., Marajofsky, A., Pirard, J.-P., Gommes, C.: Two methods of random seed generation to avoid over-segmentation with stochastic watershed: application to nuclear fuel micrographs. J. Microsc. **236**(1), 79–86 (2009)
85. Tonti, S., Di Cataldo, S., Bottino, A., Ficarra, E.: An automated approach to the segmentation of hep-2 cells for the indirect immunofluorescence ana test. Comput. Med. Imaging Graph. **40**, 62–69 (2015)
86. Tung, C.-P.: Climate change impacts on water resources of the tsengwen creek watershed in Taiwan. J. Am. Water Resour. Assoc. **37**(1), 167–176 (2001)
87. Uchida, S.: Image processing and recognition for biological images. Dev. Growth Differ. **55**(4), 523–549 (2013)
88. Van Neerbos, J., Najman, L., Wilkinson, M.: Towards a parallel topological watershed: First results. In: Lecture Notes in Computer Science Lecture Notes in Artificial Intelligence and Lecture Notes in Bioinformatics, vol. 6671, pp. 248–259, LNCS (2011)
89. Vibha, L., Harshavardhan, G., Pranaw, K., Shenoy, P., Venugopal, K., Patnaik, L.: Lesion detection using segmentation and classification of mammograms, pp. 311–316(2007)
90. Vincent, L., Soille, P.: Watersheds in digital spaces: an efficient algorithm based on immersion simulations. IEEE Trans. Pattern Anal. Mach. Intell. **13**(6), 583–598 (1991)
91. Wagner, B., Dinges, A., Müller, P., Haase, G.: Parallel volume image segmentation with watershed transformation. In: Lecture Notes in Computer Science. Lecture Notes in Artificial Intelligence and Lecture Notes in Bioinformatics, vol. 5575, pp. 420–429, LNCS (2009)
92. Wang, W., Shi, H., Wang, A.: Analysis on the future runoff changes in shiyang river basin based on genetic algorithm models (2012)
93. Wu, S., Hu, Y.: Parallelization research watershed algorithm **2012**, 1524–1527 (2012)
94. Xu, G., Zhang, D., Liu, X.: Road extraction in high resolution images from google earth (2009)
95. Yang, F., Li, J., Xu, S.-H., Pan, G.-F.: The research of video segmentation algorithm based on image fusion in the wavelet domain, vol. 7659 (2010)

96. Yu, P.-Y., Zhang, G.-P., Yan, J.-W., Liu, M.-S.: The application of the watershed algorithm based on line-encoded in lung ct image segmentation (2011)
97. Zhang, X., Chen, L., Pan, L., Xiong, L.: Study on the image segmentation based on ica and watershed algorithm, pp. 505–508 (2012)
98. Zhang, X., Cheng, Y., Qian, Y., Zhuang, X.: Automatic video object segmentation based on spatio-temporal information, pp. 5314–5317 (2011)
99. Zhu, H., Zhang, B., Song, A., Zhang, W.: An improved method to reduce over-segmentation of watershed transformation and its application in the contour extraction of brain image, pp. 407–412 (2009)

An Application Sample of Machine Learning Tools, Such as SVM and ANN, for Data Editing and Imputation

Esther-Lydia Silva-Ramírez, Manuel López-Coello and Rafael Pino-Mejías

Abstract This chapter presents studies about the data imputation to estimate missing values, and the Data Editing and Imputation process to identify and correct values erroneously. Artificial Neural Networks and Support Vector Machines are trained as Machine Learning techniques on real and simulated data sets obtaining a complete data set what help to improve the quality of the variables that define the official indicators of the eight Millennium Development Goals.

1 Introduction

United Nations Member States gathered together at the start of the new millennium to shape a broad vision to combat poverty. In this year 2000, world leaders entered into the landmark commitment to spare no effort to free our fellow men, women and children from the abject and dehumanizing conditions of extreme poverty, which was translated into a framework of eight goals that have enabled people across the world to improve their lives and their future prospects. The Millennium Development Goals (MDGs) reconfigured decision-making in developed countries and developing countries equally, by putting people and their needs in a first plane. This framework, with deadline of 2015, included *goals, targets and indicators to monitor progress on extreme poverty and hunger, education, gender equality, child survival, health, environmental sustainability and global partnerships* [57], are available at http://mdgs.un.org.

E.-L. Silva-Ramírez (✉) · M. López-Coello
Department of Computer Science and Engineering, University of Cadiz,
Avda. Universidad de Cádiz, n 10, 11519 Puerto Real (cadiz), Spain
e-mail: esther.silva@uca.es

M. López-Coello
e-mail: manuel.coello@uca.es

R. Pino-Mejías
Department of Statistics and Operational Research, University of Seville,
Avda. Reina Mercedes s/n, 41012 Seville, Spain
e-mail: rafaelp@us.es

C. Cruz Corona (ed.), *Soft Computing for Sustainability Science*,
Studies in Fuzziness and Soft Computing 358, DOI 10.1007/978-3-319-62359-7_13

Experiences and evidence from the efforts to achieve the MDGs demonstrate that they know what to do. But they recognize that it is necessary to integrate the economic, social and environmental dimensions of sustainable development. Between the aims of future a set of sustainable Development Goals are included. They aim to achieve a more prosperous, sustainable and equitable world [60].

The General Assembly required a regular assessment of progress towards the MDGs, which has led to mark or identify a series of statistical indicators as appropriate for tracking the progress of the MDGs. Representatives of the international organizations, national statisticians experts and outside expert advisers work in preparing these indicators. Progress towards the eight Millennium Development Goals is measured through 21 targets and 60 official indicators.[1]

All Millennium Development Goals are not particularly related to sustainable development, but include many indicators that may be considered relevant to the topic, not only on the goal 7 corresponding to sustainability, but in some other goals.

According to Schuschny and Soto [51], the strength or weakness of any of these indicators lies in the quality of the variables that define it. Therefore, the selection of each of the variables that compose the indicator should be documented by building metadata with the characteristics of the variable, its relevance, its quality, its availability to the public domain, the frequency with which it is sampled, the sources responsible of calculating it, the units of measure with which is expressed, etc. They comment that the selection of indicators is limited by the shortage of statistical information, which limits the possibility to draw comparisons between countries for decision making. To establish a realistic comparability between countries it is necessary to set scales and work with relative measures. To contribute to develop better strategies and action plans, that allow direct the course of the countries on a path of truly sustainable development, it is necessary to have complete, consistent and comparable information. However, there are technical applicability problems due to the absence of data or availability of inconsistent data or unreliable data. Data cleaning is a fundamental task previous to decision-making, it is concerned with the data quality.

Data sets usually contain errors in the form of incomplete or incorrect values. Missing values are due to the lack of value, no data value is stored for a field in an observation. While incorrect data are produced when the value is not accurately recorded. A value for a field is not recorded correctly because of a transcription error or a data acquisition system problem. Incomplete values and incorrect data can be costly, they can lead to false conclusions. The data must be prepared and cleaned in order to be useful for the knowledge extraction process. The treatment of non-response errors and inconsistent values are fundamental steps of data cleaning, in data knowledge discovery process, to improve the information quality.

To deal with the missing values problem is performed data imputation, defined as the process to estimate incomplete data in a data set by appropriately computed

[1]The complete list of goals, targets and indicators is available at http://mdgs.un.org/unsd/mdg/Resources/Static/Products/Progress2015/Statannex.pdf.

values. In other words, data imputation is capable of filling in the gaps of the data set with errors of non-response, producing a complete data set.

Whereas, data editing is the task of identification and correction of the values that have been erroneously recorded in a data set. Data editing is defined by UN/ECE [59] as *The activity aimed at detecting and correcting errors (logical inconsistencies) in data*. As this definition shows, data editing deals with two basic forms of errors: values out of range (for example, a negative age) and inconsistencies (for instance, married and five years old). The study performed in this chapter considers all possible errors, taking thus into account other categories of errors, such as values within range but being non-true values.

Other sources [14] distinguish between the process of finding errors and, once they erased, the process of estimating the missing values, using the terms *data editing* and *imputation* respectively. In this work, the whole process of finding errors and estimating values is considered as an only process called Data Editing and Imputation, DEI, process.

The aim of this chapter is to study an automatic procedure to the missing value imputation and to the DEI process of the values erroneously recorded in a data set based on a machine learning approach. So, in this work two studies are carried out, on the one hand the data imputation to estimate missing values, and the other hand the DEI process to identify and correct values erroneously recorded. The chapter is organized as follows. In Sect. 2 the overview of DEI process and data imputation in this topic is shown. Section 3 introduce general aspects of machine learning based models, both of the models used for the data editing and imputation process as well as models used to evaluate the influence the predicted values in classification tasks. In Sect. 4 a description of databases and perturbation process are shown. Later two clearly differentiated blocks appear. The first block, Sect. 5, is devoted to missing data imputation, in which statistical models are described. The different mechanisms that generate missing values and missing data patterns are also explained. And a detailed example of the application of machine learning techniques to the data imputation problem is shown. Whereas the second part, Sect. 6, introduces the idea of data edition, in which the errors identification and correction are carried out. An detailed example of application is also proposed.

2 Data Cleansing Process Background

The dealing with missing values and data edition are nothing new, it is possible to track the first studies several decades ago. In particular, the use of imputation methods for missing values is a task with a well established background. The machine learning based models include K-nearest neighbour, genetic algorithms, multilayer perceptron, SOM, auto-associative neural network, fuzzy-neural network, support vector machine, etc.

Artificial neural networks (ANNs from now onwards) were used in some early studies [32, 38, 55, 56]. These works were provided by several official statistical institutions, as described in the methodological material used by the Statistical Committee of United Nations and the European Economic Committee [43]. The most of these works on ANN-based DEI models adopt a different training process for each single variable. In this way, a neural network is trained for each variable and, consequently, each network provides the estimation of its associated variable. In other articles [30, 42], the DEI process was performed on a single target variable, needing only one network. In Laaksonen [31], Nordbotten [41] the experiments were conduced with data sets formed by either numeric or categorical variables, but not both. In other related works such as [13, 63], MLP is trained only over complete or almost complete data set.

Over the years other works have appeared in which these processes are performed with other techniques. Kalteh and Hjorth [28] imputed missing values with SOM, multilayer perceptron, multivariate nearest neighbours, the regularised expectation maximization algorithm and multiple imputation in the context of a precipitation-runoff process database. Kaya et al. [29] carried out a comparison of the neural networks, the expectation maximization algorithm and the multiple imputation techniques, while the application of genetic algorithms was proposed in Patil and Bichkar [44]. Rahman and Islam [47] proposed two techniques for the imputation of both categorical and numerical missing values, using decision trees and forests. The missing values were imputed using similarity and correlations, and they merged segments to achieve a higher quality of imputation. García-Laencina et al. [19] presented a Multi-Task Learning (MTL) based approach using multilayer perceptron to impute missing values in classification problems. They combined classification and imputation in only one neural architecture, being classification the main task and imputation the secondary task.

Recently, hybrid methods were proposed. Aydilek and Arslan [2] used a hybrid neural network and weighted nearest neighbours to estimate missing values. The estimation system involved an auto-associative model to predict the input data, coupled with the k-nearest neighbours to approximate the missing data. Duma et al. [12] proposed hybrid multi-layered artificial immune system and genetic algorithm. Azim and Aggarwal [3] described a 2-stage hybrid model to fill missing values using fuzzy c-means clustering and multilayer perceptrons. And Gautam and Ravi [21] proposed two novel methods viz., counterpropagation auto-associative neural network (CPAANN) and grey system theory hybridised with CPAANN.

Support vector machines (SVMs hereinafter called) were also used in DEI, Feng et al. [15] presented a SVM-based algorithm to eliminate the inconsistent examples. As in previously cited ANN articles, training was also performed with consistent examples, ignoring the remaining records. Song et al. [54] proposed a one-class SVM-based DEI model and a classification method that trained a model on the previously cleaned data set (outliers removed in the training data). In other works [33, 34], the authors conducted experiments to test the performance of SVM classifiers on distorted data according to several perturbation schemas. The results were analyzed to identify the perturbation mechanisms SVMs were less sensitive to.

The present work performs an empirical study, perturbing both real and simulated data sets, in which both Artificial Neural Networks and Support Vector Machines are trained as ML techniques to data imputation and DEI process.

The selection of the models has been guided by the good results obtained with ANN and SVM in the reviewed works on, for both theoretical and empiric evidences, considering only the group that represents the *approximate models*, based on the classification made in Luengo et al. [37]. According to Liu et al. [36], these two methods (ANN and SVM) are both frequently used as intelligent algorithms, and they may be used interchangeably in many practical cases. However, each of them has advantages and shortcomings which lead to not being completely equivalent. They recommend studying of the relationships and the differences between these two methods to select appropriate algorithms for a certain specific problem.

MLP and SVM have been proposed already to be both imputation methods for missing values and inconsistencies detectors in the literature, but the sole purpose of this chapter is to show an example of machine learning based methods application to carry out the data editing and imputation processes, obtaining a complete data set what help to improve the quality of the variables that define the official indicators of the eight Millennium Development Goals.

3 Machine Learning Based Models

In this study, two data mining techniques are tested for their applicability in missing value imputation and DEI process: artificial neural networks and support vector machines. The two methods are compared in terms of their predictive accuracy in respective sections using other classifiers such as Naïve Bayes, k-NN and decision trees. All these techniques are described briefly in this section.

3.1 Artificial Neural Networks

A three-layered perceptron is considered in this study: an input layer with p nodes which receives values corresponding to explanatory variables, an output layer with q outputs which provides approximations to q target variables, and a hidden layer with H neurons. Each node of the input layer is connected to all neurons in the hidden layer, and each hidden neuron is connected to all neurons in the output layer. Each individual connection has an associated parameter called synaptic weight. Denoting by $\{v_{ih}, i = 0, 1, 2, \ldots, p, h = 1, 2, \ldots, H\}$ the synaptic weights for the connections between the p-sized input and the hidden layer, $\{w_{hj}, h = 0, 1, 2, \ldots, H, j = 1, 2, \ldots, q\}$ the synaptic weights for the connections between the hidden layer and the q-sized output layer, the outputs of the neural network are:

$$o_j = w_{0j} + \sum_{h=1}^{H} w_{hj} g\left(v_{0h} + \sum_{i=1}^{p} v_{ih} x_i \right), \quad j = 1, 2, \ldots, q \qquad (1)$$

Hyperbolic tangent is considered as the activation function $g(u) = (e^u - e^{-u})/(e^u + e^{-u})$ in the hidden layer and identity function as the activation function in the output layer.

The input layer in a MLP only accepts numeric values. However, data files often contain quantitative variables and categorical variables, so an appropriate numerical coding is needed, following a mechanism based on binary variables, as explained in Sect. 4.

Therefore, the number p_t of inputs to the MLP is usually larger than the number p of variables in the data file. The number of outputs is equal to the number of inputs, and thus the network size is (p_t, H, p_t). Given p_t inputs, the set of the p_t outputs must be converted to the set of estimated values. For a particular variable X_j, k predictions $F_1, \ldots, F_k$ will be available, one for each binary variable E_i. The value associated to the maximum prediction defines the estimated value for X_j.

A learning algorithm is required to assign appropriate values to the synaptic weights. However, there is no known procedure to ensure a global solution and frequently one of the many possible local minima is obtained at the most. As shown in Table 1, seven learning algorithms, representative of the main learning algorithms, were considered.

3.2 *Support Vector Machines*

The SVMs have also been considered over the same data sets. Given a training set of pairs $(X_i, y_i), i = 1, \ldots, n$, where n is the number of examples, X_i vectors in $\mathbb{R}^p$, p the number of attributes, and $y_i \in \mathbb{R}$, the SVM algorithm solves the optimization problem below [61], with $C > 0$ (the penalty parameter of the error term), w the gradient and $\langle \cdot \rangle$ dot product:

Table 1 Learning algorithms

Abbr.	Algorithm
GD	Gradient Descent
GDM	Gradient Descent with Momentum
BA	Gradient Descent with Adaptive Learning Rate
BE	Resilient Backpropagation (RProp)
GC	Conjugate Gradient Fletcher-Reeves Update
QN	BFGS Quasi-Newton
LM	Levenberg–Marquardt

$$\frac{1}{2}\langle w \cdot w \rangle + C\left(\sum_{i=1}^{n} (\xi_i^* + \xi_t)\right) \quad (2)$$

which is minimized subject to the constraints:

$$\begin{gathered} y_i - \langle w \cdot X_i \rangle - b \le \varepsilon + \xi_i^*, i = 1, \ldots, n \\ \langle w \cdot X_i \rangle + b - y_i \le \varepsilon + \xi_i, i = 1, \ldots, n \\ \xi_i, \xi_i^* \ge 0, i = 1, \ldots, n \end{gathered}$$

being ξ_i and ξ_i^* slack variables which measure the error on each training point X_i according to the loss function L, defined as:

$$L(y_i, \hat{y}_i) = | y_i - \hat{y}_i |_\varepsilon = \begin{cases} 0 & \text{if } | y_i - \hat{y}_i | \le \varepsilon \\ | y_i - \hat{y}_i | - \varepsilon & \text{otherwise} \end{cases}$$

where y_i is the target value and $\hat{y}_i$ is the predicted value. The absolute loss function can be made more robust by fixing some tolerance limit (or insensitivity zone $\varepsilon > 0$) so that less than ε errors will not be punished. If $| y_i - \hat{y}_i |$ is less than ε the loss is zero; otherwise, the loss increases linearly.

The slash variable ξ_i is non-zero if the point lies above ε. The slash variable ξ_i^* is non-zero if it is below ε. The algorithm finds the flattest function (minimizing the norm of w) which passes within ε distance of the training examples.

The parameters C and ε control the evolution of the algorithm, so their choice affects the performance. The user needs to carefully select them to obtain the best performances. Hsu et al. [25] suggested to explore different parameter configurations. Hence, seven different kernels and several parameter values were analyzed in the present research, considering certain guidelines to narrow the range of values in which the parameters can vary, as described in Sect. 5.4.1. Different kernel function types were used [23], as shown in Table 2.

For each binary variable 0-1 a different SVM model is required. As in the MLP case, the estimated value of a categorical variable is obtained from the largest predicted binary variable which provides the associated category as the prediction. In

Table 2 Kernel functions

Abbr.	Function
LIN	Linear
SPL	Splines
ANS	ANOVA Splines
POL	Polynomial
RBF	Radial Basis Function
SIG	Sigmoid
ERBF	Exponential Radial Basis Function

general, this SVM regression procedure presents better results than the SVM classification model, which also requires some multiclass mechanisms like one-to-one or one-against-all. Moreover, it is also necessary to mention that while a unique MLP is trained to produce an estimated record, SVMs and other machine learning algorithms require one model for each variable in the record to be estimated.

3.3 Other Classifiers Used to Evaluate Predicted Data Sets

A desirable characteristic for data imputation process and DEI method is to improve the accuracy of classification results. In most of the analyzed literature on these processes, scholars point out the convenience of editing and imputing values with different algorithms, particularly for classification tasks. Some works as García-Laencina et al. [18] evaluate the methods performance with classification algorithms fitting in the predicted data set. Likewise, in the present work the predicted data sets were analyzed with different classifiers to evaluate the influence of the predicted values in classification tasks. A classification problem is defined for each one of the data sets analyzed in this study. The number Cl of classes is shown in Table 4.

Three well-known classifiers, and different from used to impute, were selected as a set of popular representative classifiers in the data mining community: Naïve Bayes, K-Nearest Neighbour (k-NN from now onwards) and C4.5 Decision Tree. These models are available in the WEKA System [62] used in the present study. Table 3 contains the parameters configuration for each algorithm ($\text{k} = 1$ for k-NN was the result of 10-fold cross validation in all the experiments). The parameter values were chosen according to ad-hoc experiments.

A brief description of the classification methods is as follows:

Naïve Bayes The Naïve Bayes Classifier is a simple probabilistic classifier based on applying Bayes' theorem supposing the independence of the variables and considering a simple form for the density or probability function for each component. Despite its simplicity and unrealistic assumption, the naïve bayes classifier is surprisingly effective in practice, often competing with more sophisticated classification methods.
The naïve bayes classification method uses all attributes and allows them to make contributions to the decision, equally important and independent from one another given the class [40]. Training is very easy and fast, involving the estimation of the probability function of each categorical variable and computing the posterior probability in the independence scenario. This posterior probability function is used on test records.

k-Nearest Neighbour k-Nearest Neighbours is a simple algorithm that predicts a target variable from all available cases. Given a record R to be classified, this algorithm identifies its k-nearest cases in the training data set, and R is assigned to the class most common among these k-nearest neighbours. The two main decisions governing the algorithm are the distance measure between records

Table 3 Parameters used by the classification methods

Method	Acronym	Parameters
Naive-Bayes	NB	–
k-Nearest Neighbour	k-NN	k = 1
		Distance function = Euclidean
C4.5	C4.5	Prune = true
		Confidence threshold = 0.05
		Instances per leaf = 5

and the value of k. The euclidean distance between the variables was used, previously codified with auxiliary 0-1 variables. Regarding the parameter k, the range of integer values 1 to 15 was explored through a 10-fold cross-validation mechanism, but $k = 1$ was selected in all data files.

C4.5 Decision tree learning is one of the most widely used methods applied to classification problems. These methods are robust to noisy data. A decision tree is a set of logical "if-then" rules which drive each case to a final class assignation (in classification problems). These rules can easily be plotted in order to aid to the understanding of the model. Among the different existing proposals, C4.5 algorithm [46] was selected: the tree is "grown" by binary recursive partitioning, choosing splits from the set of predictor variables. At each node of the tree, C4.5 chooses the attribute of the data with the highest normalized information gain to make a new split, and the process is repeated. J48 was used, an open source Java implementation of the C4.5 algorithm in the Weka data mining tool.

4 Data in Experiments

This section shows a description of the data sets used in this study and the perturbation procedure applied on these data sets to perform the experiments.

The processes deal with a data set S, resulting from a data collection process, where S is a nxp matrix, being p the variable measured for each of the n records. However, it is assumed that certain S cells have been missed during the data collecting process and a certain number of erroneously recorded values may be present; thus a procedure to identify the true data matrix T, with all the values completely recorded, would be useful. In this way, a data imputation model and a DEI tool is defined by a set of rules and procedures aimed at obtaining an approximation T^* to T, working on the available data set S. T^* is supposed to be a corrected and improved version of S, but the desired matrix T is not guaranteed to be achieved. In traditional DEI models, the construction and management of edits usually involves subject-matter experts, making the DEI process non-automatic, slow and time-consuming.

An alternative is to reduce the set of records to be reviewed by the experts. For example, a habitual DEI procedure can be run on a subset S^1 of S, obtaining an estimated version T^{*1}. Taking as inputs the records of S^1 and defining the records in T^{*1} as targets, a supervised machine learning model can be trained. The resulting model can be applied to $S^2 = S \backslash S^1$, providing a correction T^{*2}.

A similar scenario would be defined by the availability of an assumed correct data set T. So, a randomly perturbed version S is generated and a supervised ML model is trained working with the pairs in (S, T). In this case, the resulting model is useful to estimate new records for the data set, for instance when new survey questionnaires are received or when a panel study advances to the next time period. So, in this work, the data set T was assumed to be complete and correct, setting artificially the missing or erroneous values.

4.1 Description of the Data

Table 4 presents a brief description of the data sets used in this evaluation. These data sets were extracted from the UCI-Repository [17]. They cover different fields as social surveys, business or medicine and include different types of variables: quantitative, ordinal categorical, nominal categorical and dichotomous variables.

Table 4 Data sets used in the experiments

Name	n	Cl	p_t	p_q	p_c
Glass	214	5	9	9	0
Iris	150	3	4	4	0
Pima	768	2	8	8	0
Yeast	1484	10	9	9	0
Abalone	4177	29	10	7	1
Cleveland	303	5	25	6	7
Contraceptive	1473	3	24	2	7
Flag	194	8	77	10	18
Heart	270	2	25	6	7
Zoo	101	7	31	1	15
Breast	683	2	89	0	9
Hayes-Roth	132	3	15	0	4
Lung-cancer	32	3	155	0	56
Lymphography	148	4	60	0	18
Mushroom	8124	2	93	0	21
Solar	1389	3	16	0	7
Soybean	47	4	95	0	35

Table 4 contains the following columns for each database. Name: name of the data set; n: number of records; Cl: number of classes for the classification problem; p_t: total number of inputs and outputs for the machine learning models once the binary variables are coded; p_q: number of quantitative variables; p_c: number of categorical variables; so, $p = p_q + p_c$.

The first group corresponds to databases with qualitative and quantitative variables, the second group represents databases with only quantitative variables and the third group of databases contains only categorical variables.

ANN training requires several data preparation tasks [5, 50], and the same considerations for ANN are also applied to SVMs [25]. In this way, a variable that have the same value in all records is not informative, and hence it was ignored. The quantitative variables were normalized, computing for the quantitative variable X_i the following value for the record j:

$$\S_{ij} = \frac{x_{ij} - x_{i,min}}{x_{i,max} - x_{i,min}}$$

And, as mentioned in Sect. 3.1, this work follows a mechanism based on binary variables to code categorical variables, where given a qualitative variable X_j and being $classes = \{c_1, \ldots, c_k\}$ the finite set of possible values that can take, X_j is codified with k binary variables 0-1, $E_1, \ldots, E_k$, such that the ith variable takes the value 1 if the variable value X_j is equal to the ith category and 0 otherwise, with $i = 1, \ldots, k$:

$$E_i = \begin{cases} 1 \text{ if } X_j = c_i \\ 0 \text{ otherwise} \end{cases}.$$

4.2 Perturbation Procedure

A perturbation experiment over the data sets in Table 4 was performed. The idea of generating errors in correct data sets to evaluate data imputation methods and DEI techniques was already presented in other works such as the study by Fessant and Midenet [16].

Each available data set T in Table 4 was assumed to be complete and correct, setting artificially erroneous values in each one. Random errors of value change and non-response were introduced, obtaining a perturbed version T^d. So for each original variable X_j, a perturbed variable X_j^d was defined, considering 0.05 as the error probability. This error rate could be considered high, as in the case of a data set with 10 variables, approximately the 40% of the perturbed data records would be incorrect. Thus, the number of inconsistent values and number of missing values in a record was restricted to not more than half of the variables, whereas the number of records that may have errors is not limited. Using a realistic probability of missing value was tried. Thus, several data sets obtained from surveys were analyzed and the majority of missing value rates was not greater than 5%. Nonetheless, some preliminary studies with 1, 5 and 10% did not reveal differences between the models.

Since the quantitative variables were normalized, all variables values ranged from 0 to 1, therefore the non-response was reflected by assigning the value -1 to the disturbed variable:

$$P[Y^d = -1] = 0.05 \wedge P[Y^d = Y] = 0.95 \tag{3}$$

For a qualitative variable X_j, the non-response was reflected by assigning the value 0 to all the dummy variables used in the coding, making 0 the value of that variable.

For the non-monotone pattern, any variable in any observation was randomly set to missing value. For the monotone pattern, a set of randomly generated variables and records are set to missing value as it is more deeply described in Sect. 5.1. In particular, the missing value mechanism was assumed to be missing completely at random (MCAR).

In the DEI process, for a categorical variable X_j with K categories $c_1, c_2, \ldots, c_k$, its associated perturbed variable was defined as follows:

$$X_j^d = \begin{cases} U \sim DU(c_1, c_2, \ldots, c_K - X_j) & 0.05 \\ X_j & 0.95 \end{cases} \quad p \tag{4}$$

DU denotes the discrete uniform distribution.

5 Dealing with Missing Data

In this section, the different mechanisms that generate missing values are shown and the missing data patterns are described. Moreover, in Sect. 5.3, the three classical single imputation procedures (mean/mode imputation, regression and hot-deck), implemented to compare the efficiency of the proposed methods, are explained and the multiple imputation method, used in the proposed approach, is described. Ending the section with a detailed example that shows a serie of experiments carried out with the studied models.

5.1 Missing Data Patterns

Two types of missing data patterns are generally distinguished: monotone and non-monotone. In monotone pattern, the lack of response is observed for the same records and variables, as shown in Table 5, while in the non-monotone pattern, any variable for any record can present a missing value, as represented in Table 6.

Table 5 Monotone missing data pattern

Records	Variables					
	X_1	...	X_f	X_{f+1}	...	X_p
1						
⋮						
m						
$m+1$				?	?	?
⋮				?	?	?
n				?	?	?

Table 6 Non-monotone missing data pattern

Records	Variables					
	X_1	...	X_f	X_{f+1}	...	X_p
1	?					?
⋮			?			
m						?
$m+1$	?			?		
⋮					?	
n	?		?			

Table 7 illustrates the monotone missing data pattern in the experimental situation of this study. From the set S, a p-sized variable $X = (X^o, X^m)$ can be considered for n records, where f variables X^o are completely observed, and $p - f$ variables X^m are missing in a set of m records. The set represented by the matrix Z_4, the shaded cell in Table 7, is missing, being the sets Z_1, Z_2 and Z_3 completely observed. Hence, O denotes the set with observed values, $n - m$ records for the p variables, and M is the set of $p - f$ variables with incomplete data in m records. This pattern of missing values suggests to fit some prediction model where Z_1 provides the inputs and Z_2 defines the set of outputs. Once the model has been trained, it is applied to Z_3, obtaining an imputed set Z_4.

Let us denote:

$X = (X_1, X_2, \ldots, X_p)$ as the nxp sample with p variables and n records

$X_j = (X_{1j}, X_{2j}, \ldots, X_{nj})$ $j = 1, 2, \ldots, p$ as the variable values jth for all the records

Table 7 Monotone missing data pattern in an experimental situation

Records	Variables		
	X^o (f variables)	X^m ($p-f$ variables)	Sets
$n-m$	Z_1	Z_2	O
m	Z_3	Z_4	M

$\underline{X}_i = (X_{i1}, X_{i2}, \ldots, X_{ip})$ $j = 1, 2, \ldots, n$ as the record values ith for all the variables

x_{ij} as the value of the variable j for the record i

X_j^o as the set of records with observed values for the variable jth

X_j^m as the set of records with missing values for the variable jth

$\underline{X}_i^o$ as the variable values of the set X^o for the record ith

$\underline{X}_i^m$ as the variable values of the set X^m for the record ith.

5.2 *Missing Data Mechanisms*

Missing data can arise by different mechanisms. Little and Rubin [35] and Rubin [48] define three types of missing data mechanisms:

- MCAR (missing completely at random). The probability that the value of a variable X_j is observed or missing for any individual does not depend on any variable:

$$P[X_j = mis \mid X_1 \ldots X_p] = P[X_j = mis]$$

- MAR (missing at random). The probability that the value of a attribute X_j is observed for any record does not depend on the variable itself, but depends on the value of the other variables:

$$P[X_j = mis \mid X_1 \ldots X_p] =$$
$$= P[X_j = mis \mid X_1 \ldots X_{j-1} X_{j+1} \ldots X_p]$$

- NMAR (not missing at random). The probability that the value of a feature X_j is observed for any instance depends on the value of that variable itself, being this value unknown.

$$P[X_j = mis \mid X_1 \ldots X_p] =$$
$$= P[X_j = mis \mid X_j]$$

In most studies the response mechanism is not specified, so it is assumed that data are missing completely at random (MCAR).

5.3 Data Imputation Methods Based on Statistical Analysis

A wide range of methods and tools for data imputation may be used. Little and Rubin [35] describe extensive methods of treatment to incomplete data, many of which are intended for continuous and normally distributed data. Some methods, for example listwise, casewise and pairwise data deletion, try to make a maximum use of the available information, omitting all those records that contain missing values for some variables, depending on the population parameters to be estimated. Other methods are suitable for computing values that replace the missing data.

Little and Rubin [35] classify the methods of handling missing data according to their degree of complexity: listwise and pairwise data deletion and mean/mode are inferior, regression methods are somewhat better, but not as good as hot-deck or procedures based on multiple imputation. So, some of these methods representative are described below: mean/mode substitution, regression imputation, hot-deck and multiple imputation.

- **Mean/mode imputation**: It is a simple method where any missing value is replaced by the mean of the observed values for that attribute if it is a quantitative variable. So, if a variable presents several missing values for different records, all of them are imputed with the same value. If the variable is qualitative, the incomplete values are replaced by the mode. Since there may be variables of any type in a data set, numerical and categorical, distinguishing between types of variables is necessary. As aforementioned in Sect. 5.1, X_j^o denotes the set of records with values observed for the attribute j and X_j^m the set of records with missing values for the variable j, with $j = 1, \ldots, p$. The set of estimated values is denoted by $\widehat{X}_j^m$.
  Therefore, if the variable is quantitative, missing values for that variable are imputed with the mean of observed value:

$$\widehat{X}_j^m = E[X_j^0]$$

  On the contrary, if the variable is qualitative, missing attributes for that variable are imputed with the category that have the most of individuals with observed values, this is, with the mode:

$$\widehat{X}_j^m = \overline{X}_j^0$$

- **Regression models**: A regression model is fitted to predict missing values for a variable from the predictors. The fitted model provides a prediction for the initial incomplete values of the variable. The process is repeated for each variable with missing data.
  Three multiple regression procedures are described: Multiple Linear Regression for quantitative variables, Logistic Regression when the dependent variable is dichotomous, and Multinomial Logistic Regression used to handle categorical variables with more than two categories.

Multiple Linear Regression. For a dependent variable or response X_j with missing values, a population model $X_j = \beta_0 + \beta_1 X_1 + \beta_2 X_2 + \cdots + \beta_f X_f + \varepsilon$ is assumed using records with observed data for the variable X_j and the independent or predictor variables $X_1, X_2, \ldots, X_f$, being $f > 1$. ε denotes a random disturbance or error representing the absence of an exact relationship and being $\beta_0, \beta_1, \ldots, \beta_f$ unknown coefficients or parameters that define the regression hyperplane $\beta_0 + \beta_1 X_1 + \beta_2 X_2 + \cdots + \beta_f X_f$.
If a qualitative variable involves c categories, $c - 1$ dichotomous variables are added to the model:

$$d_{j1} \begin{cases} 0 \text{ if } j \notin \text{category } 1 \\ 1 \text{ if } j \in \text{category } 1 \end{cases}$$

$$d_{j2} \begin{cases} 0 \text{ if } j \notin \text{category } 2 \\ 1 \text{ if } j \in \text{category } 2 \end{cases}$$

$$\vdots$$

$$d_{j,c-1} \begin{cases} 0 \text{ if } j \notin \text{category } c-1 \\ 1 \text{ if } j \in \text{category } c-1 \end{cases}$$

The category c is the base category. None of the variables for this category is defined, but all observations having value 0 for the other $c - 1$ variables. For example, if X_f is qualitative and $X_1 \ldots X_{f-1}$ are quantitative, the multiple linear regression model would be:

$$x_{ij} = \beta_0 + \beta_1 x_{i1} + \beta_2 x_{i2} + \cdots + \beta_{f-1} x_{i,f-1} + \alpha_1 d_{i1} + \\ + \alpha_2 d_{i2} + \cdots + \alpha_{c-1} d_{i,c-1} + \varepsilon_i$$

Logistic Regression. This method is applied when the dependent variable X_j is dichotomous, being a Bernoulli random variable whose probability parameter (its mean) is given by a function $\mu(\underline{X}_i)$, the mean and variance of X_j depend on the value of the predictors vector. $E[X_j|X = \underline{X}_i] = \mu(\underline{X}_i)$, $V[X_j|X = \underline{X}_i] = \mu(\underline{X}_i)[1 - \mu(\underline{X}_i)]$.
The logistic function is one of the most used models to express this relationship:

$\mu(\underline{X}_i, \beta) = P[X_j = 1|\underline{X}_i] = \frac{e^{\beta_0 + \beta_1 x_{i1} + \cdots + \beta_p x_{if}}}{1 + e^{\beta_0 + \beta_1 x_{i1} + \cdots + \beta_p x_{if}}}$.

Multinomial Logistic Regression. This method is used when the dependent variable X_j is a categorical variable with more than two categories, being a generalization or extension of the previous one. Assuming $X_j \equiv 1, \ldots, C$, the log odds ratio between categories c and C (base category) is defined as $\theta(c|\underline{X}_i) = \log \frac{P[X_j = c|\underline{X}_i]}{P[X_j = C|\underline{X}_i]}$, $c = 1, \ldots, C$. This model assumes $\theta(c|\underline{X}_i) = \beta_{c0} + \beta_{c1} x_{i1} + \cdots + \beta_{cf} x_{if}$. Thus, $P[X_j = c|\underline{X}_i] = \frac{e^{\theta(c|\underline{X}_i)}}{e^{\theta(1|\underline{X}_i) + \cdots + \theta(C|\underline{X}_i)}}$.

- **Hot-deck**: This method performs the missing values estimations on the incomplete records from values of similar complete records belonging to the same data set.

Andridge and Little [1] found that no consensus exists as to the best way to apply the hot deck and obtain inferences from the completed data set. They review different forms of the hot deck and existing research on its statistical properties. The hot deck method does not rely on model fitting for the variable to be imputed, being thus potentially less sensitive to model misspecification than an imputation method based on a parametric model, such as regression imputation.
Hot deck imputation replace missing values of one or more variables for a record, non-respondent or recipient, with observed values from other record, respondent or donor, similar to the non-respondent with respect to features observed by both cases. Different ways in which donors can be identified exist. One of them is the nearest neighbour based on some distance metric: Euclidean, Mahalanobis, Manhattan, Chebyshev, etc.
In this study, records containing mixed data types are considered, usual situation in any data set. Thus a good alternative is to use the nearest neighbours technique 1-NN with Gower's general similarity coefficient [22] that allows to measure the proximity between records containing data types qualitative and quantitative.
The potential donor records are found in the complete records, choosing the most similar record to the case with missing values. The proximity between donor records and the receptor records is calculated by Gower's general similarity coefficient. Therefore, Hot-deck imputation method allows to estimate the missing value x_{ij} from the value of the variable X_j of the complete record set, which makes maximum Gower's general similarity coefficient G:

$$\hat{x}_{ij} = x_{tj} | G(\underline{X}_i, \underline{X}_t) = \max_{\underline{X}_\gamma \in X_j^o} G(\underline{X}_i, \underline{X}_\gamma)$$

For the sample with n records and p variables, Gower's general similarity coefficient s_{ij} is defined as:

$$s_{ij} = \frac{\sum_{k=1}^{p} w_{ijk} s_{ijk}}{\sum_{k=1}^{p} w_{ijk}}$$

representing by w_{ijk} the number of variables which have observed values for both records:

$$w_{ijk} = \begin{cases} 1 \text{ if } X_k \text{ is knowed in } i, j \\ 0 \text{ otherwise} \end{cases}$$

and s_{ijk} is the contribution provided by the kth variable, distinguishing between different data types. For continuous and ordinal variables:

$$s_{ijk} = 1 - \frac{|x_{ik} - s_{jk}|}{r_k}$$

denoting r_k the range of values for the kth variable and x_{ik} the value of the record i for the variable k. For nominal variables, if both records i and j present the same category for the variable k, the value s_{ijk} is equal to 1 and 0 otherwise:

$$s_{ijk} = \begin{cases} 1 \text{ if } x_{ik} = x_{jk} \\ 0 \text{ if } x_{ik} \neq x_{jk} \end{cases}$$

- **Multiple imputation**: A vector of $M_I > 1$ imputed values for each incomplete data is generated with multiple imputation method. These estimated values are alternative values to complete the missing data. From the imputation vectors, M_I sets of complete data are obtained. According to Rubin [49], the first element of its imputation vector replaces each incomplete value, generating the first complete data set. Then, the second complete data set is generated replacing each incomplete data by the second element of its imputation vector, and so on. Each obtained complete data set is analyzed with the some statistical method used for complete data and then the results of such individual analysis are suitably combined.
These M_I different values can be generated in different ways [35, 49]. Following the publication of these works, this imputation method has been studied in several articles. Various proposals are suggested, such as carrying out the multiple imputation by classification trees [45, 58]. One common method of missing value imputation for some record R is to use the k-nearest neighbours algorithm. So, the k most similar completely observed records to the incomplete record R are identified according to a similarity measure. Each missing value in R can be estimated from k possible values, and a final imputation is obtained applying some aggregation criterion, for example: the average, the median or the mode. In this approach, the use of a model, MIMLP model, to estimate these M_I values is proposed, in which the Multilayer Perceptron and k-nearest neighbours are combined. The built MIMLP-based model is a user-friendly and flexible method. This model is also free from assumptions on distributions and allows the process automation, being possible to calculate the variability introduced in the imputation process.

5.4 *Detailed Example of Data Imputation Using ML Techniques for Non-monotone and Monotone Pattern*

With so many methods in use, we set out to compare a subset of these methods using both real and artificial data set. For non-monotone pattern, automated single imputation approaches MLP-based and SVM-based were developed and compared with three classical single imputation procedures: mean/mode imputation, regression and hot-deck. For monotone patterns, a multiple imputation approach based on the combination of multilayer perceptron and k-nearest neighbours was developed and the three classical single imputation procedures were also considered.

The results, considering different performance measures, are shown in Sect. 5.4.3. These results demonstrate that, in comparison with traditional tools, the models proposed improve the automation level and data quality offering a satisfactory performance.

5.4.1 Empirical Experiments

- **Data imputation for non-monotone pattern**

An appropriate procedure to estimate the performance of the proposed procedure, for data imputation following non-monotone pattern and DEI process, was needed. Therefore, each correct data set T and its associated perturbed version T^d were randomly split into two separate files, a training subset (70%) and a test (30%) subset. Thus, two files, T^1 and T^2, were obtained with associated perturbed sets, T^{d1} and T^{d2}, respectively. The rows of T^{d1} defined the inputs to the MLP and SVM, while the rows of T^1 were the target records. This partitioning was randomly carried out 10 times to take into account the existent inherent variability in the random split.

As explained below, $M = 357$ different parameter configurations for the MLP and $M = 575$ models for the SVM were considered, so a 10-fold validation procedure was followed to select the best configuration. The rows of T^{d2} were fed to the fit MLP and SVM and the output records of the MLP and SVM -contained in T^{*2}- were compared with the true records contained in T^2. To avoid the dependence of the results on the performance of a single DEI or imputation process, 15 different perturbed data sets were generated for each database, repeating the DEI or imputation process for each of them. Thus, 15×10 trials were carried out on the data. Figure 1a displays the whole procedure for a fixed data set T.

The term "predicted data set" will be used to denote a data set resulting from a imputation or DEI process based on this ML approach. Similarly, the term

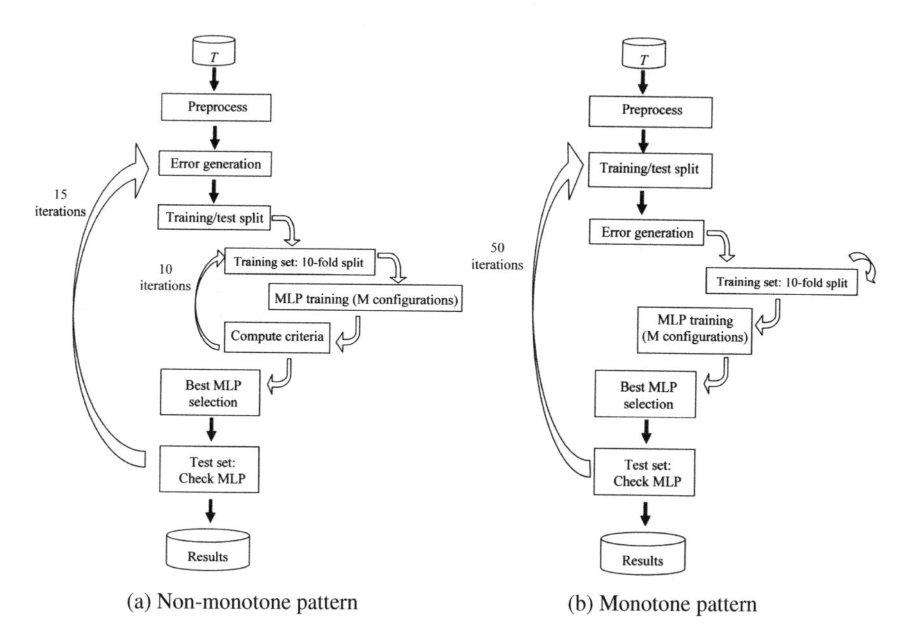

Fig. 1 Strategies used in the process of imputation for each data set T

"estimated" will denote a value or record obtained from the same ML-based imputation or DEI process.

It is well known that a proper configuration of the MLP and SVM parameters is necessary in any application. Moreover, the performance of both models is strongly dependent on their main characteristics. Following these concerns, a wide study with a great number of MLP and SVM configurations was designed. The analysis of the results obtained provided some insights into the influence of the particular parameters on the resulting performance.

Working with MLP involves several decisions to consider such as the learning algorithm, the random initialization of the MLP weights, the number of hidden units or the number of iterations (epochs) of the learning algorithm. Thus, different alternatives were considered, as discussed below.

For each of the 15 perturbed data sets generated, according to the process described in Sect. 4.2, the training algorithm was run 5 times from 5 random initial weight vectors to minimize the associated uncertainty, selecting the network with the minimum mean squared error (MSE). This is due to the fact that the initial random configuration of weights led to very different solutions.

Given the great number of configurations in this empirical study, an initial training to define appropriate ranges of values for the hidden layer size and the number of epochs was performed. The results of this preliminary phase helped us to refine the search set. All databases provided similar results. A first fact observed in this study was that the use of more than 15 hidden nodes produced a clear increase of mean squared error. Thus, three hidden layer sizes (5, 10 and 15) were evaluated. Similar performances were obtained with a number of epochs in the range (300, 3000), increasing the execution time. Consequently, the following ranges for the number training epochs were defined as follows: from 5 to 30 with an increase of 5, and from 50 to 300 with an increase of 25.

The total number of weights for each MLP architecture was $(2p_t + 1)H + p_t$. For example, for the Contraceptive database with 7 qualitative attributes and 2 quantitative, $p_t = 24$ when the categorical variables were codified with binary variables, considering $H = 15$, the ANN model comprised 759 weights.

For each perturbed data set $7 \times 3 \times 17 = 357$ different MLP parameter configurations were generated. 10 values of the GCD criterion (explained in Sect. 5.4.2) were available for each of these configurations, computing their mean value. Thereupon, assuming 15 perturbed data sets, 10-fold validation splits, 7 learning algorithms, 3 sizes of hidden layers, 17 values for the epochs and 5 different initial weights, the total number of different MLP architectures studied for each database was $15 \times 10 \times 7 \times 3 \times 17 \times 5 = 267.750$. Figure 2a shows the schema that explains the whole of tests performed.

For every perturbed data set, a grid search for the SVM parameters was also conducted and the seven different kernel function types were used. The first four functions are the most used ones, according to the description of the software LIBSVM [6]. Furthermore, some of these functions require additional parameters that must be also fit to obtain the lowest potential error.

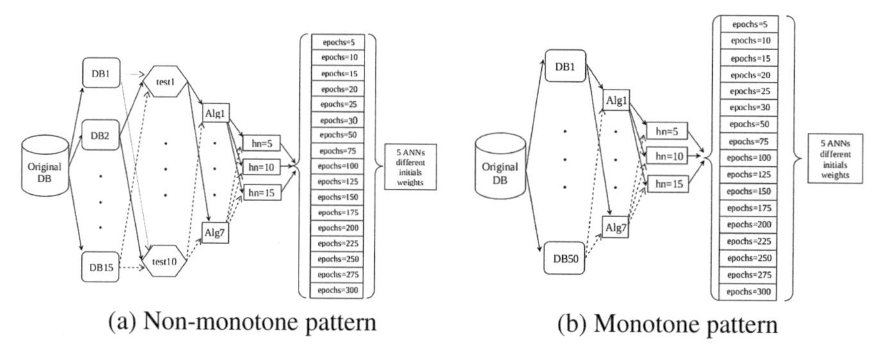

Fig. 2 Schema of the tests performed in model MLP

Linear, Splines and ANOVA Splines kernels need to choose two parameters for the algorithm. ε (the width of "tube") was searched in a reasonable range of values, from 0.02 to 0.82 with an increase of 0.2 and C varied from 2^{-5} to 2^{3}, following a geometric progression with common ratio 2^{2}. However, for the rest of kernels (i.e. Polynomial, RBF, Sigmoid and ERBF) three parameters are necessary: ε, C and σ. The first two varied in the same way previously described; as regards the third parameter (the kernel parameter) the search was performed in the range $[\frac{d}{4}, d]$, where d is the number of input variables. The performance generalization for all possible combinations of two parameters was evaluated. Thus, these ranges were selected according to ad-hoc experiments, choosing the best values for the kernel parameters.

With regard to the support vector coefficients and kernel parameters, a high number of combinations was also considered. For kernels with three parameters, the total number of different SVM architectures for each data set was $15 \times 10 \times 4 \times 5 \times 5 \times 5 = 75.000$, that is, 15 perturbed data sets, 10-fold validation splits, 4 kernels, 5 values for ε, 5 values for C and 5 for σ. For kernels with two parameters, the total number of different architectures for each data set was $15 \times 10 \times 3 \times 5 \times 5 = 11.250$, that is, 15 perturbed data sets, 10-fold validation splits, 3 kernels, 5 values for ε and 5 values for C. The total sum was 86.250 architectures.

In this way, for each perturbed data set $M = 4 \times 5 \times 5 \times 5 + 3 \times 5 \times 5 = 575$ different models were generated. For each of these models, 10 values of the GCD criterion (discussed in Sect. 5.4.2) were available, computing their mean value. Figure 3 shows the schema that explain the whole of tests performed by the kernel.

Both imputation and DEI process, in its entirety, were performed for two error probabilities: 0.05 and 0.08. An evaluation of the results obtained about performance variation with these two levels of the noise were carried out, but no major differences. Since the results are similar for both, in the experiments presented here has only considered as error probability of 0.05.

- **Data imputation for monotone pattern**

The process for the data imputation experiments following a monotone pattern is shown in Fig. 1b. Each data set T, assumed correct, was split into training set

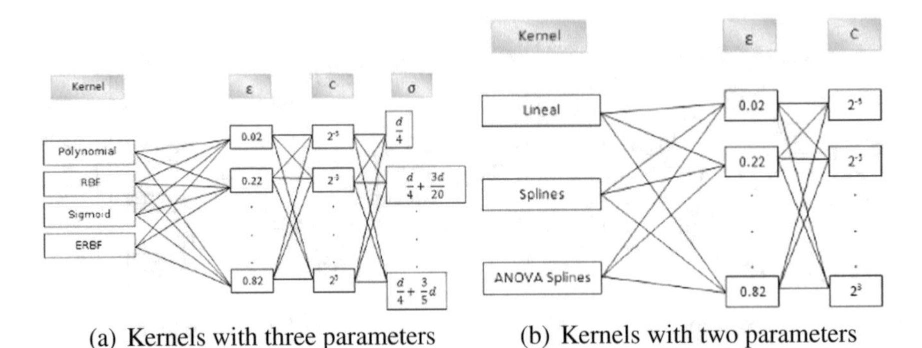

(a) Kernels with three parameters (b) Kernels with two parameters

Fig. 3 Schema of the tests performed in model SVM by the kernel

and test set, 70% and 30% respectively, obtaining the files T^1 and T^2. A 30% of variables, $p-f$ variables in Table 7, were randomly selected obtaining X^m, being T^{m2} the perturbed version from T^2 by setting all the values of all the variables in X^m to missing values. Representing by X^o the f remaining variables not included in X^m, as described above. An imputation model to predict X^m from X^o could be fitted on T^1. The rows of T^{m2} are fed to the fitted model and the output records of the model, contained in T^{*2}, were compared with the true records contained in T^2.

To avoid that the results obtained depended simply on the performance of a single imputation process, this same perturbation pattern was repeated 50 times, thus 50 perturbed files were obtained for each data set, repeating the imputation process for each one of them, that is, the five imputation methods were applied on each one of the 50 perturbed data sets. Figure 2b displays the whole procedure for a fixed data set T.

For monotone patterns the total number of different MLP architectures which have been studied for each database is: $50 \times 10 \times 7 \times 3 \times 17 \times 5 = 892.500$, that is, 50 perturbed data sets, 10 fold validation splits, 7 training algorithms, 3 sizes of hidden layers, 17 values for the epochs and 5 different initial weight sets. As in the non-monotone case, $M = 7 \times 3 \times 17 = 357$ different MLP configurations were generated. For each one of these configurations, 10 values of the GCD criterion (explained in Sect. 5.4.2) were available through the 10-fold procedure, and their mean value was also computed. The minimum mean GCD guides us to the selected configuration.

In the MIMLP model was carried out a previous step in the aforementioned process of applying the k-nearest neighbours algorithm: the training of a MLP on the complete records $O = (Z_1, Z_2)$. Thus, for a incomplete instance R in the set $M = (Z_3, Z_4)$, a completed version R' was computed handling the perceptron. The similarity function G now involves the whole set of variables, defined as a weighted sum of two similarities, one was computed with the observed values, while the other similarity was computed with the imputed values. In O set were identified the k nearest instances to R', obtaining k possible imputations. As described above, a similar work is available in Tusell [58], which can only be applied on continuous variables. In this approach, the imputation of missing values was carried out on quantitative and/or qualitative variables. MIMLP model has a single hidden layer with the Levenberg–Marquart learning algorithm. Ad-hoc studies were realized to achieve the greater MLP generalization capacity. The optimal values for these parameters and the algorithm implemented can be consulted in Silva-Ramírez et al. [53].

As previously mentioned, this chapter aim for an example of application of methods for data imputation and DEI processes, enabling greater automation of these processes, not test the influence of the parameters in the models. Thus, it is not necessary to stop in more details on the study of the parameters of models.

The source code employed in this work was written by the authors in Matlab R2012a, using Neural Network Toolbox [11] for the construction of the MLPs and the function *quadprog* of the optimization package for the SVMs [7].

5.4.2 Imputation Methods Efficiency

To evaluate the models, several measures were computed depending on the variable type.

For quantitative variables, the average of the squared linear correlation coefficient R^2, expressed as a percentage between 0 and 100%, was computed. And for each variable, where m values were set to non-response errors, R^2 was calculate measuring the association between the m real values and the m imputed values.

For qualitative variables other measures were adopted. As a first step, Euredit [13] recommends a Wald-type statistic W to analyze whether a data imputation process preserves the marginal distribution of a qualitative variable with K categories.

Under weak assumptions the large sample size distribution of W was chi-square with $K-1$ degrees of freedom, applying a statistical test for each categorical variable, calculating the p-value as the right tail probability of a chi-square distribution with $K-1$ degrees of freedom computed for the observed value W.

If the hypothesis of marginal distribution preservation was accepted, a second step was to evaluate the preservation of the true value of the categorical variable.

In the case of an ordinal categorical variable, the preservation of the true value order was studied. For more details refer Silva-Ramírez et al. [52].

The coefficient of preservation CPR was calculated as an overall measure based on these statistics. This coefficient was defined as the categorical variables percentage which preserves the marginal distribution and the true value (for nominal variables) or the true value order (for ordinal variables):

$$CPR = \frac{NCVN + NCVO}{P_c}$$

NCVN denotes the number of nominal qualitative variables which preserves the marginal distribution and true value. Let *NCVO* be the number of ordinal categorical variables which preserves the marginal distribution and true value order, and P_c represents the number of qualitative variables, Table 4.

A global criterion for the whole data imputation process GCD was computed as the mean of R^2 (when available) and CPR (when available).

5.4.3 Results

This section presents an evaluation of the obtained results, comparing the different methods for missing data imputation on the different data sets. Aforementioned, the results were evaluated by the numerical criteria described in previous section, as well as by performing of the three classifiers explained in Sect. 3.3 on the imputed data sets.

- **Data imputation evaluation for non-monotone pattern**

Table 8 shows the GCD values obtained by the different imputation models for non-monotone patterns considered over 12 databases, appearing grouped by the type of variables. In general, the values for GCD are high for all models. MLP offers the highest GCD values in six of the eight data sets including categorical variables, being the second option with values very similar to the preferred models, Regression for contraceptive and Hot-deck for Soybean. The five models provide similar GCD values in the four data sets with only quantitative variables. MLP offers the highest values in two of these data sets, in the other two datasets, regression model shows the best values.

The performance of several classification rules on the imputed data sets was also analyzed. Tables 9, 10 and 11 show the results of the different classifiers for non-monotone patterns applied to each data set, original or imputed. The results for

Table 8 Mean test GCD, non-monotone patterns

DB	Mean/mode	Regression	Hot-deck	MLP	SVM
Glass	99.3	99.4	99.2	**99.9**	**99.9**
Iris	96.4	**97.9**	97.3	94.3	99.5
Pima	98.5	**99.0**	98.7	98.5	91.4
Yeast	67.8	68.6	36.1	**67.7**	38.0
Abalone	57.1	58.8	26.7	**59.0**	48.0
Contraceptive	74.9	**79.8**	66.2	75.7	72.0
Heart	73.9	72.4	80.0	**96.7**	92.2
Zoo	83.7	79.0	93.4	**97.1**	96.9
Breast	32.6	50.7	44.7	**80.0**	49.5
Mushroom	33.9	31.5	44.0	**66.7**	58.8
Solar	55.4	69.0	73.9	**78.4**	66.4
Soybean	63.3	62.0	**83.2**	82.2	82.1

Table 9 Mean of percentage of correctly classified instances with NB

Data set	Original	Mean/mode	Regression	Hot-deck	MLP	SVM
Glass	50.0	51.9	50.9	50.5	51.0	51.0
Iris	94.7	94.0	93.3	94.0	88.0	88.0
Pima	75.1	75.4	75.1	75.0	73.3	73.3
Yeast	58.2	54.9	55.8	56.1	49.2	49.7
Abalone	19.5	17.7	17.4	15.6	20.6	20.6
Contraceptive	50.7	49.2	49.3	48.9	46.0	47.8
Heart	84.8	84.1	83.0	84.1	84.4	81.5
Zoo	95.0	93.0	93.0	94.0	95.0	78.0
Breast	97.5	97.5	97.5	97.4	86.3	65.1
Mushroom	97.3	97.3	97.2	97.3	86.0	61.8
Solar	83.5	83.4	83.2	83.2	84.3	84.3
Soybean	100.0	100.0	100.0	100.0	81.0	98.0

Table 10 Mean of percentage of correctly classified instances with *k*-NN

Data set	Original	Mean/mode	Regression	Hot-deck	MLP	SVM
Glass	66.2	68.6	67.1	68.1	65.2	63.8
Iris	92.0	93.3	92.0	92.0	93.0	92.0
Pima	69.9	69.4	71.0	67.6	67.7	67.7
Yeast	52.3	49.3	49.3	47.4	50.0	50.0
Abalone	18.7	17.7	18.2	17.4	19.5	21.5
Contraceptive	43.0	43.2	43.7	43.4	50.0	53.4
Heart	77.0	75.2	75.9	76.7	76.0	76.3
Zoo	96.0	95.0	96.0	96.0	94.0	82.0
Breast	95.9	95.6	95.7	95.9	86.3	65.1
Mushroom	100.0	100.0	100.0	100.0	89.4	61.8
Solar	83.7	83.7	84.0	84.1	84.3	84.3
Soybean	100.0	100.0	100.0	100.0	77.0	96.0

original data set are only illustrative, because in practice these data would not be available. These values of the original database were compared with the obtained results by the classifiers on classical models-based data imputed and MLP or SVM-based data imputed, offering an idea of how good the estimations performed by the data imputation models are.

The values of these tables correspond to the percentage (%) of correctly classified instances. The first column belong to the obtained results by the classifiers on original data sets, the remaining columns show the obtained results by the classifiers on different models-based imputed data sets: mean/mode, regression, hot-deck, MLP and SVM.

Table 11 Mean of percentage of correctly classified instances with C4.5

Data set	Original	Mean/mode	Regression	Hot-deck	MLP	SVM
Glass	64.8	72.9	73.8	73.3	63.0	62.4
Iris	94.7	94.7	94.7	94.0	94.7	94.7
Pima	72.4	75.1	76.0	75.5	74.5	75.4
Yeast	57.2	57.6	56.2	57.7	48.9	48.7
Abalone	23.8	22.4	23.1	21.6	23.4	24.0
Contraceptive	54.1	54.1	53.7	52.5	53.2	51.6
Heart	74.8	75.9	79.0	75.6	74.4	79.3
Zoo	89.0	90.0	89.0	91.0	88.0	79.0
Breast	92.8	92.6	92.2	94.1	85.3	65.1
Mushroom	100.0	100.0	100.0	100.0	88.6	61.8
Solar	84.3	84.3	84.3	84.3	84.3	84.3
Soybean	98.0	98.0	98.0	96.0	73.0	92.0

Broadly, the obtained precision with original data sets is slightly higher than the obtained with imputed data sets. But with much accuracy, for some databases the MLP or SVM techniques-based-imputed data sets show values higher than the obtained with original data sets.

If we focus on the comparison of imputed data sets considering the data types, as well as we did with the previous evaluation criteria in terms of the global criterion GCD, the results are as follows. For databases with only quantitative variables, there is not a clear difference between the classifiers. For databases with both variable types, qualitative and quantitative, the highest percentages of correctly classified cases correspond to MLP or SVM classifiers. And for databases with only qualitative, neither there is a clear difference between models.

- **Data imputation evaluation for monotone pattern**

Table 12 describes the results for the different methods for missing values imputation considering monotone patterns over the same databases, grouped by type of variables too. This table shows that MLP offers the highest GDC values in seven of the eight data sets with categorical variables, but in data sets with only quantitative variables do not show a so clear winner. MIMLP model offer the highest GCD value in all the data sets, competing with regression in three of them. MLP is the best method only in the Iris data set, but in this case with the same value as MIMLP and regression model. As occurs in the non-monotone patterns, the models provide very similar values in database with only quantitative variables.

Therefore, it is concluded that for data sets with only quantitative variables any of the presented models in this study produce good and similar results, while for data sets with categorical variables the best results are clearly obtained with the MLP-based model.

Table 12 Mean test GCD, monotone patterns

DB	Mean/mode	Regression	Hot-deck	MLP	MIMLP
Glass	98.8	99.4	99.3	99.1	**99.5**
Iris	98.3	**99.5**	99.3	**99.5**	**99.5**
Pima	99.5	**99.6**	99.3	99.5	**99.6**
Yeast	91.3	**93.3**	88.4	93.1	**93.3**
Abalone	94.8	95.2	91.3	89.1	**95.3**
Contraceptive	41.9	53.1	29.6	**88.4**	50.5
Heart	67.0	67.4	78.2	**87.1**	83.0
Zoo	81.4	73.1	92.2	**99.2**	98.0
Breast	59.4	52.4	56.2	**83.8**	66.8
Mushroom	50.4	54.1	57.6	**69.6**	67.1
Solar	92.5	84.0	90.9	**93.5**	**93.5**
Soybean	63.3	61.2	80.5	**90.0**	85.0

Table 13 Mean of percentage of correctly classified instances with NB

Data set	Original	Mean/mode	Regression	Hot-deck	MLP	MIMLP
Glass	48.5	50.0	50.9	50.9	49.5	50.0
Iris	95.3	94.0	88.7	88.7	88.7	88.0
Pima	76.3	76.6	76.2	76.2	75.4	75.5
Yeast	57.7	58.0	58.4	58.6	58.0	58.4
Abalone	20.6	20.7	20.3	20.4	20.9	20.3
Contraceptive	49.1	49.7	50.0	50.0	45.6	49.5
Heart	85.2	83.7	83.7	83.7	84.1	82.6
Zoo	95.0	95.0	94.1	94.1	96.0	96.0
Breast	97.6	97.4	97.5	97.5	97.6	97.5
Mushroom	97.4	97.3	97.5	97.3	97.4	97.4
Solar	83.5	83.3	83.0	83.0	83.1	82.9
Soybean	100.0	100.0	100.0	100.0	100.0	100.0

Tables 13, 14 and 15 reflect the results for monotone patterns obtained from performance of classification rules in the predicted data sets. For each of the classifiers, the values between the data sets originating from different models (original, mean/mode, regression, hot-deck, MLP and MIMLP) are similar to each other. Moreover, the obtained success rate is also similar between the classifiers. In general, a high percentage of missing values is estimated.

If we focus on the comparison of imputed data sets considering the data types, as well as we did with the previous evaluation criteria in terms of the global criterion GCD, the results are as follows. For databases with only quantitative variables, the results in the three classifiers are very similar for all the models, offering the MLP

Table 14 Mean of percentage of correctly classified instances with k-NN

Data set	Original	Mean/mode	Regression	Hot-deck	MLP	MIMLP
Glass	67.3	66.0	66.0	66.4	64.5	66.0
Iris	92.0	89.3	90.7	90.7	91.3	89.3
Pima	70.4	70.0	70.1	69.5	69.4	71.0
Yeast	51.5	50.5	51.2	50.7	52.2	52.7
Abalone	18.4	17.1	17.0	17.5	18.2	17.8
Contraceptive	43.9	43.9	45.0	43.7	51.4	44.3
Heart	78.1	65.8	60.0	50.0	76.7	68.6
Zoo	97.0	77.0	90.0	51.7	97.0	64.6
Breast	96.0	96.0	96.0	96.2	95.6	96.1
Mushroom	100.0	100.0	100.0	100.0	100.0	100.0
Solar	83.9	83.9	83.8	83.7	83.9	83.6
Soybean	100.0	70.0	60.5	60.0	100.0	54.8

Table 15 Mean of percentage of correctly classified instances with C4.5

Data set	Original	Mean/mode	Regression	Hot-deck	MLP	MIMLP
Glass	68.2	66.3	70.6	69.6	72.4	67.8
Iris	95.7	94.0	93.3	94.7	95.3	92.7
Pima	72.9	73.6	72.5	75.3	72.7	73.8
Yeast	59.5	59.1	58.5	58.3	58.6	57.9
Abalone	23.7	22.5	20.7	22.6	22.6	23.3
Contraceptive	51.4	52.1	50.7	50.5	54.9	51.8
Heart	77.4	79.0	78.1	78.5	78.1	72.2
Zoo	92.1	92.1	92.1	92.1	92.1	92.1
Breast	92.4	92.9	92.1	93.1	95.4	92.6
Mushroom	100.0	99.9	100.0	100.0	100.0	100.0
Solar	84.3	84.3	84.3	84.3	84.3	84.3
Soybean	97.9	97.9	97.9	97.9	97.9	97.9

or MIMLP-based imputed data sets the highest percentages. For databases with both variable types, qualitative and quantitative, the highest percentages of correctly classified cases correspond to MLP model. And for databases with only qualitative the MLP model also offers the highest values for monotone patterns.

In general, the results indicate that in terms of prediction and in terms of classification, the ML-based imputed data sets offer the best values.

6 Dealing with Inconsistent Values

This section focuses on the dealing with inconsistent data, the identification and the correction of the values that have been erroneously recorded in a data set. DEI process is a task usually undertaken by data editing and imputation techniques, mostly based on the definition of sets of logical rules to be verified by the records. In traditional DEI models, the construction and management of edits usually involves subject-matter experts, making the DEI process non-automatic, slow and time-consuming. This section describes a detailed example of how machine learning tools can be used to make more automatic the DEI process.

In Sect. 6.1 detection and correction of inconsistency is explained and Sect. 6.2 describes a detailed example of a serie of experiments carried out with the explained models.

6.1 Detection and Correction of Inconsistencies

A traditional approach to DEI is based on the specification of a set of edits to be satisfied by the data. Each edit is defined by a set of logical rules: a record not satisfying all edits is inconsistent. Inconsistent records can be removed from the file, but a better data exploitation can be achieved if an inconsistent record is converted into a valid one [43]. De Waal's paper [8] provided a wide introduction to data editing, while De Waal and Coutinho [9, 10] offered an overview of editing techniques.

Fellegi and Holt [14] proposed a valuable methodological framework to identify the minimal set of values in a record which need to be changed to satisfy all edits. This approach also involved the generation of all rules logically implied by the initial set of rules. However, it suffers from a number of shortcomings, as showed in several works [4, 39]. A first shortcoming of this methodology is the difficulty of application when a very high number of edits is defined, as the generation of implicit edits becomes computationally unfeasible. Secondly, the imputation process can sometimes lead to rare combinations of modified and unmodified responses for the corrected records (implausible responses). Moreover, this procedure works either for categorical data (where logical edits are used) or for numerical data (they require arithmetic edits) because algorithms for generating logical implied edits differ from the algorithms for generating arithmetic implied edits. Therefore, categorical and numerical variables cannot be simultaneously handled using the Fellegi-Holt approach. Another shortcoming is that theory is not worked out for all cases, for example, for some kinds of nonlinear edits such as conditional or mixed edits. Finally, the Fellegi-Holt approach requires the intervention of subject matter experts to define the edits, turning them into slow tasks. According to Manzari [39], the Fellegi-Holt error localization approach can increase the data quality when it is applied to random errors [20] although it is not well adapted to treat systematic errors correctly [4].

The construction and management of these rules usually involves subject matter experts, turning the data editing procedure into a non-automatic process that may be slow and time-consuming. This work offers a viable alternative to existing approaches by being an independent or complementary procedure, and reducing the intervention of subject-matter experts required by rule-based tools.

Usually error correction is realized in a two-step process. First, the erroneous records are identified, and secondly, an imputation phase lets assign appropriate values for one or more fields in those records. We follow a more automatic schema where both steps, error detection and imputation, are included as a unique phase in a machine learning model, solving both subproblems simultaneously, not independently.

This work follows a supervised learning process, where the training process is carried out only once containing both correct and inconsistent values and different variable types are considered. In contrast to what appears in the literature, in which the model is trained on consistent data and the training is performed step by step for each single variable, this approach follows a more automatic process to train MLPs and SVMs. The records that contain errors are not discarded, performing close-to-the-real-world experiments. Not perform the model training neither step by step for each single variable nor on consistent data, as in other work on data imputation [26]. The results show a clear trend towards improving the quality of the perturbed data sets, and reveal the effectiveness of machine learning approaches in the data cleaning process.

6.2 Detailed Example

This section describes a comparison between the methods explained which were applied to detection and correction of inconsistent values. The development of an automated DEI approach ML-based was carried out, applying MLP and SVM methods to the inconsistent sample, using both real and artificial data set.

6.2.1 Empirical Experiments in the Presence of Inconsistent Data

This section proposes a machine learning approaches to improve the level of automation DEI process. An empirical study of the efficiency of this approach is performed, perturbing both real and simulated data sets, and both ANNs and SVMs are trained as machine learning techniques. Given an assumed correct data set, a randomly perturbed version was generated, Sect. 4.2, and a machine learning algorithm was fit on the pairs of inputs and outputs defined by the perturbed and correct data sets. The resulting model was applied on new records for the data set, defining an automatic procedure of error correction. The perturbed data set was the input to the models and the complete and correct data set was the target to be predicted by the models. A non-estimated version and the corresponding correct data set were needed. An empirical test was conduced on ten public data sets using multilayer perceptron and support vector machines, measuring its performance with several evaluation criteria. One of these criteria is based on the study of the improvement of a supervised learning algorithm when it is applied on the predicted data set. Three classification rules were considered for this criterion.

DEI process followed the same scheme of the experiments explained in Sect. 5.4.1, 15 perturbed data sets were generated and for each one a great number of configurations were studied. The same MLP and SVM configurations were tested, 10 values of GCD criterion were available for each of these configurations, computing their mean value. As described above, 15 perturbed data sets, 10-fold validation splits, 7 learning algorithms, 3 sizes of hidden layers, 17 values for the epochs and 5 different initial weights were the total number of different MLP architectures studied for each data set, Fig. 2a.

With regard to SVM, different kernel function types with different values of their parameters were carried out for each perturbed database. 15 perturbed data sets, 10-fold validation splits, 7 kernels, 5 values for ε, 5 values for C and 5 for σ were the total number of different architectures of each database, Fig. 3.

6.2.2 Evaluation Criteria

Several performance measures were computed to evaluate the performance of the DEI models on the previously presented data sets. A first criterion was the percentage of inconsistency errors that were corrected, namely COR, defined as described below.

For a given variable X_j, let x_{ij} be its correct value for the record i, x^d_{ij} the perturbed value and x^e_{ij} the estimated value that is generated by the DEI model, being $i = 1, \ldots, n$ and $j = 1, \ldots, p$.

Let n_A be the number of values in the perturbed data set which are inconsistent:

$$\begin{aligned} A &= \{(i,j) : x_{ij} \neq x^d_{ij}\} \\ n_A &= |A| \end{aligned} \tag{5}$$

and n_B the number of values of A which are properly corrected by the DEI model:

$$\begin{aligned} B &= \{(i,j) \in A : x_{ij} = x^e_{ij}\} \\ n_B &= |B| \end{aligned} \tag{6}$$

COR is defined as the percentage of inconsistency errors corrected by the DEI model:

$$COR = 100 \cdot n_B / n_A \tag{7}$$

However, the DEI process could increase the number of inconsistency errors, so the percentage of inconsistency errors before (BEF) and after (AFT) models application was also computed, with regard to the total number of items in the data set, i.e. the number of records multiplied by the number of variables. Being $I(z) = 1$ if z is true and 0 otherwise, the reduction of the percentage of inconsistency errors was computed as a percentage, namely RPI:

$$\begin{aligned} BEF(X_j) &= \tfrac{1}{n} \textstyle\sum_{i=1}^{n} I(x_{ij} \neq x_{ij}^{d}) \cdot 100 \\ BEF &= \tfrac{1}{p} \textstyle\sum_{j=1}^{p} BEF(X_j) \end{aligned} \tag{8}$$

$$\begin{aligned} AFT(X_j) &= \tfrac{1}{n} \textstyle\sum_{i=1}^{n} I(x_{ij} \neq x_{ij}^{e}) \cdot 100 \\ AFT &= \tfrac{1}{p} \textstyle\sum_{j=1}^{p} AFT(X_j) \end{aligned} \tag{9}$$

$$RPI = 100 \cdot (1 - AFT/BEF) \tag{10}$$

Finally, a global criterion for the whole DEI process (GCD) was computed as the mean of COR and RPI. For each perturbed data set, GCD was the measure used in the 10-fold validation procedure to select the best parameter configuration of both models, such as learning algorithm, number of hidden units and number of epochs for MLP, and ε, C and σ for SVM.

6.2.3 Results and Discussion

This section presents an evaluation of the obtained results. Besides the numerical criteria described in previous section, the performance of the three classifiers explained in Sect. 3.3 on the predicted data sets has also been measured.

Evaluation criteria values for the DEI models Table 16 shows the mean of each criterion for the MLP model. This value was computed over the 15 perturbations realized over each test set, and each value in the table is the mean of the 150 numbers.

Table 16 Mean values of the criteria for the test sets in MLP model

Data set	COR	BEF	AFT	RPI	GCD
Cleveland	58.6	1.4	0.6	57.8	**58.2**
Contraceptive	51.9	0.9	0.4	55.6	**53.8**
Flag	59.2	1.2	0.5	59.2	**59.2**
Heart	58.5	1.3	0.5	58.2	58.4
Zoo	86.1	0.8	0.1	86.6	86.3
HayesRoth	40.5	0.7	0.4	41.6	**41.1**
Lung-cancer	55.5	1.0	0.4	60.0	57.8
Lymphography	67.0	0.7	0.2	68.5	**67.7**
Solar	59.9	0.7	0.3	57.1	**58.5**
Soybean	69.3	0.5	0.2	70.0	69.6

In Table 16 it can be observed that the reduction of the inconsistency errors percentage (RPI) ranges from 41.6% to 86.6%. The columns BEF and AFT show a clear trend towards improving the quality of the perturbed data sets. Moreover, a McNemar test was applied to perform a statistical comparison between the probability of inconsistency errors before and after the application of the MLP-based model. The mean of the significance value over the different repetitions was computed, being for the ten data sets lower than 0.01. Therefore, a statistically significant reduction of the number of inconsistency errors was achieved. The last column in Table 16 shows the global criterion GCD, noticing that it is greater than 53.8% for most of the data sets, except for the HayesRoth data set.

The same data sets were also used to train the SVM model. Table 17 contains the test mean values of the criteria. As observed in this table, the reduction of the percentage of inconsistency errors RPI ranges from 28.6% to 87.5%. A McNemar test was also applied to perform a comparison between the probability of inconsistency errors before and after the application of the SVM-based model. The columns BEF and AFT show a clear trend toward improving the quality of the perturbed data sets, reducing significantly the number of inconsistency errors. The last column of Table 17 presents the global criterion GCD, which was greater than 50% for seven of the ten data sets; the HayesRoth data set generated the lowest values, as occurred in the MLP model and in other works [27].

The comparison of both methods, bolding the best DEI method in Tables 16 and 17, suggests that SVM can offer similar results to MLP. Each method was chosen twice in data sets with mixed types of data: Heart and Zoo for SVM model, and Cleveland and Contraceptive for the MLP model, showing both models the same result for the Flag database. Although the results are very similar, we can establish that the SVM model excelled in two of the data sets with only qualitative variables, Led7 and Soybean, doing likewise the MLP model in three of them.

Performance of classification rules in the predicted data sets As reported in Sect. 5.4.1, 10 test average accuracies were computed for each predicted data

Table 17 Mean values of the criteria for the test sets in SVM model

Data set	COR	BEF	AFT	RPI	GCD
Cleveland	53.4	1.3	1.2	57.7	55.6
Contraceptive	47.9	0.9	0.4	55.6	51.8
Flag	60.0	1.2	0.5	58.3	**59.2**
Heart	65.1	1.3	0.4	69.2	**67.2**
Zoo	86.1	0.8	0.1	87.5	**86.8**
HayesRoth	36.9	0.7	0.5	28.6	32.8
Lung-cancer	66.3	1.0	0.3	70.0	**68.2**
Lymphography	52.2	0.7	0.4	42.9	47.6
Solar	39.5	0.7	0.4	42.9	41.2
Soybean	77.7	0.5	0.1	80.0	**78.9**

set, each average resulting from the 15 perturbed versions. The comparison of the 10 pairs of measures provides some insight into the capacity of a predicted data set to offer a classification performance similar to the one of the original data.
The study of all the classifiers for each data set, comparing the results between the different databases, are shown in Tables 18, 19 and 20. The values correspond to the percentage (%) of correctly classified instances. The first column shows the obtained results by the classifiers on original data sets; the second and third columns show the obtained values by the classifiers on MLP-based predicted and SVM-based predicted data sets, respectively. The difference between columns means the grade in which the editing process is able to recover errors.
The results for original databases are only useful for illustrative purposes, because in practice these data would not be available. These results are compared with the obtained results by the classifiers on machine learning-based predicted data. So, an idea of how good the estimations performed by the data editing techniques is achieved. It is observed that for each of the classifiers, the values between the different data sets (original, MLP and SVM) are similar to each other. Moreover, the obtained success rate is also similar between the classifiers. In general, a high percentage of errors is corrected. The machine learning techniques-based predicted data sets show values higher than the ones obtained with original data sets for some databases. Even though it is true that in other databases the opposite occurs, the obtained precision with original data sets is slightly higher than the one obtained with predicted data sets.
To facilitate the comparisons, Table 21 reproduces the obtained results, previously shown in Tables 18, 19 and 20.
The NB classifier offers the same results for two SVM-based and MLP-based predicted data sets, being the best choice for eight SVM-based predicted databases. For nine SVM-based predicted data sets, k-NN classifier is the best method with the highest values, compared to one database for the MLP model. Finally, the C4.5 classifier is the best solution in three MLP-based edited data sets and in six for the ones SVM edited, offering the same values than SVM in only one database.
If we focus on the comparison of DEI methods considering data types, as well as we did with the previous evaluation criteria in terms of the global criterion GCD, the results are as follows. In the five databases with both variable types, qualitative and quantitative, the highest differences is found in the Zoo database. The SVM model the best method in databases Zoo and Heart, with the same results in Flag. But now k-NN and C4.5 offer better results for MLP in Heart, being very similar for NB classifier. According to GCD criterion, the MLP model is the best method in the other two databases, Cleveland and Contraceptive, which coincides with the C4.5 classifier. But for the k-NN classifier, SVM offers the highest percentage in both databases, being the best model in database Contraceptive for the NB classifier and with equal results than MLP in Cleveland.

The best result in terms of prediction, in the databases with only qualitative variables, is obtained with the SVM model in Lung-cancer and Soybean. Likewise, in terms of classification, the best results are obtained for the classifiers in the

Table 18 Mean of percentage of correctly classified instances with NB

Data set	Original	MLP	SVM
Cleveland	54.4	55.5	55.5
Contraceptive	49.5	45.7	50.1
Flag	46.8	31.6	42.6
Heart	85.2	83.3	83.7
Zoo	95.0	88.2	97.0
HayesRoth	80.3	53.0	79.5
Lung-cancer	55.9	33.0	48.0
Lymphography	85.3	60.7	83.0
Solar	83.5	84.3	84.3
Soybean	100.0	87.2	100.0

Table 19 Mean pf percentage of correctly classified instances with *k*-NN

Data set	Original	MLP	SVM
Cleveland	55.0	50.2	52.1
Contraceptive	43.7	51.5	53.7
Flag	51.0	36.0	51.5
Heart	76.3	78.9	77.4
Zoo	96.0	83.0	98.0
HayesRoth	63.8	59.2	62.3
Led7	70.4	68.8	8.6
Lung-cancer	43.3	51.9	60.0
Lymphography	81.8	60.8	78.5
Solar	83.9	82.7	85.0
Soybean	100.0	85.0	100.0

Table 20 Mean of percentage of correctly classified instances with C4.5

Data set	Original	MLP	SVM
Cleveland	52.5	57.9	52.9
Contraceptive	54.1	54.0	50.1
Flag	62.1	40.0	54.7
Heart	78.5	78.9	77.0
Zoo	91.0	82.0	91.0
HayesRoth	64.6	47.7	67.7
Lung-cancer	53.7	25.0	64.8
Lymphography	72.7	53.3	83.0
Solar	84.3	84.3	84.3
Soybean	98.0	77.0	98.0

Table 21 Mean of percentage of correctly classified instances with all the classifiers

	NB		*k*-NN		C4.5	
	MLP	SVM	MLP	SVM	MLP	SVM
Cleveland	55.5	55.5	50.2	52.1	57.9	52.9
Contraceptive	45.7	50.1	51.5	53.7	54.0	50.1
Flag	31.6	42.6	36.0	51.5	40.0	54.7
Heart	83.3	83.7	78.9	77.4	78.9	77.0
Zoo	88.2	97.0	83.0	98.0	82.0	91.0
HayesRoth	53.0	79.5	59.2	62.3	47.7	67.7
Lung-cancer	33.0	48.0	51.9	60.0	25.0	64.8
Lymphography	60.7	83.0	60.8	78.5	53.3	83.0
Solar	84.3	84.3	82.7	85.0	84.3	84.3
Soybean	87.2	100.0	85.0	100.0	77.0	98.0

SVM-based predicted data sets. MLP provides the best results, in terms of prediction, in HayesRoth, Lymphography and Solar databases. For all the classifiers, the SVM model is the best option in HayesRoth and Lymphography, while SVM and MLP show the same values for NB and C4.5 classifier in database Solar, and SVM-based predicted data set show better results for *k*-NN.

In general, Tables 18, 19 and 20 agree with the results of [24] research, in which a better prediction does not imply a better classification rule, according to one or more measures. This fact supports the necessity of considering a wide set of criteria to measure the quality of DEI techniques.

7 Conclusions

A data editing and imputation methodology based on machine learning models was proposed in this study. The method performance was empirically assessed considering multilayer perceptron and support vector machines for data editing, and for data imputation was used multilayer perceptron and a multiple imputation technique combining multilayer perceptron with $k-$nearest neighbours. These techniques were proved especially effective, improving data quality, as alternative automatic procedures to the other methods, for example imputation classical models or Fellegi-Holt-based methods. Overall, this study was remarkably productive achieving the proposed objectives: filling missing values, correcting inconsistency errors, using categorical and numerical variables simultaneously, reducing the intervention of the subject-matter experts, studying the influence of the parameters with a considerable number of architectures and parameter configurations, and comparing different models, solving the two steps of error detection and imputation, training machine learning models only once with both correct and inconsistent values, no discarding records with errors in inputs.

Several data sets were exposed to a perturbation experiment, and an extensive range of parameter configurations for the machine learning models was explored. Several measures were computed to evaluate the performance of the data editing and imputation models. Machine learning models offer a good enhancement in this more automatic data editing and imputation process. These machine learning models present similar results for the evaluation criteria described, although the main advantage of MLP is that only one model is needed for each data set to be predicted, while SVM requires a model for each variable. Regarding the MLP architecture, simpler MLPs were preferred with a reduced number of hidden units and a small number of training epochs; no learning algorithm can be recommended for network training, although a quasi-newton procedure has little prevalence.

The analyzed data sets could be used to build classification rules. Therefore, it was also measured whether predicted data sets could be used to derive classification models with a similar performance to the one in the original and correct data set. This comparison evidences that the machine learning models show a clear trend to maintain the performance of the classifiers, providing a clear advantage over the MLP model.

Both models make few assumptions about data, are flexible and resilient to noise, although a careful exercise of parameter configurations has to be appropriately followed.

Future works include hybrid models, trying to focus all efforts on the imputation phase, both single and multiple imputation. Another future research topic would be the treatment of missing values in databases varying in time, for examples panel surveys or econometric time series.

References

1. Andridge, R., Little, R.: A review of hot deck imputation for survey non-response. Int. Stat. Rev. **78**(1), 40–64 (2010). doi:10.1111/j.1751-5823.2010.00103.x
2. Aydilek, I., Arslan, A.: A novel hybrid approach to estimating missing values in databases using k-nearest neighbors and neural networks. Int. J. Innov. Comput. Inf. Control **8**(7A), 4705–4717 (2012)
3. Azim, S., Aggarwal, S.: Hybrid model for data imputation: Using fuzzy c means and multi layer perceptron. In: IEEE International Advance Computing Conference, Gurgaon, New Delhi (India) (2014)
4. Barcaroli, G., Venturi, M.: The probabilistic approach to automatic edit and imputation: Improvements of the Fellegi-Holt methodology. In: Quaderni di Ricerca 4/1997. Istituto Nazionale di Statistica, Rome (Italy) (1997)
5. Bishop, C.: Neural Networks for Pattern Recognition. Oxford University Press, Oxford (1995)
6. Chang, C., Lin, C.: LIBSVM: a library for support vector machines. http://www.csie.ntu.edu.tw/~cjlin/libsvm (2001)
7. Coleman, T., Zhang, Y.: Optimization Toolbox 4. http://www.mathworks.com/ (2008)
8. De Waal, T.: Processing of Erroneous and Unsafe Data. Ph.D. thesis, Erasmus University Rotterdam (2003)

9. De Waal, T.: An Overview of Statistical Data Editing. Statistics Netherlands (2008)
10. De Waal, T., Coutinho, W.: Automatic editing for business surveys: an assessment of selected algorithms. Int. Stat. Rev. **73**(1), 73–102 (2005)
11. Demuth, H., Beale, M.: Neural Network TOOLBOX for Use with Matlab. User's Guide. The Math Works Inc. http://www.mathworks.com (1997)
12. Duma, M., Marwala, T., Twala, B., Nelwamondo, F.: Partial imputation of unseen records to improve classification using a hybrid multi-layered artificial immune system and genetic algorithm. Appl. Soft Comput. **13**(12), 4461–4480 (2013). ISSN 1568-4946. doi:10.1016/j.asoc.2013.08.005. http://www.sciencedirect.com/science/article/pii/S156849461300269X
13. Euredit. Interim Report on Evaluation Criteria for Statistical Editing and Imputation. http://www.cs.york.ac.uk/euredi (2005)
14. Fellegi, I., Holt, D.: A systematic approach to automatic edit and imputation. J. Am. Stat. Assoc. **71**(353), 17–35 (1976)
15. Feng, H., Liao, M., Chen, G., Yang, B., Chen, Y.: SVM and reduction-based two algorithms for examining and eliminating mistakes in inconsistent examples. In: Proceedings of 2004 International Conference on Machine Learning and Cybernetics, 2004, vol. 4, pp. 2189–2192 (2004). doi:10.1109/ICMLC.2004.1382161
16. Fessant, F., Midenet, S.: Self-organising map for data imputation and correction in surveys. Neural Comput. Appl. **10**(4), 300–310 (2002)
17. Frank, A., Asuncion, A.: UCI machine learning repository. http://archive.ics.uci.edu/ml (2016)
18. García-Laencina, P., Sancho-Gómez, J., Figueiras-Vidal, A., Verleysen, M.: *k* nearest neighbours with mutual information for simultaneous classification and missing data imputation. Neurocomputing **72**(7-9), 1483–1493 (2009). ISSN 0925-2312. doi:10.1016/j.neucom.2008.11.026
19. García-Laencina, P., Sancho-Gómez, J., Figueiras-Vidal, A.: Classifying patterns with missing values using multi-task learning perceptrons. Expert Syst. Appl. **40**(4), 1333–1341 (2013). ISSN 0957-4174. doi:10.1016/j.eswa.2012.08.057
20. García-Rubio, E., Peirats, V.: Evaluation of data editing procedures: results of a simulation approach. In: Statistical Data Editing Methods and Techniques Volume No. 1, Conference of European Statisticians, Statistical Standards and Studies No 44, New York (USA) and Geneva (Switzerland), 1994. United Nations Statistical Commission and Economic Commission for Europe
21. Gautam, C., Ravi, V.: Counter propagation auto-associative neural network based data imputation. Inf. Sci. **325**, 288–299 (2015). ISSN 0020-0255. doi:10.1016/j.ins.2015.07.016. http://www.sciencedirect.com/science/article/pii/S0020025515005083
22. Gower, J.: A general coefficient of similarity and some of its properties. Biometrics **27**(4), 857–871 (1971)
23. Gunn, S.: Support vector machines for classification and regression. Technical report, University of Southampton, England (United Kingdom) (1998)
24. Hruschka, E., Hruschka, E., Ebecken, N.: Bayesian networks for imputation in classification problems. J. Intell. Inf. Syst. **29**(3), 231–252 (2007). ISSN 0925-9902. doi:10.1007/s10844-006-0016-x
25. Hsu, C., Chang, C., Lin, C.: A practical guide to support vector classification. Technical report, Department of Computer Science, National Taiwan University, Taiwan (2010)
26. Jerez, J.M., Molina, I., Garca-Laencina, P.J., Alba, E., Ribelles, N., Martn, M., Franco, L.: Missing data imputation using statistical and machine learning methods in a real breast cancer problem. Artif. Intell. Med. **50**(2), 105–115 (2010). ISSN 0933-3657. doi:10.1016/j.artmed.2010.05.002. http://www.sciencedirect.com/science/article/pii/S0933365710000679
27. Jiang, Y., Zhou, Z.: Editing training data for knn classifiers with neural network ensemble. In: Lecture Notes in Computer Science, vol. 3173, pp. 356–361. Springer (2004)
28. Kalteh, A., Hjorth, P.: Imputation of missing values in a precipitation-runoff process database. Hydrol. Res. **40**(4), 420–432 (2009)
29. Kaya, Y., Yesilova, A., Almali, M.: An application of expectation and maximization, multiple imputation and neural network methods for missing value. World Appl. Sci. J. **9**(5), 561–566 (2010). ISSN 1818-4952

30. Koikkalainen, P.: Neural networks for editing and imputation. In: DataClean 2002 Conference, Jyväskylä (Finland) (2002)
31. Laaksonen, S.: Traditional and new techniques for imputation. J. Stat. Transit. **5**(6), 1013–1035 (2002)
32. Larsen, B., Madsen, B.: Error identification and imputations with neural networks. In: Working Paper No 26, UN/ECE Work Session on Statistical Data Editing. Conference of European Statistics, Rome (Italy) (1999)
33. Lin, P., Thapa, N., Omer, I., Liu, L., Zhang, J.: Feature selection: a preprocess for data perturbation. IAENG Int. J. Comput. Sci. **38**, 168–175 (2011)
34. Lin, P., Zhang, J., Omer, I., Wang, H., Wang, J.: A comparative study on data perturbation with feature selection. In: Lecture Notes in Engineering and Computer Science: Proceedings of the International MultiConference of Engineers and Computer Scientist 2011, IMECS, vol. 1, pp. 454–459 (2011)
35. Little, R., Rubin, D.: Statistical Analysis with Missing Data. Wiley, New York (1987)
36. Liu, X., Gao, C., Li, P.: A comparative analysis of support vector machines and extreme learning machines. Neural Netw. **33**(0), 58–66 (2012). ISSN 0893-6080. doi:10.1016/j.neunet.2012.04.002
37. Luengo, J., García, S., Herrera, F.: On the choice of the best imputation methods for missing values considering three groups of classification methods. Knowl. Inf. Syst. pp. 1–32 (2011). doi:10.1007/s10115-011-0424-2
38. Madsen, B.: Data editing and imputation at statistics denmark. In: Working Paper No 26, UN/ECE Work Session on Statistical Data Editing. Conference of European Statisticians, Prague (Czech Republic) (1997)
39. Manzari, A.: Combining editing and imputation methods: an experimental application on population census data. J. R. Stat. Soc. Ser. A **167**(2), 295–307 (2004)
40. Mitchell, T.: Machine Learning. Computer Science Series. McGraw-Hill International Editions, New York (1997)
41. Nordbotten, S.: Editing statistical records by neural networks. J. Off. Stat. **11**(4), 391–411 (1995)
42. Nordbotten, S.: New methods of editing and imputation. In: Washington D.C. International Statistical Institute, editor, Agriculture Statistics 2000, The Haag (Netherlands), (1998)
43. Nordbotten, S.: Evaluating efficiency of statistical data editing: general framework. In: Conference of European Statisticians, Methodological Material, Geneva (Switzerland), United Nations Statistical Commission and Economic Commission for Europe (2000)
44. Patil, D., Bichkar, R.: Multiple imputation of missing data with genetic algorithm based techniques. IJCA Spec. Issue Evol. Comput. (2), 74–78 (2010). Published by Foundation of Computer Science
45. Puerta, A.: Imputación basada en Árboles de Clasificación. Cuadernos Técnicos, EUSTAT (2001)
46. Quinlan, J.: C4.5: Programs for Machine Learning. Morgan Kaufmann Publishers, Inc., San Mateo (1993)
47. Rahman, M., Islam, M.: Missing value imputation using decision trees and decision forests by splitting and merging records: two novel techniques. Knowl.-Based Syst. **53**, 51–65 (2013). ISSN 0950-7051. doi:10.1016/j.knosys.2013.08.023
48. Rubin, D.: Inference and missing data. Biometrika **63**(3), 581–592 (1976)
49. Rubin, D.: Multiple Imputation for Nonresponse in Surveys. Wiley, New York (1987)
50. Sarle, W.: Neural network FAQ. Periodic posting to the usenet newsgroup comp.ai.neural-nets ftp://ftp.sas.com/pub/neural/FAQ.html (2002)
51. Schuschny, A., Soto, H.: Guía metodológica: diseño de indicadores compuestos de desarrollo sostenible. Naciones Unidas, Comisión Económica para América Latina y el Caribe (CEPAL), Santiago de Chile (Chile) (2009)
52. Silva-Ramírez, E., Pino-Mejías, R., López-Coello, M., Cubiles-de-la-Vega, M.: Missing value imputation on missing completely at random data using multilayer perceptrons. Neural Netw. **24**(1), 121–129 (2011)

53. Silva-Ramírez, E.L., Pino-Mejías, R., López-Coello, M.: Single imputation with multilayer perceptron and multiple imputation combining multilayer perceptron and k-nearest neighbours for monotone patterns. Appl. Soft Comput. J. **29**, 65–74 (2015). doi:10.1016/j.asoc.2014.09.052. http://www.scopus.com/inward/record.url?eid=2-s2.0-84920696294&partnerID=40&md5=e1ceb18f460914bce29133f3dbc4abd0
54. Song, X., Fan, G., Rao, M.: SVM-based data editing for enhanced one-class classification of remotely sensed imagery. IEEE Geosci. Remote Sens. Lett. **5**(2) (2008)
55. Sonnberger, H., Maine, N.: Editing and imputation in Eurostat. In: Working Paper No 21, UN/ECE Work Session on Statistical Data Editing. Conference of European Statisticians, Cardiff, England (United Kingdom) (2000)
56. Statistics Norway. Data editing with artificial neural networks. In: Working Paper No 10, UN/ECE Work Session on Statistical Data Editing. Conference of European Statisticians, Geneva (Switzerland) (1997)
57. The Official United Nations site for the MDG Indicators. Millennium development goals indicators. http://mdgs.un.org/ (2015)
58. Tusell, F.: Neural networks and predictive matching for flexible imputation. In: Proceedings of DataClean 2002 Conference, Jyväskylä (Finland) (2002)
59. UN/ECE. Glossary of terms on statistical data editing. In: Conference of European Statisticians, Methodological Material, Geneva (Switzerland), United Nations Statistical Commission and Economic Commission for Europe (2000)
60. United Nations: The millennium development goals report 2015. Technical report, United Nations, New York (USA) (2015)
61. Vapnik, V.: The Nature of Statistical Learning Theory. Springer, New York (1995)
62. Witten, I., Frank, E.: Data Mining: Practical Machine Learning Tools and Techniques with Java Implementations. Morgan Kaufmann, San Francisco (2000)
63. Yoon, S., Lee, S.: Training algorithm with incomplete data for feed-forward neural networks. Neural Process. Lett. **10**(3), 171–179 (1999)

Multimodal Transport Network Problem: Classical and Innovative Approaches

Juliana Verga, Ricardo C. Silva and Akebo Yamakami

Abstract This work shows a review about the multimodal transport network problem. This kind of problem has been studied for several researchers who look for solutions to the large numbers of problems relating on the transport systems like: traffic jam, pollution, delays, among others. In this work are presented a standard mathematical formulation for this problem and some other variations, which make the problem more complex and harder to be solved. There are many approaches to solve it that are found in the literature and they are divided according to classical methods and soft computing methodologies, which combine approximate reasoning as fuzzy logic and functional as metaheuristics and neural networks. Each approach has its advantages and disadvantages that are also shown. A novel approach to solve the multimodal transport network problem in fuzzy environment is developed and this approach is also applied in a theoretical problem to illustrate its effectiveness.

Keywords Decision support system · Fuzzy mathematical optimization · Soft computing · Transport system · Fuzzy logic

1 Introduction

Although many policies in metropolitan areas have been created to improve the quality of life of their citizens, they have not kept up with the population growth of these areas yet. The quality of life is an index created to measure the life conditions

J. Verga · A. Yamakami
Department of Telematics, School of Electrical and Computer Engineering,
University of Campinas, Av. Albert Einstein, 400, Campinas, SP 13083-852, Brazil
e-mail: juverga@dt.fee.unicamp.br

A. Yamakami
e-mail: akebo@dt.fee.unicamp.br

R.C. Silva (✉)
Institute of Science and Technology, Federal University of São Paulo,
Rua Talim, 330, Vila Nair, São José dos Campos, SP 12231-280, Brazil
e-mail: ricardo.coelho@unifesp.br; rcoel-hos@dema.ufc.br

C. Cruz Corona (ed.), *Soft Computing for Sustainability Science*,
Studies in Fuzziness and Soft Computing 358, DOI 10.1007/978-3-319-62359-7_14

of a human being, which encompass welfare, social relationships, health, education, purchasing power, housing, sanitation, transport, among others. Nowadays the people try to get a job near from their houses because they do not want to spend a lot of time in the traffic. However, this situation is not usual and many times people need to take more than one transport mode, which belongs to a transport network that can be described as transport by land, maritime and air transports. These transport modes can be used to shop among other things. When a transport network have many modes it is defined as a multimodal transport network.

There are many works in the literature that deal with the multimodal transport network problems and related ones. Some formulations and methods are proposed to solve this kind of problem, which has become a challenge to the administrative agencies that control the traffic in the cities. Not only has the population growth helped to worsen the conditions of the public transport, which do not have sufficient vehicles to serve the citizens, but the amount of private vehicles become the flow on the streets and roads more complicated and stressful. For this reason, finding a solution to multimodal and monomodal transport networks problems and related ones is so important and necessary in order to improve the growth planning of the cities. This growth has to allow a welfare for the people that use both public transports and private vehicles.

Therefore, the main objective of this paper is to show some approaches used to solve the multimodal transport network problem which are found in the literature. There are many other works solving this problem and its variations, however, the selected papers use different formulations and methodologies (e.g. classical methods, methaheuristics, fuzzy methods and hybrid methods) to solve this problem. Another objective is to develop an approach that solves the multimodal transport network problem with uncertain parameters.

This paper is organized as follows: Sect. 2 shows the standard mathematical formulation of the multimodal transport network problem and others variations, which have different constraints that turn this problem more realistic and complex to solve it. Section 3 shows three different type of methods to solve this kind of problem: classical methods, metaheuristics, and fuzzy methods. Section 5 presents a novel approach which combine graph formulation, metaheuristic and fuzzy set theory. To clarify the above development, a theoretical numerical example is solved in Sect. 6. Finally, conclusions are presented in Sect. 7.

2 Formulation of the Problem

Urban transport systems can be formulated mathematically by complex topology and constraints. For this reason, there are many formulations that are harder to be modeled [36].

Up against various ways of modeling the transport systems, we can use a basic formulation based on a network, where each network is related to a transport mode and these networks are connected together to modal transfers. This system can be viewed as a multimodal graph, in which some node represents the places where the consumer chooses between continuing in the current mode or change the mode. Each edge represents a connection between nodes and they can be distinguished in two types: transport or transfer. The transport edges link two nodes in the same mode while the transfer edges link one mode to another.

Following this section a basic formulation for this problem is presented. Some others factors can be included in this basic formulation which transform it into a formulation further difficult to solve.

2.1 Standard Formulation

Let $G = (N, E, M)$ be a graph representing the multimodal transport network, where N is the set of the nodes, E is the set of edges, and M is the set of the transport modes considered. The multimodal transport problem can be formulated mathematically as a linear programming problem as follows:

$$
\begin{aligned}
&\min \quad z = \sum_{m=1}^{M} \sum_{(i,j)\in E} c_{ij}^m x_{ij}^m \\
&\text{s.t} \begin{cases}
\displaystyle\sum_{j:(i,j)\in E} x_{ij}^m - \sum_{j:(j,i)\in E} x_{ji}^m = b_i^m, \forall i \in N,\ \forall m = 1, \dots, M \\
\displaystyle\sum_{m=1}^{M} x_{ij}^m \le u_{ij},\quad \forall (i,j) \in E \\
x_{ij}^m \ge 0,\quad \forall (i,j) \in E,\ m = 1, \dots, M.
\end{cases}
\end{aligned}
\tag{1}
$$

where

- c_{ij}^m is the cost of the edge (i, j) in the mode m;
- u_{ij} is the capacity of the edge (i, j);
- x_{ij}^m is the flow of the edge (i, j) in the mode m;

A modeling often found in the literature is: G can be considered as the union of subgraphs, representing the transport modes considered. For example, if $M = \{\text{car, bus, metro}\}$ then $G = (N, E)$ is such that $G = G_{car} \cup G_{bus} \cup G_{metro}$, where $G_{car} = (N_{car}, E_{car})$ is the private network, $G_{bus} = (N_{bus}, E_{bus})$ is the bus network and $G_{metro} = (N_{metro}, E_{metro})$ is the metro network. Noting that we can have transfer between modes, then $E = E_{car} \cup E_{bus} \cup E_{metro} \cup T$, where T represents the modal transfers.

The cost of each edge is constant in Formulation (1), but the situation of it can also be considered depending on its flow. In this case, the costs are modeled by functions that represent the travel time and there are many works in the literature which deal with this case as follows.

$$\min \ z = \sum_{w \in W} \sum_{(i,j) \in E} t_{ij}(x_{ij}) x_{ij}^w$$

$$\text{s.t} \begin{cases} \sum_{j:(i,j) \in E} x_{ij}^w - \sum_{j:(j,i) \in E} x_{ji}^w = b_i^w, \forall i \in N, \quad \forall w \in W \\ \sum_{w \in W} x_{ij}^w \leq u_{ij}, \quad \forall (i,j) \in E \\ x_{ij}^w \geq 0, \quad \forall (i,j) \in E, \ w \in W. \end{cases} \tag{2}$$

where:

- W is the set of containing all the origin/destination pairs;
- w is a origin/destination pair (O-D);
- x_{ij}^w is the flow of passengers in the edge (i, j) of the pair w;
- $t_{ij}(x_{ij})$ is the travel time in the edge (i, j) depending on the flow (x_{ij}) in the edge;
- u_{ij} is the capacity of the edge (i, j).

According to the travel time function chosen, the formulation above can be non-linear programming problem.

In traffic flow, there are many ways to model travel time as a function. The first travel time function was proposed in 1964 by Bureau of Public Roads (BPR) in the USA. Nowadays, this function is still used to design the transport network [60]. The function BPR is given by:

$$t = t_0 \left[1 + \rho \left(\frac{x}{c} \right)^{\lambda} \right] \tag{3}$$

where

- t is the travel time;
- t_0 is the free flow travel time;
- c is the capacity of the edge;
- x is the flow of the edge;
- ρ and λ are the parameters of the model (normally $\rho = 0.15$, $\lambda = 4$).

According to [60], there are a wide variety of the values to the parameters ρ and λ.

Other function used to estimate the travel time was proposed by [18] and it is given by:

$$t = t_0 \left[1 + \left(\frac{J_D y}{1 - y} \right) \right] = t_0 \left[1 \left(\frac{J_D x}{c - x} \right) \right] \tag{4}$$

where

- t is the travel time;
- t_0 is the free flow travel time;

- J_D is a delay parameter that represents the features of the road and the environmental conditions;
- $y = \frac{x}{c}$, where x is the flow in the edge and c is the capacity of the edge;

The Davidson's function has a disadvantage when we compute the term y because the travel time function can be infinity if the flow is similar to the capacity. For this reason, a modification in the Davidson's function was proposed by [3] as follows

$$t = t_0\left(1 + 0.25r_f\left[z + \left(z^2 + 8J_D\frac{x}{r_f}\right)^{0.5}\right]\right) \tag{5}$$

where

- $r_f = \frac{T}{t_0}$ is ratio of flow (analysis) period to the free flow travel time (T and t_0 must be in the same units);
- J_D is a delay parameter that represents the features of the road and the environmental conditions;
- $z = x - 1$;
- $x = \frac{q}{Q}$ is the degree of saturation;
- q is the demand (arrival) flow rate (in veh/h);
- Q is the capacity (in veh/h).

It is easy to see that Function (5) does not have the same disadvantage than the original Davidson's function but this modification become it more complex. In [3, 22] some travel time function are compared.

2.2 Formulation with Limited Modal Transfer

Some constraints can be inserted in the basic formulation described in Problem (2). One of them is the maximum number of the modal transfers that are permitted for the user. The modal transfers include a set of subjective factors that can vary strongly among different users and the behavior of these users towards the modal transfers can be unpredictable. Under the same conditions, the best path for a specific user cannot be for the other.

In practice, the number of modal transfers that the users are willing to do during a trip is low. In [36, 37], the maximum number of the modal transfers are chosen by the users. Thus, the best path depends on the two criteria: cost and number of modal transfers.

It is clear that the difficulty increases when this constraint is included in the basic formulation of the multimodal transport problem. In this situation the problem has two criteria that must be analyzed to reach the efficient path. Thus, this formulation is more realistic because the users do not want ordinarily to change the mode, but they have to change it due to the lack of other options.

In Problem (2) this constraint can be included in as follows:

$$\text{number of the modal transfers} \leq k \tag{6}$$

This constraint permits that the number of the modal transfers in the path is not greater than k, where k is defined by the users.

2.3 Formulation with Constraints Considering the Order of the Different Modes

The constraint that considers the order of the different modes can be analyzed in the multimodal transport network problem, as described in [11, 36, 37, 41]. Some works takes into account this kind of constraint introducing the viable path conception, which do not violate the set of constraints that defines the sequence of the transport modes used. The available transport modes define this set of constraints that can represent bike, bus, car, metro and others types of transport modes.

For instance, if a person uses his/her car and changes to the bus and/or metro he/she cannot use it to continue the trip again, which is another constraint. This constraint is often utilized to limit the sequence of modes that can be used between private and public transport.

It is easy to observe that this type of constraint depend on the transport modes considered. In some places, like in Brazil, it makes sense to consider that the user cannot change the transport mode if he/she uses a private vehicle. In [36] the multimodal viable path problem with these constraints is formulated.

2.4 Formulation with Time Window

Another constraint that transforms this problem more realistic is when the concept of the time is used whereas the parameters of the problem depend on the time. For example, the waiting time when the user is at the bus stop depends on the timetable and the number of buses in the line. The frequency of the buses in some lines is shorter in the rush hour.

The time constraints can be formulated by several ways because the total cost of the objective function depends on the time. In the same way the constraint related to the timetable of the bus, van, train, and metro can be introduced. Other important constraint is the time window, which restricts the users trip time. These constraints can be easily included in a basic formulation of the multimodal transport network problem.

There are many works in the literature that consider time window. Multiple time windows are considered in [38], which are described in starting and stopping nodes for each commodity.

In [71] the itinerary planning problem is expressed as a shortest path problem in a multimodal time-schedule network with time windows and time-dependent travel times. A dynamic programming-based algorithm has been developed for the solution of the emerging problem. The special case of the problem involving a mandatory visit at an intermediate stop within a given time window is formulated as two nested itinerary planning problems which are solved by the foregoing algorithm.

In [53], the problem of bus service from house to workplace in a metropolitan area is analysed. It is considered the equilibrium among conflicting criteria such as efficiency and equity. Therefore, the authors proposed a multiobjective approach looking upon the equity among time windows which combine the arrival time at the bus stop. The time windows can have other applications such as ensuring the synchronization of the service with other transport modes.

The time window constraints can be inserted in Problem (2) in the following way:

$$a \leq \sum_{(i,j)\in E} t_{ij}(x_{ij}^{w}) \leq b \quad \forall w \in W \tag{7}$$

where an interval of the time window is described as $[a, b]$.

2.5 Formulation with Uncertain Decision Variables

Another factor that can be regarded is the uncertainty in the decision variables. Usually, some parameters that are associated to real problems are uncertain and they can be found in the capacity, cost and demand, travel time, among others [5, 23, 24, 56]. When uncertain parameters are included in the optimization problem, it becomes more realistic and the obtained solution is closer to reality.

The uncertainty in the decision variables and others parts of the formulation can be treated statistically or through the fuzzy set theory. In [61], the supply and the demand of the network are regarded using stochastic variable based on Poisson distribution. According to the authors, these uncertainties represent mainly the adverse climatic conditions with different degrees of impact in the transport modes considered. Other works in the literature also deal with the uncertain using the statistical modeling, as described in [14, 49, 52, 66].

3 Resolution Methods

This research presents a brief description of some methods found in the literature to solve the multimodal transport network problem. The resolution methods can be classified in three groups: (i) classical methods; (ii) metaheuristics and (iii) fuzzy methods. Some advantages and disadvantages for each method are presented.

3.1 Classical Methods

Some classical methods shown in this sub-section are based on classical shortest path algorithm like Dijkstra. Usually, the multimodal transport network problem is modeled by using graph theory that helps to represent the topology of this problem. Thus, some traditional algorithms based on graph theory can be adapted to the multimodal transport network problem.

3.1.1 Methods Based on Graph Theory

In [39], the multimodal shortest path problem in urban transport network is studied. The tasks are to minimize the global cost, travel time and user discomfort, which are conflicting goals. An approach based on classical shortest path problem is presented. It is applied in a network representing multimodal urban transport system, modeling private and public modes and pedestrian. The main idea is to use a utility function that weights the cost and time. The proposed approach was applied in an Italian city.

The transport modes considered are:

- **Private mode**: own car;
- **Public mode**: bus, train, etc.;
- **Pedestrian mode**: walk at the street.

The urban transport network was modeled as graph $G = (N, E)$, where N is the set of n nodes that represent relevant places in the urban area (shoppings, workplace and others) and E is the set of m edges (i, j) linking the i-th node with the j-th node. Each edge have a pair of associated weights (cost and time). Each transport mode is represented by a subgraph, then G is a union of three subgraphs.

The total time is formed by the time that the user spend to travel with the own car between two nodes, the time spent when the user ride on a bus, metro and other public transport modes, the average waiting time for the public transport in a specific node, and the time that the user have to walk. The total cost is formed by cost to travel by own car, which includes the fuel cost, parking cost, and the public transport cost to each public mode.

In [6], the authors focused on determining what nodes can be transformed into an attractive point to change the transport mode seeking new services like parking lot, bus stop and informative panels. In this approach, a path, intermodal or monomodal, is selected by the user that tries to balance the estimated travel cost and some subjective elements like discomfort, walking and economic factors. Tests with some data in the central area of Genoa, Italy, are made. The modelling of this problem is made by using the graph $G = (N, E, C, I, D, M, R)$ with the following specifications:

- N is the set of nodes;
- E is the set of edges;
- C is the set of weights (travel times) associated to each edge $(i, j) \in E$;
- I is the set of weights (costs of mode transfer) associated to each node;

- D is the set of the associated qualities with the nodes of modal transfer;
- M is the set of enabled transport models in the considered networks, which can be divided in two different classes:
  - Restricted model like private vehicles (car and motorcycle), which can be used only by decision maker;
  - Collective model like bus and train, which transport many users with one only vehicle.
- R is the set of possible decision criteria that evaluate the total cost for whole trip in a multimodal network.

The proposed algorithm is an heuristic approach that obtains good multimodal routes in urban transport network problem. This algorithm identifies the nodes that belong to the best transition mode and compute the minimum cost based on the classical Dijkstra's algorithm [19].

In [12], some important issue are presented that help to develop a dynamic approach, which reach the minimum travel time in multimodal routes combining feasible transport modes. This approach computes the estimated travel time and it can be used to obtain a solution multimodal routing problem. In this context, the best solution obtained includes the round trip and it is not equivalent to the best sequence that is reached to one way.

Other type of modeling is to use a directed graph $G = (N, E)$ with associated cost function for each edge $(i, j) \in E$ which can describe the travel time, distance, and others. The multimodal network was represented by a structure with some layers. Each layer features a transport mode and the layers are connected through a transfer edges. To solve such problem, a proposed algorithm, is based on classical ones that set labels, like Bellman [9] and Dijkstra [19]. This proposal can compute the multimodal minimum time for one way trip under a dynamic environment. It was also extended to compute the solution for round trip in the same environment. The proposed approach is tested in a urban transport network of Lyon, France.

In [33], highway and railway networks are based on scanned real maps from Denmark, Hungary, New Zealand, Norway, and Spain. These maps was obtained by GIS (Geographical Information System) and these networks are modeled through colored graphs. The cardinality of the obtained set is analyzed and it was concluded that the connectivity of the nodes and the shape of network affect quite the total number of optimal paths. The authors use graph coloring to represent specific attributes of the transport network such as a transport mode. A colored graph is define as $G = (N, E, \omega, \lambda)$, where (N, E) is a directed graph, $\omega : E \rightarrow \mathbb{R}^+$ is a weight function of the edges, and $\lambda : E \rightarrow M$ is the color function in the edges. In this case, M is a finite set of colors with $K = |M|$.

In [13], a dynamic algorithm of label setting is developed obtaining a multimodal shortest path to one way trip. This algorithm is an adaptation of classical Dijkstra's algorithm. Another algorithm, which is based on this dynamic version, was also proposed and used a strategy to solve the shortest path problem to a round trip. This last proposed strategy is important because the optimal solution to the multimodal transport problem to a round trip is not equal to obtain two optimal solutions for each way.

In [30], a clustering technique is described improving the performance of the conventional computation of shortest k-paths in multimodal transport networks. This network is transformed into an acyclic one and each cycle is identified and clustered. The shortest k-paths are obtained applying the generalized Floyd's algorithm, which computes the shortest path for each pair of nodes in a graph, in the clusters. According to some numerical experiments, this proposal improves significantly its performance in comparison with conventional algorithms, specially when the number of shortest k-paths increases.

Dijkstra's algorithm solves the shortest path problem in a directed graph or not but the costs have not to be negative. This algorithm is easy to be implemented and used. The approaches based on it keep the same features, then its simplicity of implementation is an advantage. Another advantage is that in this approach it is easy to interpret the results and then they can be compared with other models. Although these approaches are simple they do not treat the subjective features that can be found in the real life which is a disadvantage.

3.1.2 Methods Based on Optimization Techniques

In [35], the multimodal network problem is transformed into a multicommodity optimization problem and an algorithm is proposed, which is a nonlinear structure with two layers. The objective function minimizes the costs of transport for each user and the environmental impact produced from utilizing inefficient transport modes. The proposed algorithm is an heuristic based on column generation techniques and decomposition procedures, which enable to solve large scale problems in reasonable time. This heuristic is divided in main problems and subproblems using the column generation. So, this problem can be described as a multicriteria problem with two objective functions: (i) minimizing the transport cost and (ii) minimizing the necessary investment cost for the selected routes. The subproblems represent a submodel that build the new routes and they are considered in the main problem. This submodels are solved using the modified approach from Dijkstra's algorithm [19]. Some network optimization problems can be highlighted (like shortest path, traveling salesman, vehicle routing, facility location, traffic balance, among others) because they can be included in the multimodal network transport problem.

In [31, 32], the objective is to reduce the delay of the trips or improve the social benefits. For this reason, the authors propose an optimization model and it can be formulated as follows:

$$\min_{v,x,\beta,\rho} \quad t(v)^T v$$

$$\text{s.t} \begin{cases} v = \sum_{w} x^w \\ Ax^w = E^w d_w \quad \forall w \\ t_{ij}(v) + \beta_{ij} \geq \rho_i^w - \rho_j^w \quad \forall w, \ (i,j) \in L \\ x_{ij}^w (t_{ij}(v) + \beta_{ij} - \rho_i^w + \rho_j^w) = 0 \quad \forall w, \ (i,j) \in L \\ \rho_{o(w)}^w - \rho_{d(w)}^w \leq c_w^{UE} \quad \forall w \\ 0 \leq x_{ij}^w \leq x_{max}, \ 0 \leq \beta_{ij} \leq \beta_{max}, \quad \forall (i,j) \in L, \ \forall w. \end{cases} \tag{8}$$

where: UE is user equilibrium and c_w^{UE} is the travel time under a user equilibrium for users of OD pair w. $E^w \in R^m$ is input-output vector, e.g., a vector that has exactly two non-zero components, has a value 1 in the component corresponding the origin node and the other has a value 1 in the component for the destination. d_w denotes the travel demand for OD pair w. β is the toll vector, each of which is measured in units of time. ρ_i^w is the KKT multiplier associated with the flow balance constraint at node i and $t_{ij}(\cdot)$ represents the travel time function for link (i, j).

The goal is to minimize the total travel or delay time in the system. The first constraint represents the flow, the second guarantees that the graph is balanced, the third and the fourth constraints are the KKT optimality conditions associated with the problem satisfying the equation $(t(v) + \beta)^\top (u - v) \geq 0 \ \forall u \in V$, where V is a set which satisfies the first and second constraints and its elements are nonnegative. The fifth is related to the path cost, and the last constraint is the nonnegative of the flow. The variables are bounded by the intervals $[0, \beta_{max}]$ and $[0, x_{max}]$, where β_{max} is a sufficiently large positive constant (e.g., $\beta_{max} = \sum_w c_w^{UE}$) and x_{max} is sufficiently large, e.g., $x_{max} = \sum_w d_w + \epsilon, \ \epsilon > 0$. As stated above, ρ_i^w is unrestricted.

A new algorithm is proposed and converges for a stationary solution in a finite number of operations. This algorithm can be apply in problems where the demand is fixed or variable. The efficiency of this approach was proved computationally in some large scale networks.

In [65], it is proposed, a strategy to decrease the traffic jam in multimodal transport networks that includes traffic services, tolls, high occupation in the road, among others. In this scenario, the price regime refers to a strategy to toll in roads and highways and setting the rates in several transit lines. In addition, this regime tries to maximize the social benefit without increasing outlays related to trips that include transport authority, transit passengers, etc. In this work it is regarded three transport modes: single occupancy vehicle (SOV), high occupancy vehicle (HOV), and traffic. The general multimodal network consists in two subnetworks: vehicles and traffic networks. These networks form a multimodal network $G = (N, E)$, where N is the set of nodes and stops, and E is a set of edges and traffic lines when they are integrated. Each edge is associated to a delay time and a travel time in vehicle. In this study the problem was transformed into the optimization problems with complementary constraints and was solved by the algorithm developed in [31].

There are many other techniques based on optimization that are used to solve transport network problems, like linear, nonlinear, integer and continuous programming, complementary problems, column generation, among others. These techniques can model clearly the problem to be solved. They allow a consistent analysis of the obtained solutions, which is an advantage. However, these approaches have the same disadvantage than the approaches based on graph theory because they do not treat the subjective features that can be found in real life.

3.1.3 Other Classical Methods

In [36], an approach that uses labels to find the viable shortest path in multimodal transport network is proposed. A path is viable if the sequence found do not violate at least one function in the set of constraints. In this work, a modified algorithm based on Chronological algorithm which solves the multimodal viable shortest path problem developed in [50] is presented. The obtained results are in a set of nondominated solutions and the best path depend on preferences of the user towards the cost and number of modes. The modes that was considered are: bus, metro, private vehicles (car and motorcycle), and walk. Nevertheless, the metro and private vehicle modes are subject to limitations. In this case, the task is to find the shortest path towards the total cost, which is formed by the edge costs and cost of changing in mode, for each time there is a change in mode it is not great than k, which is the maximum number of modal transfer that the user is willing to do. This way, the two criteria used to choose the best path are the cost and the number of changing in mode.

In [37], the concepts of multimodal hypergraph and viable hyperpath are presented, and this work generalizes the proposal shown in [36]. The viable shortest hyperpath problem for multimodal transport network is defined and the obtained solutions belong to a set of Pareto-optimal shortest hyperpaths. These solutions represent the personal preferences for each user towards expected travel time and the maximum number of modal transfer, which the user is willing to do. A hypergraph is a generalization of a graph in which an edge can connect any positive number of nodes. The hypergraph is a pair $H = (N, E)$, where N is the set of nodes and E is the set of h-edges. A h-edge $e = (t(e), h(e))$ is identified by its border $t(e) \in N$ and its roof $h(e) \subset N/t(e)$. If $|h(e)| = 1$ the h-edge is equivalent to an edge $e = (i, j)$. The objective is to obtain the viable hyperpaths with the minimal expected travel time where the user does not have to do more than k modal transfers. The proposed algorithm to solve this problem is based on what was the developed in [36].

In [11], a multimodal transport system is designed to meet with the necessities of a range of supply/demand. In urban trips, the passengers have to ride a bus, metro, and private vehicle while intercity trip they can use bus, private vehicle, and train. An algorithm was developed to reach a solution to planning problem to long term. This proposal recognizes the set of constraints formed by the delay time and the sequence of used modes in a trip. The objective is provide a tool to detect the facilities in using different transport modes in a trip. Geographic Information Systems are essential to obtain the effective cost and validation of the sequence of used modes and selected path. This proposed algorithm also deals with the time constraint which represents the discontinuity time and the delays in transfer points. A software with graphical interface is presented in this study and the user can obtain some information as routes, timetable, among others. The traffic network is modeled through a directed graph $G(N, E)$, where N represents objects of the network (bus stop, parking,...) and E represents the edge that connects two nodes in N. A modified shortest k-path algorithm is proposed and it composes multimodal viable paths and time constraint. The authors define a set of dominated solutions, that compare time and number of changes among modes, and this set contains the first k solutions that satisfy these objectives.

In [41], a method that uses data from GIS (Geographic Information Systems) tool is proposed. This method handles information to work with multimodal routes between an origin and a destination. The objective is to provide a trip model that helps the user making a decision, which involves different combinations of transport mode, and choosing the best route for him/her. The user can access information about the available public transport modes in a specific area, find route faster and safer for the destination. In this case, the transport problem, which also consider the time constraint is formulated by subgraphs where each subgraph represents a transport mode and the total graph is an union of all subgraphs. The proposed algorithm is based on shortest k-path one, which is modified to solve viable multimodal shortest path problems.

In [71], a new mathematical formulation and an algorithm to solve itinerary planning problems are proposed. The itinerary is to optimize a set of criteria (total travel time, number of modal transfer, total walking time, and total waiting time) that has to satisfy a determined time window between an origin and a destination. This formulation is based on the shortest path problem in a multimodal network with time window.

The authors assume that N is the set of nodes (vertices) that denotes the stops of the urban public transport network. Each service is defined by a sequence of nodes $\{v_1, v_2, \ldots, v_k\}$ which specifies the routes $R_s = \{(v_1, v_2), (v_2, v_3), \ldots, (v_{k-1}, v_k)\}$, whereas the departure from each node v_i is allowed at specified points in time $ST^s_{v_i} = \{\tau^1_{v_i}, \tau^2_{v_i}, \ldots, \tau^{\gamma}_{v_i}\}$ within a time horizon $[0, T]$. The travel time is time dependent, e.g., it depends on the departure time τ from the upstream node v_i and is denoted by $t^{\tau}_s(v_i, v_{i+1})$. The set of arcs formed by the routes R_s,$s \in S$ is denoted with A. In addition, assume $A^{'}$ as the set of walking arcs (v_i, v_j),$v_i \in s_1, v_j \in s_2, s_1 \neq s_2$ which denote the transfer between any two services s_1, s_2.

The departure for traversing any interchange arc $(v_i, v_j) \in A^{'}$ may occur at any point in time, whereas the corresponding walking time that is denoted by $t^{\tau}_w(v_i, v_j)$ is also assumed to be dependent on the departure time from the upstream node v_i. Thus, an urban public transport network can be modeled by a multimodal time-schedule network, which is denoted by $G(N, A, A^{'}, S, ST)$, where $ST = \bigcup_{s \in S} \bigcup_{v_i \in N} ST^s_{v_i}$. It should be clarified that the term multimodal is used in the sense of multiple fixed scheduled transport services. Any itinerary may be written as a sequence of arcs enhanced with the associated departure times and the services used to traverse them, e.g.:

$$p^{\tau}(v_0, v_n) = \{[(v_0, v_1); \tau_{v_0}; s_0], [(v_1, v_2); \tau_{v_1}; s_1], \ldots, [(v_{n-1}, v_n); \tau_{v_{n-1}}; s_{n-1}]\}. \quad (9)$$

The itinerary planning problem relates to the determination of the itinerary that satisfies the following scheduling constraints while optimizing a set of criteria including the total travel time c_1, the number of transfers c_2, and the total time transfer time, e.g., walking and waiting time c_3:

$$\tau_{v_i} + t_s^{\tau_{v_i}}(v_i, v_{i+1}) \leq \tau_{v_{i+1}}; \quad (v_i, v_{i+1}) \in p^t(v_0, v_n); \quad \tau_{v_i} \in ST_{v_i}^s \quad (10)$$

$$d_{v_0}^e \leq \tau_{v_0} \leq d_{v_0}^l \quad (11)$$

$$a_{v_n}^e \leq \tau_{v_{n-1}} + t_s^{\tau_{v_{q-1}}}(v_{n-1}, v_n) \leq a_{v_0}^l \quad (12)$$

Constraint (10) implies that the departure from node v_{i+1} of the itinerary should occur after the arrival from the preceding node v_i. Constraint (11) implies that the departure time from the origin should occur within the time window $[d_{v_0}^e, d_{v_0}^l]$, where $d_{v_0}^e$ and $d_{v_0}^l$ denote the earliest and latest departure times, respectively. On the other hand, constraint (12) expresses that the arrival at the destination node should occur within the time window $[a_{v_n}^e, a_{v_0}^l]$, where $a_{v_n}^e$ and $a_{v_0}^l$ denote the earliest and latest arrival times, respectively. It should be noted that the total time of transfers (waiting and walking time) constitutes part of the travel time of the itinerary, and therefore, these two criteria cannot be considered totally conflicting. The algorithm is based on dynamic programming and solves a special case where there is a required visit in an intermediate stop within a determined time window. This algorithm is integrated with a travel planning system based on web to show a route in the urban public transport in Athens, Greece.

In [63], the multimodal transport networks are modeled by an hierarchical structure. The relationship among different levels is described in details while techniques of dynamic segmentation and linear referencing are used to solve overlapping problems in multimodal networks. Moreover, a shortest path algorithm is proposed to obtain a solution to the transferring problem with many public vehicle modes. Some numerical examples reached from GIS were used to show the efficiency of this proposed approach.

In [42], the transport network system users can choose some modes to travel from any origin to any destination. The objective is to obtain the best route that satisfies the constraints about expected travel time, delays in the modes and the changing points, feasibility sequence of used modes, and the number of modal transfer. The system is described by a multimodal graph and each mode is modeled by a subgraph. The computational complexity of this proposal was proved and an operator of multimodal path based on the proposed algorithm was developed.

In [44], the shortest hyperpath problem is presented. This problem is a extension of the classical shortest path problem and it can be applied in some fields. The algorithm proposed in this work is better than a shortest k-hyperpath problem in a acyclic hypergraph because it improves the computational complexity of the worst case. This is possible due to application of new optimization techniques to the shortest hyperpaths. This algorithm was applied in some practical examples that show its efficiency. These examples are based on stochastic networks dependent on time to find the k best strategies to solve bi-criteria problems, as described in [45, 46].

In [28], it is shown algorithms based on defining labels to find optimal k-paths between two predefined points in a multimodal transport network. These algorithms keep a list of labels for each node. These labels define the paths and the iterative way is updated. The solution depends on the preferences of the users towards distance, travel time and number of modal transfer. The developed project in this work is applied in a multimodal transport network in Mashhad, Iran. This proposal is included in a software called ArcMap.

In [4], a spacial analysis is proposed to find the optimal route between two specific places in a network where the traffic jam moves continually. In this proposal some heuristic functions, which are extracted of graphic features, are used to find a solution in each partition. Lastly, a model to collect traffic data is introduced, then data in real time in different places and schedules can be obtained. Furthermore, this model can helps the users to reach the best route in a urban trip using GIS.

In [69], an algorithm to solve an intermodal optimal path problem dependent on time is proposed. This problem models the multimodal transport network subject to delays in modes and in transfer points. This work shows the convergence proof and computational complexity of this algorithm. A simple representation of the transfer options mode by mode is proposed and it improve substantially its efficiency. Some numerical tests are made in network based on real data and they reach good solutions. The problem is formulated by a graph $G = (N, E, T, M)$, where N is the set of nodes, E is the set of edges, T is the period of discrete time, and M is the set of transport modes.

Some approaches that use hypergraph, geographical information systems, dynamic programming are also used to solve transport network problems. Usually, a set of optimal solutions can be shortest viable path, shortest viable hyperpath, solutions to itinerary planning, among others. This set of solutions is shown to the users and they can choose them according to determined criteria, which is an advantage. A disadvantage is the computational complexity of these algorithms that increase when more constraints and decision variables are considered. Another disadvantage is towards the modeling and the obtained solutions that do not represent actually what happen in the reality due to the uncertain data.

3.2 *Metaheuristic Methods*

In this section some methods based on metaheuristics are presented. These methods can be divided in artificial neural networks, genetic and hybrid algorithms that solve multimodal transport network problems.

3.2.1 Methods Based on Artificial Neural Networks

The artificial neural networks (ANN) are computational systems based on some synaptic connections and join nodes (also called neurons, processors or units) that form a network. Each neuron is responsible for solving a particular part of the problem and then an aggregation operator is used to combine them. The obtained combination is a solution to the whole problem and this methodology guarantees a good approximation for any problem. The original inspiration for this method is based on studies about the brain structure, in particular of the exam of neurons.

In [47], an approach was proposed to compare the descriptive and predictive power of two model classes to estimate flows of multimodal network: the family of models of discrete choice (e.g., logit and probit models), and neural network models. This approach was applied in a large set of data in Europe to two type of products (foods and chemical products). Some computational experiments were done after political and methodological aspects, and modeling problems are exposed. The results showed that predictive models reach better solution than discrete models.

The artificial neural networks are based on the human nervous system, which has an enormous capability to learn. This feature is an advantage because an ANN is flexible with respect to incomplete and noisy data. Before the computational tests are done, the ANN has to pass for the learning phase, which reaches the best values to the parameters for each neuron according to test data. After test phase has finished, the same ANN is applied in new data of the problem to show its efficiency, which is called validation phase. The capability of learning and forecasting results turns the ANN an important option to obtain a solution to many kind of problems, mainly problems with unknown structure. Another advantage has a parallel feature where a solution in real time can be obtained.

Although the idea behind an ANN is easy, the best values of the parameters for each neuron is difficult to set up. Another disadvantage is to choose what the best architecture of ANN to solve a specific problem. The ANNs are compared as a black box during the process of the resolution and they do not allow a simple analysis about the obtained results.

3.2.2 Methods Based on Metaheuristics

Genetic algorithms are a special class of evolutionary algorithms that are inspired in the nature. They are based on Darwin's theory and has among their characteristics the natural selection, heredity, crossover, and mutation. This approach performs a computational simulation to search the best solution through population that represents some points in decision space of the problem. Usually the first generation of the population is generated randomly and some individuals of the population are selected to crossover, mutation, and/or a new population. The new population is used as instance to the next generation and this procedure occurs until a stop criterion is satisfied.

Actually, the genetic algorithms differ of the traditional optimization methods in four aspects:

(i) a set of solutions is obtained in the last generation;
(ii) each individual represents a point in decision space;
(iii) they are based on a stochastic search instead of deterministic rules;
(iv) they do not need to know about the problem but they need a way to evaluate it.

In [1], the multimodal shortest path problem, which is dependent of time, in large and complex urban areas is proposed. An adapted evolutionary algorithm was applied to solve this problem. It has individuals with variable sizes and evolutionary steps. The evolution consists in computing the multimodal shortest path among hundreds pairs of selected origin-destination nodes randomly with different distances. The proposed solution was tested in a set of data in Tehran. The computational testes were done by considering taxi, van, micro-bus, bus and metro as transport modes. In this case, there are three types of paths: monomodal, dual-modal and multimodal. Each solution depends on the time that the user begins the trip because each edge of the network is dynamic.

In [2], the authors applied the genetic algorithm proposed in [1] in a network with 250 pairs of origin-destination nodes, which were selected randomly with different distances. The transport modes considered in this case were bus, metro, and walking. In addition, the paths can be classified as monomodal, dual-modal and multimodal.

In [67], a model to strategic planning was proposed, specially in development of interregional transport network of freight. The model determines a satisfied set of actions, such as improving the existent infrastructure or proposing to build highways, railways, sea routes, and load terminals. The authors model this problem as a two-level programming problem where a technique of multimodal traffic assignment is introduced in the low-level problem. The upper-level problem determines the best combination of actions so the ratio freight-cost is maximized. An heuristic approach based on genetic local search is applied in the upper-level programming problem, which describes a combinatorial optimization. Empirical results show that this proposal obtains the good performance when it is compared with other heuristics. The proposal was applied to a intermodal regional load transport network of large scale in Philippines, where the planning of a load transport network is necessary to increases the use of others modes of transport.

In [7], an approach to the shortest path problem in multimodal time-dependent networks was presented. This approach is based on a strategy called Ant Colony Optimization (ACO), which solved a problem modeled as a graph that is a easy structure without time-dependent. The time-dependent problem is solved with an adaptation of Dijkstra's algorithm and it is used to reach the shortest path to the new structure.

Even though the most of the works presented in this paper were focused on academic environment there are initiatives that were in the financial one. For instance, an algorithm called TIMIPLAN, which combines Operational Research techniques with computational intelligence, was developed to solve multimodal transport problems, described in [21]. This program was applied in a large scale problem with 300 containers, trucks, among other services. It obtains good quality solutions because the objective is to reach a plan that minimizes the service cost of all daily requests.

In [40], an approach that uses two or more modes in the project of multimodal transport networks was presented. This proposal tries to minimize both the cost of the users and of the operators. For users, the objective is to minimize the time walking or waiting for specific transport mode, while the objective for operators is to reduce the total costs to invest more in extending the multimodal transport system. This total cost of operators are described by initial cost service, operation cost, and maintenance cost. Initially, the bus mode was used for the authors and then the van mode was included to analyze two modes. This approach uses ant colony optimizations to obtain the routes, besides using ACO to choose terminal points.

In [43], the transport problem of emergency supplies by using many transport modes was studied. When any type of disaster occurs, there are many places affected and need for a vast amount of emergency supplies. In addition, the transport modes has limited capacity, so many modes must be used to provide these supplies in the affected areas. This problem can be formulated as

$$\min\ T = \sum_{k=1}^{K}\sum_{i=1}^{m}\sum_{j=1}^{n} T_{i,j}^{k}\frac{X_{i,j}^{k}}{b_j}$$

$$\text{s.t}\begin{cases}\displaystyle\sum_{k=1}^{K}\sum_{i=1}^{m} X_{i,j}^{k} = b_j \quad j = 1,\ldots,n;\ \ k = 1,2,\ldots,K \\ \displaystyle\sum_{i=1}^{m} a_i \geq \sum_{j=1}^{n} b_j \quad i = 1,\ldots,m;\ \ j = 1,\ldots,n \\ \displaystyle\sum_{k=1}^{K} S_k \geq \sum_{j=1}^{n} b_j \\ X_{i,j}^{k} \geq 0. \end{cases} \tag{13}$$

where: $a_i,\ i = 1,2,\ldots,m$ is supply quantity, $b_j,\ j = 1,2,\ldots,n$ is demand quantity, $S_k,\ k = 1,2,\ldots,K$ is the maximum transport capacity of mode, X_{ij}^{k} is the actual transport quantity from i to j by mode k, T_{ij}^{k} is time from i to j by mode k.

The objective function has to minimize the weighted time subject to four kind of constraints. The first constraint guarantees that the sum of demands for each transport mode is equal to the demand of the node. The second guarantees that the sum of supply

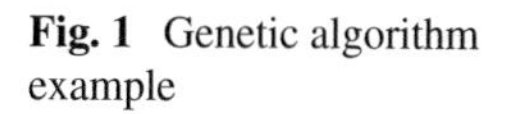

Fig. 1 Genetic algorithm example

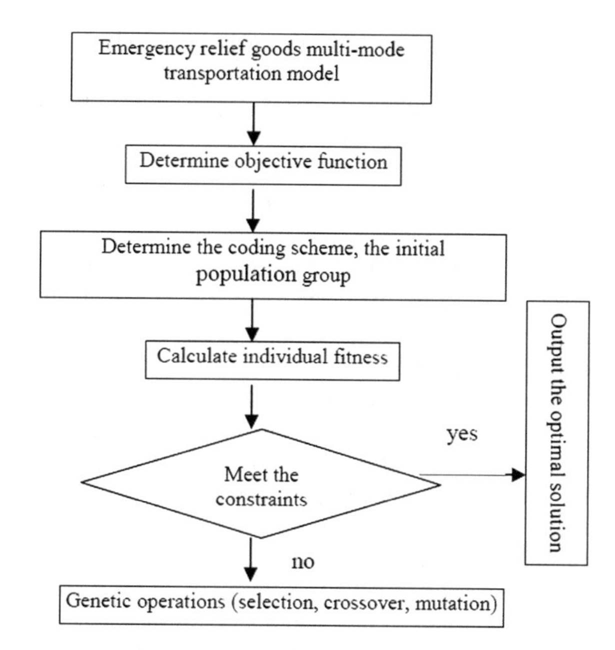

in each origin node is not less than the total demand in each destination node. The third constraint is that the quality of transported products in each mode must not exceed its capacity. The last constraint guarantees the flows is nonnegative. Due to this problem a NP-complete, an optimal global solution or a satisfactory solution is hard to find and then the authors use genetic algorithms to solve this problem. This proposal is applied in a practical example to show its efficiency. Figure 1 shows a design of the proposed algorithm.

In [17], it is presented a method to solve multimodal traffic network problems where the urban demands are elastic. The mathematical formulation considers the relationship among division of modal demand, level of traffic services, and extern transports. The proposal is a combination of heuristics that include a routine to generate routes, a genetic algorithm to find a set of suboptimal routes with associated frequency and many practical rules to improve the solutions. The network performance is estimated for a probabilistic model that simulates the behavior of users when they choose the modes, a model of traffic assignment that simulates behavior of users in relation to choice of bus line, and a model of deterministic user balance that estimates an intersection between the behavior of drivers when they choose the routes and the traffic jam in the edges.

In [64], a genetic algorithm to solve planning problems of multimodal routes is developed. The individuals have variable sizes that are divided in many parts, which represent a type of transport mode. Two new operators are proposed to be used in intermodal way and they are called hiper-crossover and hiper-mutation. A multicriteria evaluation method that uses an p-dimensional vector to represent multiple criteria is inserted in the fitness function to choose the best solutions. The problem was modeled by graph theory, where each transport mode is a subgraph and the hypergraph is the union of all subgraphs. Each edge $e = (i, j)$ has an associated weight $w(e)$ and an p-dimensional vector of criteria:

$$C(e) = (C_1(e), C_2(e), \ldots, C_p(e)), \quad \text{with } e \in R(s,t)$$

where $R(s,t)$ is one route from node s to node t. The value The value of any criterion $k \in (1, \ldots, p)$ for the given route $R(s,t)$ is defined as $C_k^{R(s,t)} = \sum_{e \in R(s,t)} C_k^e$. So the optimal problem can be stated as min $C^{R(s,t)}$. Criteria such as time, transfer, fare and others can be described together in multi-criteria problem. Some computational tests was done with data in Beijing and the proposal reached satisfactory solutions.

In [34], the freight transport problem is formulated by a multimodal transport network. In this context, the commodity transport cost, time and transport quality are influenced when the transport modes are chosen. A virtual network of transports is made, where the original problem is transformed into a shortest path with time and capacity constraints. A genetic algorithm is proposed to solve this problem and it can reduce substantially the complexity of the network through an example, which shows the efficiency and viability of this method.

Genetic algorithms have a great advantage to solve transport problems due to the evaluations of optimal solutions based on stochastic data. By starting off a known set that contains the best solutions it is easier to find the solutions more adapted to the context. However, genetic algorithms need more runtime and/or more computational power which consist in are some disadvantages. They depend on the problem and instance size before the convergence, in order to search for solutions, which can be premature.

3.3 *Fuzzy Methods*

In this section some works that use fuzzy set theory to model the uncertain data of multimodal transport network problems are presented. The uncertainty can be found in the costs of edges, capacity of the edges, supply/demand of a network, and others parameters of the model.

3.3.1 Uncertain Costs in the Edges

In [24], it is presented an algorithm based on uncertain shortest dioid k-paths where is, uncertainty, in the costs of the edges that are described by fuzzy discrete numbers. This proposal obtains a solution to shortest path problem in multimodal transport networks. It considers the number of mode transfer, the correct order of the used modes and the modeling round trip. The modeling uses graph theory, which is composed of many subgraphs and each subgraph represents a transport mode and dioids. The transport modes considered in this work are: bus, metro, private vehicle and taxi. The private vehicle is a restricted mode because the mode is not possible to be changed. In each subgraph the corrected order and availability of the chosen modes are considered. Each one has a level that is defined by the minimum number of transfer edges used from origin to subgraph. In this formulation, the number and variety of the offered services is not limited. The authors are concerned about the number of mode transfer that turns the problem more expensive if the user chooses more than three or four changes.

In [56], a process to choose the routes when each travel time in the edges is a fuzzy number was proposed to solve a fuzzy shortest path problem (FSPA). This uncertain time is called perceived travel time (PTT) and the goal is to find the minimal path with respect to the travel time perceived about saturation level over the routes using traffic jam. The results obtained by this proposal were compared with real data.

In [57], an algorithm that evaluates the perception of the driver with respect to the travel time interferes the route chosen by the user is presented. The proposal that uses a incremental strategy solves traffic assignment problems with uncertain data, which is described by fuzzy set theory to model the perceived travel time by the users. This algorithm is tested in a real network from Mashhad, Iran, and a fuzzy balance is suggested to define the flow in networks. The traffic flow was obtained by FITA (Fuzzy Incremental Traffic Assignment) algorithm and it was compared with other conventional algorithms in real instances. The results showed that FITA was more precise to estimate the traffic flow.

The use of the fuzzy set theory helps to deal with the uncertainty in the real data that is an advantage to solve the real-world problems. In this case, a set of nondominated solutions is obtained because each solution has a satisfaction level that is chosen by the user whereas the best solution depends on the his/her well-being. The complexity is a disadvantage because the problem become more real when the uncertainty is inserted in the formulation. Another disadvantage is how to model this subjectivity as each user has preferences and they are hard to compute and analyzed like traffic, climate conditions, among others.

3.3.2 Uncertain Coefficients of the Demand in the Network

In [23], the minimal fuzzy cost flow problem is presented. The uncertainty can be in the supply or demand of the nodes and also in the cost or capacity of the edges of the network. In this case of supply/demand, the uncertainty represents incomplete stochastic data or obtained by simulations. Three models are presented with different uncertain ways in the formulation: supply/demand, costs and combination of them. The latter is solved by an exact method and heuristic methods. The models were applied in a practical problem made to plan the bus network.

In [51], a new method that find the optimal solution to integer transport problems called separation method is proposed. The uncertainty is represented by an interval that is used to model the transport costs, supply and demand. This method can be an important tool to solve many types of logistical problems with interval parameters. After, this method is extended to use fuzzy numbers instead of intervals and it solves transport problems with fuzzy parameters.

In [5], the objective is to find efficient and established solutions of multiobjective transport problems with fuzzy coefficients $\tilde{c}^r_{ij} \in \tilde{c}^r$ and/or fuzzy supply $\tilde{a}_i$ and/or fuzzy demand $\tilde{b}_j$. The concept of efficient α-fuzzy was introduced in which the ordinary efficient solution is extended based on α-cut of fuzzy numbers. A necessary and sufficient condition to obtain this kind of solution is also established. A parametric analysis is used to feature the parametric optimal solutions to auxiliary problems. An algorithm to determined the established set is developed and it is applied in some numerical examples.

An advantage of the methods presented here is to deal with uncertainty in the supply/demand of the network by using fuzzy set theory. This approach formulates the problems more realistic and robust because it is hard to know exactly the number of the users/commodities/products to be transported among determined points in the practice. In the same way as before, a disadvantage is in increasing the complexity of the problem.

3.3.3 Other Kind of Uncertainties

In [59], the transport planning problem of cement in Taiwan is solved by using fuzzy linear programming methods. Three fuzzy linear methods are used to determined the considerable amount to transport and the capacity of new installations. The goal is to formulate a transport planning problem subject to constraints as port capacity, traffic jam, and demand accomplishment. The used methods in this work are: Zimmermann approach [70], Chanas approach [16], and Julien approach [27]. The obtained results can be used in planning the general infrastructure and the logistic to the cement companies in Taiwan.

In [29], controllers based on fuzzy logic are used to minimize the energy costs in Montreal metro, in Canada. Two fuzzy controllers are presented. The first does not reach practically improvement while the second reaches six per cent of the economy in average and minimizes the average travel time. Many computational tests were done to analyze the performance of two proposed controllers and the results are satisfactory.

In [62], some different transport modes of commodities for a company in Turkey are examined. There are many qualitative and quantitative criteria that are conflicting to evaluate the alternatives of transport modes. Quantitative criteria has often ambiguity and imprecise and they are treated by using fuzzy analytic process. Many criteria can be combined to produce the transport modes more appropriated. This evaluation was proposed to obtain solutions more precise and acceptable. In addition, the used model in this work was compared with the obtained results with the company preferences to validate it.

The fuzzy linear programming is a class of optimization problems where the parameters are not well-known, e.g., the costs of the objective function and the coefficients of the set of constraints are uncertain. The advantage is that some real-world problems are modeled as a linear programming problem and the uncertainty can be modeled by using fuzzy set theory. Again, the high complexity to solve problem more realistic is a disadvantage of this approach. One way to fix this problem is to use the defuzzification methods to obtain a classical number that represents the fuzzy number. However, these methods allow that information is lost during the process, which is a disadvantage.

4 Hybrid Methods

In [54], a hybrid multicriteria decision-making method is developed to find the best route in multimodal transport networks. This method is based on fuzzy analytic hierarchal process and artificial neural networks. When other methods obtain only one solution, this hybrid method reaches a set of nondominated solutions. This set solves the multimodal network problem that is formulated by a multiobjective programming problem with conflicting goals. Thus, the decision maker have many solutions to choose one that is more satisfactory. This proposal is tested in a multimodal network with some transport modes as highway, railway, air and water transports.

In [55], artificial neural networks are used to multicriteria decision-making in multimodal transport networks, which usually are complex and all components must be connected on an efficient way. Since the criteria are conflicting, non-linear and subjective, the transport mode is chosen by using a multicriteria optimization formulation. Some theoretical concepts of feedforward artificial neural networks to solve this problem is developed, that is, a neural network system that uses a predetermined topology for fuzzy analytic hierarchical process. The number of nodes of the artificial neural network adapts the preferences of decision makers. Empirical computational results show that the proposed method is efficient and flexible to select transport modes.

In [8], a hybrid approach is tested in time-dependent real instances that provide an appropriated balance between computational time and memory space. Thus, this proposal can be applied to solve real problems involving many cities, regions or countries. The solutions that solve time-dependent multimodal transport problems can be applied to real networks to minimize the impact of traffic jam about pollution, economy and welfare of citizens. Two previous approaches are compared in relation to theoretical point and experimental performance. This proposal is based on Dijkstra's algorithm and ant colony optimization. It uses the transfer graph modeling that maintains all transport modes in different unimodal networks. Because of this, such networks are easily and independently updated.

In [10], there is an assignment of green vehicles in an existing public transport fleet, in which the traffic network is multimodal with variable demands. Two traffic subnetworks are considered and composed in the following way: the polluted subnetwork contains for traditional vehicles that have permission only to move over roads where environmental protection must not be guaranteed; the clean subnetwork is composed of green vehicles belonging to a limited size fleet that can be driven over routes with high environmental quality and without environmental protection. The main objective of this work is to find an ideal set of traffic network to minimize the operational and user costs. They are composed of elasticity modal demand and availability of green vehicles. The users can choose any network of multimodal system: bus, car and metro. In this case, they can choose the polluted or green vehicles. This proposal includes a routine to generate routes and a genetic algorithm to find a set of suboptimal routes with associated frequency.

In [15], the vehicle routing problem in uncertain environment is solved. This uncertain environment is described by fuzzy set theory and it can deal with travel time, costs of the objective function and set of constraints. A hybrid heuristic algorithm to solve the different types of problems is proposed and it is based on GRASP (Greedy Randomized Adaptive Search Procedure) [58] and VNS (Variable Neighborhood Search) [25]. The obtained results show that the proposed hybrid algorithm is efficient for reach good solutions in a short time.

The hybrid methods combine the best part of some techniques belonging to soft computing which is a collection of metodologies that aim to exploit the tolerance for imprecision and uncertainty to reach tractability, robustness, and low solution cost, as example, fuzzy logic, neurocomputing, metaheuristics and probability it reasoning [68]. Thus, this advantage solves the deficiencies in using only one technique and they enable to build systems stronger and efficient. On the other hand, the hybrid approaches are more complex and the obtained solutions cannot be guaranteed as optimal. This occurs because the worst part of these techniques can be predominated as being a disadvantage.

5 Proposed Approach

The application of the traditional methods has not reached by the realistic solutions because the data normally are uncertain in the real life. Models that take into account the uncertainties are more realistic when these uncertainties are considered in the formulation to be optimized.

In this work, multimodal transport problem is modeled using graph theory and considering the uncertainties.

Each subgraph represents one transport mode. In addition, these subgraphs are connected and they represent the transfer edges. The total graph join all the subgraphs and the transport modes regarded are bus, vehicle, motorcycle and van. The parameters related to the cost and the time are uncertain, which are represented by fuzzy numbers.

Each node represents a place where the user chooses the mode. The options are to continue in the current mode or change. It is clear that there are two type of edges in this model: (i) transport edge, which connects two nodes in the same mode; and (ii) transfer edge, which connects two different modes.

Let be $G = (N, E)$ a graph where N is the set of nodes and E is the set of edges. This graph is divided in different types of transport modes defined in the set M, e.g., $M = \{\text{car (c), motorcycle (mt), bus (b), train (t), metro (me)}\}$. Each edge is represented by (i, j) where $i,\ j \in N$.

The change between on mode to another one is represented by transfer edges that are in the set T. In this case we have

$$E = E_c \cup E_{mt} \cup E_b \cup E_t \cup E_{me} \cup T$$

Here G is described by:

$$G = G_c \cup G_{mt} \cup G_b \cup G_v \cup G_{me}$$

where $G_c = (N_c, E_c)$, $G_{mt} = (N_{mt}, E_{mt})$, $G_b = (N_b, E_b)$, $G_t = (N_t, E_t)$ and $G_{me} = (N_{me}, E_{me})$.

When the user chooses to start the trip by using a private vehicle (car or motorcycle), no transfer edges will be considered.

The fuzzy multimodal transport network problem can be formulated as a fuzzy linear programming problem. However, it is a non-linear one when the cost of each edge depend on its flow. In this case the formulation follows as:

$$\min\ z = \sum_{w \in W} \sum_{(i,j) \in E} \tilde{t}_{ij}(x_{ij}) x_{ij}^w$$

$$\text{s.t} \begin{cases} \sum_{j:(i,j)\in A} x_{ij}^w - \sum_{j:(j,i)\in E} x_{ji}^w = b_i^w, \forall i \in N, \quad \forall w \in W \\ \sum_{w \in W} x_{ij}^w \leq u_{ij}, \quad \forall (i, j) \in A \\ x_{ij}^w \geq 0, \quad \forall (i, j) \in E, \ w \in W. \end{cases} \tag{14}$$

where

- W is the set of all origin-destination pairs;
- w is an origin-destination pair (O-D);
- x_{ij}^w is the flow of passengers of w in edge (i, j);
- $\tilde{t}_{ij}(x_{ij})$ is the uncertain travel time in the edge (i, j);
- u_{ij} is the capacity of edge (i, j).

The travel time is modeled as a cost function proposed by Bureau of Public Roads (BPR) in 1964:

$$\tilde{t}_{ij}(x_{ij}) = \tilde{t}_0 \left[1 + \rho \left(\frac{x_{ij}}{u_{ij}} \right)^{\lambda} \right]$$

where

- $\tilde{t}$ is the travel time fuzzy in edge (i, j);
- $\tilde{t}_0$ is a free flow travel time fuzzy;
- u_{ij} is the capacity of edge (i, j);
- x_{ij} is the flow in edge (i, j);
- ρ e λ are parameters of the model (usually $\rho = 0.15$, $\lambda = 4$).

Without loss of generality, the costs (travel time) are described by fuzzy triangular numbers. Each fuzzy triangular number is represented by a 3-tuple (m, α, β), where m is the modal value, α is the left spreading, and β is the right spreading. The values $(m - \alpha)$ and $(m + \beta)$ are called inferior and superior bound, respectively.

To find the nondominated paths we use the algorithm proposed by Hernandes [26], which is based on the classic Ford-Moore-Bellman algorithm [9] to find the paths nondominated networks, in which costs are fuzzy. This is an iterative algorithm, taking as stopping criteria the numbers of iterations or no alteration in costs of all paths found in the previous iteration with respect to current iteration. Thus, there are found all nondominated paths between origin and destination nodes of a given network, applying the Okada and Soper's order relation [48] to discard the paths dominated.

In the case where the capacities of the edges are fuzzy numbers can proceed as follows:

1. Calculate the possibility of each path to be minimal according to the theory of possibility [20];
2. Calculate the possibility of flow through the edges given path, we make the intersection of these possibilities and take the minimum;
3. Make the intersection between the possibility of path to be minimum, and the minimum of intersection of the possibility of flow pass in the edges of given path and take the maximum, sending flow through the corresponding path.

The flow is sending incrementally according to the ordering performed in the nondominated paths. In each iteration of algorithm, the nondominated paths are calculated, because costs of edges depend of flow, thus one path that is minimal in given iteration may leave be minimal according sending flow.

6 Numerical Example

In order to illustrate how the proposed approach works, a theoretical example which considers two transport modes (bus and metro) is presented. According to Fig. 2, the nodes in gray belong to the bus mode while the nodes in white belong to the metro mode. The dashed edge represents the transfer edge. The demand from origin O to destination D is 20. The edges that connect O to nodes 1 and 5, such as the edges that connect the nodes 4 and 8, to D has zero cost. 'O' represents the origin of trip of users,

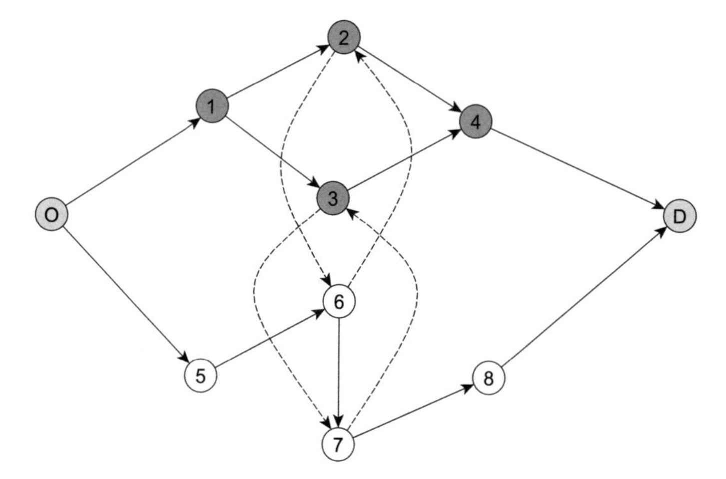

Fig. 2 Illustrative example

Table 1 Data of the network in Fig. 2

Edge	Origin → Destination	$\tilde{t}_0$	Capacities
1	1 → 2	(5, 2, 2)	7
2	1 → 3	(5, 1, 1)	5
3	2 → 4	(4, 1, 1)	5
4	3 → 4	(6, 1, 1)	3
5	2 → 6	(4, 1, 1)	30
6	6 → 2	(4, 1, 1)	30
7	5 → 6	(4, 1, 1)	10
8	6 → 7	(2, 1, 1)	12
9	7 → 8	(2, 1, 1)	15
10	3 → 7	(6, 1, 1)	20
11	7 → 3	(6, 1, 1)	20

that in this case could be made using the bus or the subway, considering is convenient to have bus stops and metro stations close to certain origin trip. 'D' represents the destination of users, that in the practice can be the workplace, recreation, studies, among others.

The capacity and the free flow travel time, t_0, of each edge are shown in Table 1.

The spreading values of the free flor travel time fuzzy are used only initially and they change according to the send flow.

The following paths from O to D are shown below:

- $O \rightarrow 5 \rightarrow 6 \rightarrow 7 \rightarrow 8 \rightarrow D$ with cost (8; 3; 3);
- $O \rightarrow 1 \rightarrow 2 \rightarrow 4 \rightarrow D$ with cost (9; 3; 3);
- $O \rightarrow 1 \rightarrow 3 \rightarrow 4 \rightarrow D$ with cost (11; 2; 2);

- $O \to 5 \to 6 \to 2 \to 4 \to D$ with cost (12; 3; 3);
- $O \to 1 \to 3 \to 7 \to 8 \to D$ with cost (13; 3; 3);
- $O \to 1 \to 2 \to 6 \to 7 \to 8 \to D$ with cost (13; 4; 4);
- $O \to 5 \to 6 \to 7 \to 3 \to 4 \to D$ with cost (18; 4; 4);

Now, the first step is to choose a path to send the flow which is $O \to 5 \to 6 \to 7 \to 8 \to D$. The capacity of edge $5 \to 6$ is 10, then the flow allowed in this path is at most 10.

After sending the flow, the costs in the edges must be updated according to the travel time function described in (3) the following was: $t_{ij} = t_{ij}(x_{ij})$, $t^l_{ij} = t_{ij}[(1 - \mu)x_{ij}]$ and $t^r_{ij} = t_{ij}[(1 + \theta)x_{ij}]$. The μ and θ values belong to the interval [0, 1] and without loss of generality they can be equal to 0.5. Then, when the costs of the edges $5 \to 6$, $6 \to 7$ and $7 \to 8$ are updated we reach: $t_{56} = 4.6$, $t^l_{56} = 4.0375$ and $t^r_{56} = 7.0375$. Thus, the updated cost of the edge $5 \to 6$ is (4.6; 0.5625; 2.4375), while the updated cost of the edge $6 \to 7$ is (2.1447; 0.1357; 0.5877) and of the edge $7 \to 8$ is (2.0593; 0.0556; 0.2407) using the same travel time function.

The following paths and their respective costs are shown below:

- $O \to 5 \to 6 \to 7 \to 8 \to D$ with cost (8.8040; 0.7538; 3.2659);
- $O \to 1 \to 2 \to 4 \to D$ with cost (9; 3; 3);
- $O \to 1 \to 3 \to 4 \to D$ with cost (11; 2; 2);
- $O \to 5 \to 6 \to 2 \to 4 \to D$ with cost (12.6; 2.5635; 4.4375);
- $O \to 1 \to 3 \to 7 \to 8 \to D$ with cost (13.0593; 2.0556; 2.2407);
- $O \to 1 \to 2 \to 6 \to 7 \to 8 \to D$ with cost (13.0593; 4.0556; 4.2407);
- $O \to 5 \to 6 \to 7 \to 3 \to 4 \to D$ with cost (18.6; 4.5625; 6.4375);

The path $O \to 5 \to 6 \to 7 \to 8 \to D$ it is with the exhausted capacity in the edge $5 \to 6$, therefore, can not be used to send flow.

Now, using the path $O \to 1 \to 2 \to 4 \to D$, the maximum flow that can pass this path is 5, due to the capacity of edge $2 \to 4$. Updating the edges costs, we have the edge $1 \to 2$ cost (5.1952; 0.1830; 0.7932) and the edge $2 \to 4$ cost (4.6; 0.5625; 2.4375). After updated of costs, we have the following paths with their respective costs:

- $O \to 5 \to 6 \to 7 \to 8 \to D$ with cost (8.8040; 0.7538; 3.2659);
- $O \to 1 \to 2 \to 4 \to D$ with cost (9.7952; 0.7455; 3.2307);
- $O \to 1 \to 3 \to 4 \to D$ with cost (11; 2; 2);
- $O \to 5 \to 6 \to 2 \to 4 \to D$ with cost (13.2; 3.1250; 6.8750);
- $O \to 1 \to 3 \to 7 \to 8 \to D$ with cost (13.0593; 2.0556; 2.2407);
- $O \to 1 \to 2 \to 6 \to 7 \to 8 \to D$ with cost (13.2545; 4.2386; 5.0339);
- $O \to 5 \to 6 \to 7 \to 3 \to 4 \to D$ with cost (18.6; 4.5625; 6.4375).

Note that the paths $O \to 5 \to 6 \to 7 \to 8 \to D$ and $O \to 1 \to 2 \to 4 \to D$ could no longer be used for sending flow. The paths $O \to 5 \to 6 \to 2 \to 4 \to D$ and $O \to 5 \to 6 \to 7 \to 3 \to 4 \to D$ also cannot be used due to the capacities of the edges $5 \to 6$ and $2 \to 4$. Using the path $O \to 1 \to 3 \to 4 \to D$, the maximum flow that can pass this path is 3, due to the capacity of edge $3 \to 4$. Updating the edges costs, we have the edge $1 \to 3$ cost (5.0972; 0.0911; 0.3948) and the edge $3 \to 4$ cost (6.9; 0.8438; 3.6562).

After sending this flow, we have the following paths with their updated costs:

- $O \to 5 \to 6 \to 7 \to 8 \to D$ with cost (8.8040; 0.7538; 3.2659);
- $O \to 1 \to 2 \to 4 \to D$ with cost (9.7952; 0.7455; 3.2307);
- $O \to 1 \to 3 \to 4 \to D$ with cost (11.9972; 0.9349; 4.0510);
- $O \to 5 \to 6 \to 2 \to 4 \to D$ with cost (13.2; 3.1250; 6.8750);
- $O \to 1 \to 3 \to 7 \to 8 \to D$ with cost (13.1565; 2.1467; 2.6355);
- $O \to 1 \to 2 \to 6 \to 7 \to 8 \to D$ with cost (13.2545; 4.2386; 5.0339);
- $O \to 5 \to 6 \to 7 \to 3 \to 4 \to D$ with cost (19.5; 5.4063; 10.0937).

Now, only the paths $O \to 1 \to 3 \to 7 \to 8 \to D$ and $O \to 1 \to 2 \to 6 \to 7 \to 8 \to D$ can be used for sending the remaining flow (2). Using the path $O \to 1 \to 3 \to 7 \to 8 \to D$, the remaining flow passes fully in this path. Table 2 shows the paths used for sending flow, the updated cost and the total cost of each path.

The demand of network in Fig. 2 was completely fulfilled using four paths according as shown in Table 2 and the overall cost is (200.7537; 15.7072; 68.1654). We note that there was a transfer on the path $O \to 1 \to 3 \to 7 \to 8 \to D$ that occurred between the nodes 3 and 7. In the other paths used for sending flow there are no transfers. The path $O \to 5 \to 6 \to 7 \to 8 \to D$ was used only in the metro mode and the paths $O \to 1 \to 2 \to 4 \to D$ and $O \to 1 \to 3 \to 4 \to D$ was used only in the bus mode.

Despite the works which were shown in this paper dealt with the multimodal transport network problem, the obtained solutions by the approaches presented in these works cannot be directly compared with the our proposed approach for three reasons: (i) these published works usually focus in reaching the shortest path but they are not interested in its flow; (ii) some of them consider changing between private and public modes, although our approach consider the Brazilian case where this changing does not usually happen; and (iii) the most of approaches did not deal with the uncertain data presented from the real problems.

Table 2 Send of flow

Path	Sending flow	Updated cost of path	Total cost
$O \to 5 \to 6 \to 7 \to 8 \to D$	10	(8.8040; 0.7538; 3.2659)	(88.04; 7.5380; 32.6590)
$O \to 1 \to 2 \to 4 \to D$	5	(9.7952; 0.7455; 3.2307)	(48.9760; 3.7275; 16.1535)
$O \to 1 \to 3 \to 4 \to D$	3	(11.9972; 0.9349; 4.0510)	(35.9919; 2.8047; 12.2539)
$O \to 1 \to 3 \to 7 \to 8 \to D$	2	(13.8729; 0.8185; 3.5495)	(27.7458; 1.6370; 7.0990)

7 Conclusion

This work has dealt with the multimodal transport network problems and different mathematical formulations were presented. Some approaches were developed to solve multimodal transport network problems and correlated problems, e.g., multimodal shortest path problem. These approaches tried to use the specific feature for each formulation and they can be classified in three groups: (i) classical methods, which use graph theory and the optimization techniques; (ii) methaheuristics, which combines a formulation by using graph with bio-inspired algorithms and/or artificial neural network; and (iii) fuzzy methods, which deal with the uncertainty presented in real life problems. In addition, some advantages and disadvantages for each type of method is also presented.

Multimodal transport network problem is increasily important nowadays. Each day, more and more cars are in the roads and streets, which promote long traffic jams, delays on product deliveries, lack of parking and other restrictions. Then, approaches found in the literature tries to find a solution to transport problem, despite different methods and different formulations. They seek the optimal solution for at least one of the following objectives: (i) minimize the travel cost; (ii) minimize transport cost; (iii) minimize the travel time; and (iv) minimize modal transfer. Besides, practical problems have some uncertainties which normally are not included in the classical mathematical formulation. For this reason, some works use fuzzy set theory, which is a way to deal with the uncertainties of the practical problems.

Finally, a novel approach that combines graph theory and fuzzy set theory is proposed. This proposal tries to formulate the multimodal transport network problem more realistically. Thus, the objective is to suggest a route with respect to the offered services and transport ways in order to improve the welfare and comfort of the users.

As it can be seen, the network demand in Sect. 6 was complied with using the four paths below:

1. $O \rightarrow 5 \rightarrow 6 \rightarrow 7 \rightarrow 8 \rightarrow D$
2. $O \rightarrow 1 \rightarrow 2 \rightarrow 4 \rightarrow D$
3. $O \rightarrow 1 \rightarrow 3 \rightarrow 4 \rightarrow D$
4. $O \rightarrow 1 \rightarrow 3 \rightarrow 7 \rightarrow 8 \rightarrow D$

The first path represents the metro mode and its flow was whole used, which is the maximum capacity of one of the its edges. The second and third paths represents the bus mode and its flow was also whole used.

Note that a path cannot be used any more when its flow reaches the maximum capacity of one of its edges. Moreover, all the paths which share these edges cannot be also used any more. It is easy to see that this strategy tries to avoid large traffic jams, which are common nowadays.

The fourth path presents a modal transfer between bus mode and metro mode.

The modal transfers are necessary when it is not possible to use only one mode to arrive in the destination. This is normal in big cities. It is obvious the people prefer to minimize these changes and they do them only when it is really necessary.

In order to obtain a mathematical formulation more realistic, the costs of all edges are described by fuzzy numbers, which can be dealt with imprecise data like traffic jam, delay caused by weather conditions, among others.

In this work, the capacity of the edges are not uncertain. However, some studies considering uncertain capacities are being done and they will be the goal in future works. Thus, the modeling become more realistic allowing to violate some restrictions about flow to achieve a better solution.

Acknowledgements The authors want to thank the Brazilian agencies CAPES and FAPESP with project number 2010/51069-2.

References

1. Abbaspour, R.A., Samadzadegan, F.: An evolutionary solution for multimodal shortest path problem in metropolises. Comput. Sci. Inf. Syst. **7**(4), 1–24 (2010)
2. Abbaspour, R.A., Samadzadegan, F.: A solution for time-dependent multimodal shortest path problem. J. Appl. Sci. **9**(21), 3804–3812 (2009)
3. Akçelik, R.: Travel time functions for transport planning purposes: Davidson's function, its time-dependent form and an alternative travel time function. Aust. Road Res. Rep. **21**(3), 49–59 (1991)
4. Alivand, M., Alesheikh, A.A., Malek, M.R.: New method for finding optimal path in dynamic networks. World Appl. Sci. J. **3**(1), 25–33 (2008)
5. Ammar, E.E., Youness, E.A.: Study on multiobjective transportation problem with fuzzy numbers. Appl. Math. Comput. **166**, 241–253 (2005)
6. Ambrosino, D., Sciomachen, A.: A shortest path algorithm in multimodal networks: a case study with time varying costs. In: Proceedings of International Network Optimization Conference, Pisa, Italy (2009)
7. Ayed, H., Galvez-Fernandez, C., Habbas, Z., Khadraoui, D.: Solving time-dependent multimodal transport problems using a transfer graph model. In: Computer & Industrial Engineering (2010, in Press)
8. Ayed, H., Galvez-Fernandez, C., Habbas, Z., Khadraoui, D.: Hybrid algorithm for solving a multimodal transport problems using a transfer graph model. In: UBIROADS Workshop, Tunisia (2009)
9. Bellman, R.E.: On a routing problem. Q. Appl. Math. **16**, 87–90 (1958)
10. Beltran, B., Carrese, S., Cipriani, E., Petrelli, M.: Transit network design with allocation of green vehicles: A genetic algorithm approach. Transp. Res. Part C **17**, 475–483 (2009)
11. Bieli, M., Boumakoul, A., Mouncif, H.: Object modeling and path computation for multimodal travel systems. Eur. J. Oper. Res. **175**, 1705–1730 (2006)
12. Bousquet, A.: Rounting strategies minimizing travel time within multimodal urban transport networks. In: ECTRI Young Researcher Seminar, Torino, Italy (2009)
13. Bousquet, A., Sophie, C., Nour-Eddin, E.F.: On the adaptation of a label-setting shortest path algorithm for one-way and two-way routing in multimodal urban transport networks. In: International Network Optimization Conference, Pisa, Italy (2009)
14. Bovy, P.H.L., Uges, R., Lanser, S.H.: Modeling route choice behavior in multimodal transport networks. In: 10th International Conference on Travel Behaviour Research, Lucerne (2003)
15. Brito, J., Martínez, F.J., Moreno, J.A., Verdegay, J.L.: Fuzzy approach for vehicle routing problems with fuzzy travel time. In: International Conference Fuzzy Systems, Barcelona, Spain (2010)
16. Chanas, S.: The use of parametric programming in fuzzy linear programming. Fuzzy Sets Syst. **11**, 243–251 (1983)

17. Cipriani, E., Petrelli, M., Fusco, G.: A multimodal transit network design procedure for urban areas. Adv. Transp. Stud. Int. J. **10**, 5–20 (2006)
18. Davidson, K.B.: A flow travel time relationship for use in transportation planning. In: Proceedings of the Australian Road Research Board, Conference **3**(1) (1966)
19. Dijkstra, E.W.: A note on two problems in conexion with graphs. Numer. Math. **1**, 269–271 (1959)
20. Dubois, H., Prade, D.: Fuzzy Sets and Systems: Theory and Applications. Academic Press, INC, New York (1980)
21. Flórez, J.E., Torralba, A., García, J., López, C.L., Olaya, A.G., Borrajo, D.: TIMIPLAN: an application to solve multimodal transportation problems. In: Association for the Advancement of Artificial Intelligence (2010)
22. Gattuso, D., Hashemi, S.M.: Estimating running speeds on urban roads. Traffic Eng. Control **45**(5), 182–186 (2004)
23. Ghatee, M., Hashemi, S.M.: Generalized minimal cost flow problem in fuzzy nature: An application in bus network planning problem. Appl. Math. Model. **32**, 2490–2508 (2008)
24. Golnarkar, A., Alesheikh, A.A., Malek, M.R.: Solving best path on multimodal transportation networks with fuzzy costs. Iran. J. Fuzzy Syst. **7**(3), 1–13 (2010)
25. Hansen, P., Mladenovic, N., Moreno, J.A.: Variable neighbourhood search: methods and applications. Q. J. Oper. Res. **6**(4), 319–360 (2008)
26. Hernandes, F.: Algorithms for fuzzy graphs problems. Ph.D. thesis. School of Electrical and Computer Engineering, State University of Campinas (2007)
27. Julien, B.: An extension to possibilistic linear programming. Fuzzy Sets Syst. **64**, 195–206 (1994)
28. Kheirikharzar, M.: Shortest path algorithm in multimodal networks for optimization of public transport. In: XXIV FIG Congress Facing the Challenges Building the Capacity, Sydnei, Australia (2010)
29. Khanbaghi, M., Malham, R.P.: Reducing travel energy costs for a subway train via fuzzy logic controls. In: International Symposium on Intelligent Control, Ohio, USA (1994)
30. Lam, S.K., Srikanthan, T.: Accelerating the K-shortest paths computation in multimodal transportation networks. In: 5th International Conference on Intelligent Transportation Systems, Singapura (2002)
31. Lawphongpanich, S., Yin, Y.: Solving the Pareto-improving toll problem via manifold suboptimization. Transp. Res. Part C **18**, 234–246 (2010)
32. Lawphongpanich, S., Yin, Y.: Pareto-improving congestion pricing for general road networks, Technical report, Department of Industrial and Systems Engineering, University of Florida, Gainesville, Florida (2007)
33. Lillo, F., Schmidt, F.: Optimal paths in real multimodal transportation Networks: An appraisal using GIS data from New Zealand and Europe. In: Proceedings of the 45th Annual Conference of the Operations Research Society of New Zealand, New Zealand (2010)
34. Liu, X., Lin, H.: Optimization model of multimodal transportation mode and its algorithm. In: International Conference on Transportation Information and Safety, pp. 1068–1075 (2011)
35. Loureiro, C.F.G.: Column generation in solving design problems of multimodal transport networks. In: XVII National Meeting of Production Engineering, Porto Alegre-RS (1997)
36. Lozano, A., Storchi, G.: Shortest viable path algorithm in multimodal networks. Transp. Res. **35**, 225–241 (2001)
37. Lozano, A., Storchi, G.: Shortest viable hyperpath in multimodal networks. Transp. Res. - Part B **36**, 853–874 (2002)
38. Moccia, L., Cordeau, J.F., Laporte, G., Ropke, S., Valentini, M.P.: Modeling and solving a multimodal transportation problem with flexible-time and scheduled services. Networks **57**(1), 53–68 (2011)
39. Modesti, P., Sciomachen, A.: A utility measure for finding multiobjective shortest paths in urban multimodal transportations networks. Eur. J. Oper. Res. **111**, 495–508 (1998)
40. Mohaymany, A.S., Gholami, A.: Multimodal feeder network design problem: Ant colony optimization approach. J. Transp. Eng. **138**(4), 323–331 (2010)

41. Mouncif, H., Boulmakoul, A., Chala, M.: Integrating GIS-technology for modelling origin-destination trip in multimodal transportation networks. Int. Arab J. Inf. Technol. **3**, 256–263 (2006)
42. Mouncif, H., Rida, M., Boulmakoul, A.: An eficient multimodal path computation integrated within location based service for transportation networks system (Multimodal path computation within LBS). J. Appl. Sci. **11**(1), 1–15 (2011)
43. Na, L., Zhi, L.: Emergency relief goods multi-mode transportation based on genetic algorithm. In: Second International Conference on Intelligent Computation Technology and Automation, pp. 181–184 (2009)
44. Nielsen, L.R., Andersen, K.A., Pretolani, D.: Finding the k shortest hyperpaths using reoptimization. Oper. Res. Lett. **34**(2), 155–164 (2006)
45. Nielsen, L.R.: Route choice in stochastic time-dependent networks. Ph.D. thesis. Department of Operations Research, University of Aarhus, Dinamarca (2004)
46. Nielsen, L.R., Andersen, K.A., Pretolani, D.: Bicriterion shortest hyperpaths in random time-dependent networks. IMA J. Manag. Math. **14**(3), 271–303 (2003)
47. Nijkamp, P., Reggiani, A., Tsang, W.F.: Comparative modelling of interregional transport flows: Applications to multimodal European freight transport. Eur. J. Oper. Res. **155**, 584–602 (2004)
48. Okada, T., Soper, S.: A shortest path problem on a network with fuzzy arc lengths. Fuzzy Sets Syst. **109**, 129–140 (2000)
49. Palma, A., Picard, N.: Route choice decision under travel time uncertainty. Transp. Res. Part A **39**, 295–324 (2005)
50. Pallottino, S., Scutell, M.G.: Shortest path algorithms in transportation models: classical and innovative aspects. In: Proceedings of the Equilibrium and Advanced Transportation Modelling Colloquium, Klumer (1997)
51. Pandian, P., Natarajan, G.: A new method for finding an optimal solution of fully interval integer transportation problems. Appl. Math. Sci. **4**(37), 1819–1830 (2010)
52. Pattanamekar, P., Park, D., Rilett, L.R., Lee, J., Lee, C.: Dynamic and stochastic shortest path in transportation networks with two components of travel time uncertainty. Transp. Res. Part C **11**, 331–354 (2003)
53. Perugia, A., Moccia, L., Cordeau, J.F., Laporte, G.: Designing a home-to-work bus service in a metropolitan area. Transp. Res. Part B **45**, 1710–1726 (2011)
54. Qu, L., Chen, Y.: A hybrid MCDM method for route selection of multimodal transportation network. In: Proceedings of the 5th international symposium on Neural Networks: Advances in Neural Networks, pp. 374–383 (2008)
55. Qu, L., Chen, Y., Mu, X.: A transport mode selection method for multimodal transportation based on an adaptive ANN System. In: Fourth International Conference on Natural Computation, pp. 436–440 (2008)
56. Ramazani, H., Shafahi, Y., Seyedabrishami, S.E.: A shortest path problem in an urban transportation network based on driver perceived travel time. Sci. Iran. A **17**(4), 285–296 (2010)
57. Ramazani, H., Shafahi, Y., Seyedabrishami, S.E.: A fuzzy traffic assignment algorithm based on driver perceived travel time of network links. Sci. Iran. A **18**(2), 190–197 (2011)
58. Resende, M.G.C., Ribeiro, C.C.: Greedy randomized adaptive search procedure. In: Handbook in Metaheuristics, pp. 219-249. Kluwer (2003)
59. Shih, L.H.: Cement transportation planning via fuzzy linear programming. Int. J. Prod. Econ. **58**, 277–287 (1999)
60. Sreelekha, M.G., Anjaneyulu, M.V.L.R.: Development of link travel time model in mixed mode environment. In: Proceedings of Inter-American Congress on Traffic and Transportation (2010)
61. Sumalee, A., Uchida, K., Lam, W.H.K.: Stochastic multi-modal transport network under demand uncertainties and adverse weather condition. Transp. Res. Part C **19**(2), 338–350 (2011)
62. Tuzkaya, U.R., Önüt, S.: A fuzzy analytic network process based approach to transportation-mode selection between Turkey and Germany: A case study. Inf. Sci. **178**, 3133–3146 (2008)
63. Xin-bo, W., Gui-jun, Z., Zhen, H., Hai-feng, G., Li, Y.: Modeling and implementing research of multimodal transportation network. In: The 1st International Conference on Information Science and Engineering, pp. 2100–2103 (2009)

64. Yu, H., Lu, F.: A multimodal route planning approach with an improved genetic algorithm. Int. Arch. Photogramm. Remote Sens. Spat. Inf. Sci. **38**(2), 343–348 (2011)
65. Wu, D., Yin, Y., Lawphongpanich, S.: Pareto-improving congestion pricing on multimodal transportation networks. Eur. J. Oper. Res. **210**, 660–669 (2011)
66. Wellman, M.P., Larson, K., Ford, M., Wurman, P.R.: Path planning under time-dependent uncertainty. In: Proceedings of the 11th Conference on Uncertainty in Artificial Intelligence (1995)
67. Yamada, T., Russ, B.F., Castro, J., Taniguchi, E.: Designing multimodal freight transport networks: A heuristic approach and applications. Transp. Sci. **43**(2), 129–143 (2009)
68. Zadeh, L.: Soft computing and fuzzy logic. IEEE Softw. **11**(6), 48–56 (1994)
69. Ziliaskopoulos, A., Wardell, W.: An intermodal optimum path algorithm for multimodal networks with dynamic arc travel times and switching delays. Eur. J. Oper. Res. **125**, 486–502 (2000)
70. Zimmermann, H.J.: Fuzzy Sets Theory and its Applications. Kluwer Academic Publishers, Boston (1991)
71. Zografos, K.G., Androutsopoulos, K.N.: Algorithms for itinerary planning in multimodal transportation networks. IEEE Trans. Intell. Transp. Syst. **9**, 175–184 (2008)

A Linguistic 2-Tuple Based Environmental Impact Assessment for Maritime Port Projects: Application to Moa Port

Yeleny Zulueta, Rosa M. Rodríguez and Luis Martínez

Abstract Maritime port operations usually comprise a spread spectrum of environmental challenges, which are often unique to each port site. For this reason, the Environmental Impact Assessment (EIA) process has been developed for evaluating the impact of port operations on the environment, including its natural, social and economic aspects. In this chapter, an EIA model based on a 2-tuple linguistic model is proposed to assess the overall environmental impact of Moa Port in Cuba by using a double matrix that represents impacts, which are characterized by multiple-criteria. The environmental impacts of factors and actions are also ranked from the most to the least risky. This EIA 2-tuple linguistic based model facilitates the handling of inherent uncertainty of criteria involved in an EIA problem by simplifying the sophisticated structure of the problem under consideration while computations are made without loss of information and provide a high interpretability of the EIA results.

1 Introduction

The proper design, construction and maintenance of coastal and marine resources are critical for successful activities of maritime trade, fishing industry and naval defence due to the fact that they heavily depend on the development of ports and harbors. Operations such as creation and deletion of materials, development of beach areas, maritime and vehicular traffic in the harbor, can cause the release of natural and

---

Y. Zulueta
University of Informatics Science, carretera San Antonio Km 2 1/2,
Havana, Cuba
e-mail: yeleny@uci.cu

R.M. Rodríguez (✉)
University of Granada, Calle Periodista Rafael Gmez Montero, 2,
18014 Granada, Spain
e-mail: rosam.rodriguez@decsai.ugr.es

L. Martínez
University of Jaen, Campus Las Lagunillas, s/n, 23071 Jaen, Spain
e-mail: martin@ujaen.es

C. Cruz Corona (ed.), *Soft Computing for Sustainability Science*,
Studies in Fuzziness and Soft Computing 358, DOI 10.1007/978-3-319-62359-7_15

anthropogenic contaminants to the environment [6]. At the same time, actions such as the construction, extension, and operation of seaports; the dredging deepening; the construction of breakwaters, channels and hydraulic fills; the beaches stabilization and coastal waterways; the artificial creation of beaches and dunes and in general, the alteration of natural waters and the construction of artificial structures affect existing water mass, ecosystems and relevant communities near ports [6].

Moa Port plays an important role in the infrastructure of the Cubaniquel Business Group in Cuba. Its main purpose is the reception and storage of products imported for industrial consumption in Nickel enterprises and the export of the final production of these industries. Moreover, it also plays an important natural, cultural and socio-economic role as it provides floral species and existing animal species with a vital environmental performance for the environment maintenance. There is also a significant creation of permanent jobs in both production and administrative areas, reflecting a significant increase in the quality of life of the working population and becoming a cornerstone for economic growth. However, the port maritime activity causes a negative impact to the coastal ecosystem, which is associated not only to the lack of treatment systems, the dredging activity and absence of a proper sewage system; but also to the synergy of the environmental impacts of all the economic activity. Therefore, it is of vital importance to assess the environmental impacts for conducting strict control on the effect of the port maritime activity in order to avoid compromising the ability of future generations to meet their own needs using these resources. To achieve this goal, different models can be used including the Environmental Impact Assessment (EIA) [2, 10].

EIA [21] is a tool for decision makers which takes into account the possible effects of a project on the environment at both construction and operation stage by means of scientific methods and techniques, and develop preventive actions. General objectives of EIA include to ensure that environmental considerations are explicitly addressed and incorporated into the development decision-making process to anticipate, avoid, minimize or offset the adverse significant biophysical, social and other relevant effects of development proposals; to protect the productivity and capacity of natural systems and the ecological processes which maintain their functions; and to promote a form of development that is sustainable and optimizes resource use and management opportunities.

The assessment of environmental impacts in order to make a decision cannot be a straightforward process because a large number of parameters may significantly affect the natural environment. It is necessary to take into account the probable several sources of data including extensive knowledge from multiple experts as well as conflicting issues among the considered criteria. Therefore, stakeholders face a Multi-Criteria Decision-Making (MCDM) problem [9, 10]. An EIA problem can be intuitively modelled as a MCDM problem in which a project is evaluated through the impacts caused by the effect of its interactions with the surrounding environmental factors, and such impacts are assessed considering multiple criteria, usually under uncertain contexts.

Environmental factors are not constant in time because ecosystems are dynamic and highly complex, so detailed measurements and surveys on environmental indicators cannot be undertaken because of areal extent, time constraints, or cost considerations. In other words, potential environmental impacts of projects cannot be quantified precisely, because of the imprecision of environmental impacts and the frequent lack of quantitative data for the model, or because data collection might be too expensive. Even when the environmental components are well defined, and the data to characterize them are established, it may not be practical to spend years measuring these components completely. The use of qualitative data can be helpful to supplement the measured data and to replace missing or incomplete data. It can also be crucial for developing a sufficient basis for supporting a technically sound and legally defensible analysis [21]. In such cases, when in presence of vagueness, subjectivity, fuzziness or incompleteness, one key issue in EIA is to represent and handle various types of uncertainty.

In general, Soft Computing exploits the tolerance for imprecision, uncertainty and partial truth to achieve tractability, robustness and low solution cost. In the case of the environmental impacts, scales for many criteria in EIA are hard to measure and it is not easy for experts to assign exact numerical values for these types of criteria; consequently, representing them with precise numbers does not give credible results. Therefore, solutions based on Soft Computing provide a useful approach to EIA [1, 5, 18, 29]. Fuzzy logic and a fuzzy linguistic approach [26–28] in fact allow the decision makers to have a broader and more realistic perception of the possible project environmental impact.

In such scenario, linguistic modelling and computational models [14, 19] are demanded to capture these linguistic terms within a mathematical framework and facilitate the Computing with Words (CW) [15, 16, 25]. Initially, two classical linguistic computational models were introduced to perform linguistic computations based on fuzzy linguistic approach.

- Linguistic computational model based on membership functions [3, 13] which uses fuzzy arithmetic to accomplish their operations. The results obtained are fuzzy values that usually do not match with any linguistic term from the initial linguistic term set.
- Symbolic linguistic computational model [4, 24] uses an ordered structure of the linguistic term set to operate. The results are numeric values which are approximated to a numerical value that indicates the index of the associated linguistic term.

Both models produce loss of information due to the necessity of an approximation process to obtain linguistic results, and hence a lack of precision in the results. In order to avoid these limitations different approaches have been proposed in the literature. Some of the symbolic linguistic computational models most widely used in linguistic decision making are the 2-tuple linguistic mode [7], the virtual linguistic model [23] and the proportional 2-tuple model [22]. However, the 2-tuple linguistic model seems the most suitable one to deal with linguistic information in decision making problems [20], because it is based on the fuzzy linguistic approach, it is considered in the CW paradigm and its computational model provides accurate and understandable results for human begins. For these reasons, the 2-tuple linguistic model is used in this proposal to solve the Moa Port EIA problem.

The rest of the chapter is set out as follows. Section 2 describes basic notions on the 2-tuple linguistic model and its extensions necessary for our proposal. Section 3 introduces Moa Port environmental problem and provides a linguistic EIA based on 2-tuple linguistic model for its resolution. Section 4 points out some concluding remarks.

2 Preliminaries

This section introduces some necessary basic knowledge regarding the fuzzy linguistic approach, the 2-tuple linguistic model and its extension of linguistic hierarchies that will be used across the proposed linguistic EIA to evaluate the Moa Port environmental impact.

2.1 *The Fuzzy Linguistic Approach*

The fuzzy linguistic approach [26–28] is commonly used to manage the uncertainty and model linguistic information by means of the concept of linguistic variable.

The use of linguistic variables demands the selection of suitable linguistic descriptors for the term set, including the analysis of their granularity of uncertainty, and their syntax and semantics. The granularity, commonly noted as $g + 1$, is determined by the level of discrimination among different counts of uncertainty modelled by the linguistic descriptors in the linguistic term set, $S = \{s_0, \ldots, s_g\}$. Therefore its definition depends on some attributes of the decision making problem. For instance, decision makers, stakeholders and experts participating in an EIA problem may have different degrees of knowledge about impacts, factors and actions involved in a project, consequently their degree of distinction for expressing their knowledge may determine different meaning of words. As a matter of fact, the selection of the syntax and suitable semantics is essential to keep the basis of fuzzy linguistic approach. Further information can be found in [14].

2.2 *2-Tuple Linguistic Representation Model*

The 2-tuple linguistic model was introduced in [7] to improve the accuracy of the CW processes and avoid the loss of information keeping the linguistic basis (semantics and syntax) [20] and providing linguistic results by a simple retranslation process [15, 25]. This model represents the linguistic information by means of a pair of values (s_i, α) [7, 12, 14]:

1. Let $s_i \in S = \{s_0, \ldots, s_g\}$ be a linguistic term whose semantics is provided by a fuzzy membership function and the syntax chosen according to the choices offered by the fuzzy linguistic approach.
2. α is a numerical value, *Symbolic Translation*, that indicates the translation of the fuzzy membership function which represents the closest term, $s_i \in \{s_0, \ldots, s_g\}$.

Definition 1 ([12]) The symbolic translation is a numerical value assessed in [−0.5, 0.5) that supports the "difference of information" between a counting of information β assessed in the interval of granularity $[0, g]$ of the linguistic term set S, and the closest value in $\{0, \ldots, g\}$ which indicates the index of the closest linguistic term in S.

This representation model defines the functions Δ and Δ^{-1} to facilitate the CW processes [7].

Definition 2 ([7]) Let $S = \{s_0, \ldots, s_g\}$ be a set of linguistic terms and $\beta \in [0, g]$ a value supporting the result of a symbolic aggregation operation. A 2-tuple linguistic value that expresses the equivalent information to β is then obtained as follows:

$$\Delta : [0, g] \longrightarrow \bar{S}$$

$$\Delta(\beta) = (s_i, \alpha), \quad \text{with} \quad \begin{cases} i = \text{round}\ (\beta), \\ \alpha = \beta - i, \end{cases} \tag{1}$$

being *round* the usual round operation, i the index of the closest label, s_i, to β, and α the value of the symbolic translation.

Δ is a bijective function [7] and $\Delta^{-1} : \bar{S} \longrightarrow [0, g]$ is defined by $\Delta^{-1}(s_i, \alpha) = i + \alpha$.

A symbolic computation on linguistic terms in S obtains a value $\beta \in \{s_0, \ldots, s_g\}$ that will be transformed into a equivalent 2-tuple linguistic value, (s_i, α) by means of the Δ function meanwhile from a 2-tuple linguistic value, the Δ^{-1} function returns its equivalent numerical value.

2.3 2-Tuple Linguistic Computing Model

The 2-tuple linguistic model defined a computational model based on the functions Δ and Δ^{-1} and introduced the comparison between two 2-tuples linguistic values and several aggregation operators [7].

Let us suppose two 2-tuple linguistic values, (s_k, α_1) and (s_l, α_2), the comparison is as follows:

- if $k < l$ then $(s_k, \alpha_1) \prec (s_l, \alpha_2)$.
- if $k = l$ then

– if $\alpha_1 = \alpha_2$ then $(s_k, \alpha_1) = (s_l, \alpha_2)$;
– if $\alpha_1 < \alpha_2$ then $(s_k, \alpha_1) \prec (s_l, \alpha_2)$;
– if $\alpha_1 > \alpha_2$ then $(s_k, \alpha_1) \succ (s_l, \alpha_2)$.

In the literature can be found different aggregation operators defined for 2-tuple linguistic values [11, 14]. Here only are revised those applied in the EIA of Moa Port.

Definition 3 ([7]) Let $x = \{(s_1, \alpha_1), \ldots, (s_m, \alpha_m)\}$ be a set of 2-tuple linguistic values, the 2-tuple arithmetic mean is the function $2TAM : \bar{S}^m \rightarrow \bar{S}$ defined as:

$$2TAM(x) = \Delta\left(\frac{1}{n}\sum_{i=1}^{m} \Delta^{-1}(s_i, \alpha_i)\right) \quad (2)$$

Definition 4 ([7]) Let $x = \{(s_1, \alpha_1), \ldots, (s_m, \alpha_m)\}$ be a set of 2-tuple linguistic values, and $W = (w_1, \ldots, w_m)$, $w_i \in [0, 1]$ be a weighting vector such that $\sum_{i=1}^{m} w_i = 1$, the 2-tuple weighted mean operator associated with W is the function $2TWM : \bar{S}^m \rightarrow \bar{S}$ defined as:

$$2TWM(x) = \Delta\left(\sum_{i=1}^{m} w_i \Delta^{-1}(s_i, \alpha_i)\right) \quad (3)$$

2.4 2-Tuple Model for Managing Linguistic Information: Linguistic Hierarchies

Sometimes, it is necessary to deal with linguistic frameworks in which the linguistic information can belong to linguistic term sets with different granularity. For instance, qualitative criteria or environmental indicators assessed in an EIA problem can have different nature or involve different types of uncertainty, in such cases the information can be elicited by means of linguistic terms belonging to different linguistic term sets with either different semantics or granularity. In [8] was presented an approach to manage multigranular linguistic information which builds a structure so-called Linguistic Hierarchy (LH), and a computational symbolic model based on the 2-tuple linguistic model is defined over it to accomplish the CW processes.

A LH is the union of all levels $t : LH = \bigcup_t l(t, n(t))$, where each level t of a LH corresponds to a linguistic term set with a granularity of uncertainty of $n(t)$ denoted as: $S^{n(t)} = \{s_0^{n(t)}, \ldots, s_{n(t)-1}^{n(t)}\}$ [8]. The construction of LH must satisfy a pair of rules, so-called LH basic rules:

1. to preserve all former modal points of the membership functions of each linguistic term from one level to the following one.
2. to make smooth transitions between consecutive levels.

The goal is to add a new linguistic term set $S^{n(t+1)}$ by adding a new linguistic term between each pair of terms belonging to the linguistic term set of the previous level t. To do so, it is necessary to reduce the support of the linguistic labels to keep place for the new one located in the middle of them. Therefore, a linguistic term set in the level $t+1$ is obtained from its predecessor as $l(t, n(t)) \longrightarrow l(t+1, 2 \cdot n(t) - 1)$.

A transformation function was defined to transform a linguistic term in level t to its correspondent linguistic term in level $t+1$ following the LH basic rules.

Definition 5 ([8]) Let $LH = \bigcup_t l(t, n(t))$ be a LH whose linguistic term sets are denoted as $S^{n(t)} = \{s_0^{n(t)}, \ldots, s_{n(t)-1}^{n(t)}\}$ and let us consider the 2-tuple linguistic representation. The transformation function $TF_t^{t'} : S^{n(t)} \rightarrow S^{n(t')}$, from a linguistic label in level t to its correspondent label in level t', satisfying the LH basic rules, is defined as:

$$TF_t^{t'}((s_i^{n(t)}, \alpha^{n(t)})) = \Delta\left(\frac{\Delta^{-1}(s_i^{n(t)}, \alpha^{n(t)}) \cdot (n(t') - 1)}{n(t) - 1}\right) \tag{4}$$

It is then remarkable that the final results of any computational process, dealing with linguistic information assessed in a LH, can be expressed in any linguistic term set of the LH by means of a retranslation process accomplished by $TF_t^{t'}$ without loss of information.

3 EIA of Moa Port Based on 2-Tuple Linguistic Models

This chapter studies the impact of the Moa Port operations from a environmental point of view. Hence, it was found that there exist uncertain information about project's impacts due to the inability to collect the required data sets on initial existing baseline conditions or a base index for the evaluation and comparison of alternative proposals. Because of this, EIA based on measurement collection and comparisons are not suitable for this problem.

This section first describes the Moa Port operations and its environmental problem based on information from [6]. Later on it presents a linguistic multi-criteria evaluation framework for the EIA which will use the 2-tuple linguistic model to capture the vagueness of the information and deal with the complexity of the problem. The environmental parameters are defined through linguistic variables assessed by 2-tuple linguistic values and the different criteria can be assessed by using different linguistic scales [8] according to experts' knowledge and nature of criteria. Therefore, the proposal deals with multi-granular linguistic information without loss of information by using the linguistic 2-tuple extension, so-called *linguistic hierarchies*, because they allow to define an evaluation framework that models linguistic variables with different granularity according to the real situation. To operate with linguistic values, CW processes are carried out by using 2-tuple linguistic computing model. Eventually, this section provides and discusses the results of applying the linguistic EIA based on the 2-tuple linguistic model to the Moa Port problem.

3.1 The Moa Port Environmental Problem

Moa municipality is a 766.33 km^2 area located in Holgun province on the island of Cuba. It has borders to the north with the Atlantic Ocean to the west with the municipalities of Frank Pas and Sagua de Tnamo, to the East and South with the municipalities of Baracoa and Yateras and the Moa Port is located in one of its 4 bays. Moa Port has facilities for receiving goods and fuels where maneuvers for moving supplies are made, including fuel oil by structures that carry it. This is mainly used for loading and unloading of international voyage, through which nickel exports and import of inputs is done such as crude fuel oil, anthracite coal, sulfur in solid state, among others; all of them critical to the industry. It also provides services of maneuvering, dredging, storing and distributing raw materials (coal, ammonia, fuel, sulfur, etc.) and mining services, among others. However, the port maritime activity causes a negative impact on the coastal ecosystem mainly due to the absence of systems to treat liquid waste and solid waste and lack of an adequate sewerage system. This is largely related to its high production activity and capacity compared to other similar in the area.

Different variables such as waste water, solid waste and industrial materials cause impacts generated by activities on ships as increasing pollutants are released by re-suspension and dispersion of sediments in the water column, suspended sediments increase drags as ocean currents prevail in the dump area and the effect on existing species in the area near the entrance of the canal coral reef [6]. Other port activities also cause the modification of the bottom relief in the dump area, due to the deposition of the extracted material, the changes in the circulation of ocean currents as a result of the modification of bathymetry, the effect on existing species in the coral reef near the entrance channel has also altered some fish species by ingestion and accumulation of heavy metals. It is affected by the activity of human origin, basically in the industrial area, extending its influence to tens of kilometers away in the prevailing wind directions, causing the presence of pollutants such as CO_2, CO, SO_2, N_2, CH_4, H_2, even in low concentrations, these substances can affect not only air quality, but the soil biota of ecosystems and the health of the population.

A small tropical forest alluvial Ombrophilous, bordering the mangrove to the east of the port, is an example of change in vegetation with only three endemic species: Sabal parviflora Becc., Ficus membranacea Wright and Bucida espinosa Jenn [6]. This is not only because of the indiscriminate cutting of tree species but also because of the activity of construction of the port facilities and the dumping of waste oil from an Oil Base, which has killed large numbers of specimens of ocuje and jcaro.

In the coastal port area, mangroves are affected due to soil contamination that caused a change in physical-chemical properties and soil erosion; altering flood and salinity lead to defoliation, habitat fragmentation, changes in species composition and soil compaction. All this has led to a decrease in associated wildlife, alteration of surface runoff, soil loss of mangroves, decreased scenic value of the landscape, mangrove mortality and the spread of unwanted species [6].

3.2 *The Moa Port MCDM Problem Defined in a Linguistic Multigranular Framework*

Once it has been described the Moa Port environmental problem, it is time to define the framework of the multi-criteria evaluation problem that will be used to solve the EIA of this site. The operation phase of the Moa Port considered for EIA includes the following actions:

- Wastewater from human activities discharge (a_1),
- wastewater from industrial activities discharge (a_2),
- pollutant emissions (a_3), and
- waste accumulation due to creation of industrials drains (a_4).

These actions interact with the following environmental factors:

- Quality of air (f_1),
- surface/groundwater (f_2),
- soil (f_3),
- vegetation (f_4),
- fauna (f_5),
- ecological relationships (f_6) and
- health and hygiene (f_7).

Impacts are assessed considering ten criteria adapted from [2] whose weights are given by $W^c = \{\frac{3}{13}, \frac{2}{13}, \frac{1}{13}, \frac{1}{13}, \frac{1}{13}, \frac{1}{13}, \frac{1}{13}, \frac{1}{13}, \frac{1}{13}, \frac{1}{13}, \frac{1}{13}\}$:

- (c_1) Intensity: The effect of the action on the factor.
- (c_2) Extension: The impact's sphere of influence of the action in relation to the site.
- (c_3) Moment: The time between the appearance of the action and the start of the effect on the factor.
- (c_4) Persistence: The time that the effect of the action would supposedly last.
- (c_5) Reversibility: The possibility of restoring the affected factor to its initial state by natural means.
- (c_6) Accumulation: The progressive increase in manifestation of the effect.
- (c_7) Probability: Likehood of occurrence of the impact.
- (c_8) Effect: How the effect of the action on an environmental factor is manifested.
- (c_9) Periodicity: The regularity in the manifestation of the effect on the environmental factor.

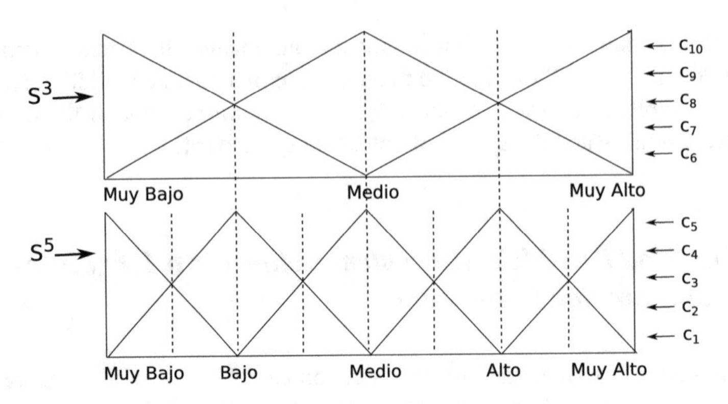

Fig. 1 LH with linguistic term sets of 3 and 5 labels and criteria assessed

- (c_{10}) Recoverability: Social perception of the possibility of artificially restoring the affected factor through human intervention.

Inspired by the use of different numerical scales to assess criteria in crisp traditional matrices and based on information from [6] for the Moa Port EIA, the linguistic information about impacts is defined using a multigranular linguistic framework. The LH is conformed by the linguistic term sets S^3 and S^5 with three and five linguistic terms, as depicted in Fig. 1, which fulfil the LH basic rules.

The granularities of the linguistic term sets have been selected based on levels of discrimination used in previous works on linguistic EIA [29] and the number of values in scales presented in conventional proposals [2] which was used in the original problem solution [6].

3.3 The 2-Tuple Based Solving Procedure for EIA

The solving procedure is based on the scheme presented in [29] and consists of four main steps:

1. Constructing the impact matrix.
2. Gathering multigranular linguistic information about impacts.
3. Unifying multigranular linguistic information.
4. Computing impacts.

The proposed procedure schematically is shown in Fig. 2.

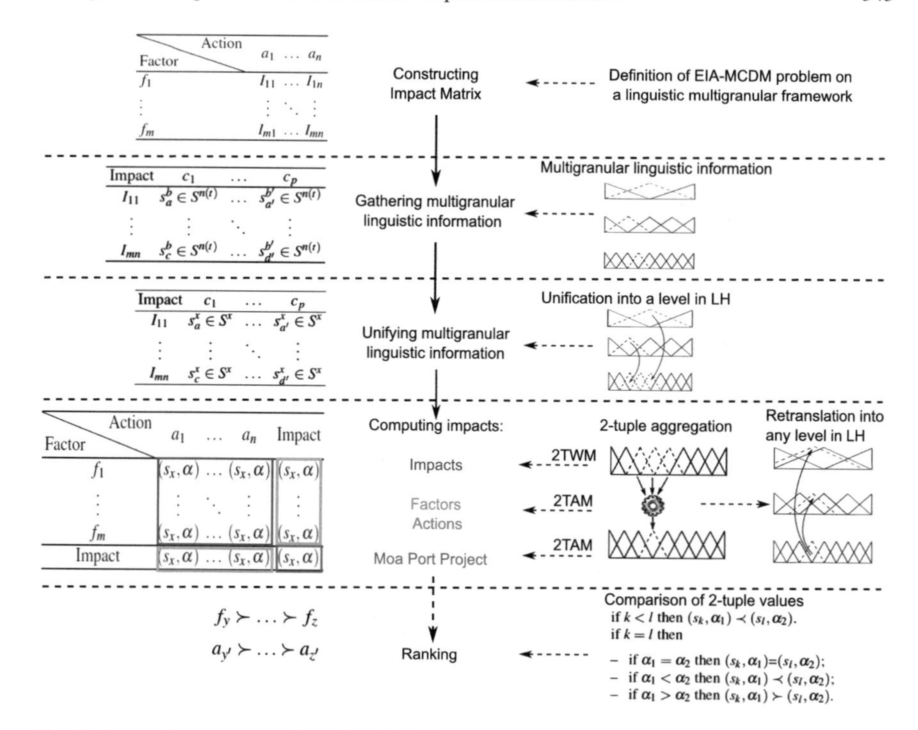

Fig. 2 Resolution procedure for EIA of the Moa Port

Table 1 Impact matrix for Moa Port

Factor\ Action	a_1	a_2	a_3	a_4
f_1	–	–	I_{13}	I_{14}
f_2	I_{21}	I_{22}	I_{23}	I_{24}
f_3	–	I_{32}	I_{33}	I_{34}
f_4	–	I_{42}	I_{43}	I_{44}
f_5	I_{51}	I_{52}	I_{53}	I_{54}
f_6	I_{61}	I_{62}	I_{63}	I_{64}
f_7	I_{71}	I_{72}	I_{73}	I_{74}

3.3.1 Constructing the Impact Matrix

The double-entry matrix relates the set of factors and the set of actions developed in Moa Port and their intersections represents the possible impacts. There are 24 impacts with negative nature under consideration as depicted in Table 1.

3.3.2 Gathering Linguistic Information About Impacts

Once the impact matrix is constructed, the linguistic information about impacts is defined by means of linguistic terms from the LH (see Table 2) considering the information provided in the original problem [6]. The linguistic terms have the syntax illustrated in Fig. 1, but in this case we shall use a normalized syntax for a better comprehensiveness of the resolution processes.

3.3.3 Unifying Multigranular Linguistic Information

The conversion of gathered linguistic information into linguistic 2-tuple values consists of adding a value 0 as symbolic translation. After that, a linguistic term set is chosen to make uniform the multigranular linguistic information. In this case, we

Table 2 Gathered criteria value for impacts

Impact	c_1	c_2	c_3	c_4	c_5	c_6	c_7	c_8	c_9	c_{10}
I_{13}	s_1^5	s_1^5	s_2^5	s_1^5	s_0^5	s_0^3	s_2^3	s_2^3	s_1^3	s_1^3
I_{14}	s_4^5	s_2^5	s_2^5	s_2^5	s_1^5	s_2^3	s_2^3	s_2^3	s_2^3	s_2^3
I_{21}	s_1^5	s_1^5	s_0^5	s_0^5	s_0^5	s_0^3	s_1^3	s_0^3	s_0^3	s_0^3
I_{22}	s_2^5	s_2^5	s_2^5	s_2^5	s_1^5	s_2^3	s_2^3	s_0^3	s_1^3	s_1^3
I_{23}	s_3^5	s_2^5	s_2^5	s_2^5	s_1^5	s_2^3	s_0^3	s_2^3	s_0^3	s_2^3
I_{24}	s_3^5	s_1^5	s_2^5	s_2^5	s_1^5	s_2^3	s_2^3	s_2^3	s_2^3	s_1^3
I_{32}	s_1^5	s_1^5	s_2^5	s_2^5	s_1^5	s_2^3	s_2^3	s_2^3	s_1^3	s_0^3
I_{33}	s_2^5	s_2^5	s_2^5	s_2^5	s_1^5	s_2^3	s_0^3	s_2^3	s_0^3	s_0^3
I_{34}	s_2^5	s_1^5	s_1^5	s_2^5	s_0^5	s_2^3	s_0^3	s_2^3	s_1^3	s_0^3
I_{42}	s_2^5	s_1^5	s_2^5	s_2^5	s_1^5	s_2^3	s_0^3	s_2^3	s_1^3	s_0^3
I_{43}	s_2^5	s_2^5	s_2^5	s_2^5	s_1^5	s_1^3	s_0^3	s_2^3	s_0^3	s_0^3
I_{44}	s_1^5	s_1^5	s_1^5	s_1^5	s_0^5	s_2^3	s_0^3	s_2^3	s_0^3	s_0^3
I_{51}	s_0^5	s_1^5	s_0^5	s_0^5	s_0^5	s_0^3	s_0^3	s_0^3	s_0^3	s_0^3
I_{52}	s_2^5	s_1^5	s_1^5	s_1^5	s_0^5	s_2^3	s_0^3	s_2^3	s_0^3	s_0^3
I_{53}	s_3^5	s_2^5	s_2^5	s_2^5	s_1^5	s_2^3	s_2^3	s_2^3	s_1^3	s_1^3
I_{54}	s_2^5	s_2^5	s_2^5	s_2^5	s_1^5	s_2^3	s_2^3	s_2^3	s_1^3	s_0^3
I_{61}	s_1^5	s_0^5	s_1^5	s_0^5	s_0^5	s_2^3	s_0^3	s_2^3	s_0^3	s_0^3
I_{62}	s_2^5	s_1^5	s_0^5	s_0^5	s_0^5	s_1^3	s_0^3	s_2^3	s_0^3	s_0^3
I_{63}	s_3^5	s_2^5	s_2^5	s_2^5	s_1^5	s_2^3	s_2^3	s_2^3	s_1^3	s_0^3
I_{64}	s_2^5	s_2^5	s_2^5	s_2^5	s_1^5	s_2^3	s_2^3	s_2^3	s_1^3	s_0^3
I_{71}	s_2^5	s_0^5	s_0^5	s_1^5	s_0^5	s_2^3	s_0^3	s_2^3	s_0^3	s_1^3
I_{72}	s_2^5	s_2^5	s_2^5	s_2^5	s_1^5	s_2^3	s_2^3	s_2^3	s_2^3	s_2^3
I_{73}	s_1^5	s_2^5	s_1^5	s_2^5	s_1^5	s_2^3	s_2^3	s_2^3	s_2^3	s_2^3
I_{74}	s_2^5	s_2^5	s_2^5	s_2^5	s_1^5	s_2^3	s_2^3	s_2^3	s_2^3	s_2^3

Table 3 2-tuple linguistic values unified on S^5

Impact	c_1	c_2	c_3	c_4	c_5	c_6	c_7	c_8	c_9	c_{10}
I_{13}	$(s^5_1, 0)$	$(s^5_1, 0)$	$(s^5_2, 0)$	$(s^5_1, 0)$	$(s^5_0, 0)$	$(s^5_0, 0)$	$(s^5_4, 0)$	$(s^5_4, 0)$	$(s^5_2, 0)$	$(s^5_2, 0)$
I_{14}	$(s^5_4, 0)$	$(s^5_2, 0)$	$(s^5_2, 0)$	$(s^5_2, 0)$	$(s^5_1, 0)$	$(s^5_4, 0)$	$(s^5_4, 0)$	$(s^5_4, 0)$	$(s^5_4, 0)$	$(s^5_4, 0)$
I_{21}	$(s^5_1, 0)$	$(s^5_1, 0)$	$(s^5_0, 0)$	$(s^5_0, 0)$	$(s^5_0, 0)$	$(s^5_0, 0)$	$(s^5_2, 0)$	$(s^5_0, 0)$	$(s^5_0, 0)$	$(s^5_0, 0)$
I_{22}	$(s^5_2, 0)$	$(s^5_2, 0)$	$(s^5_2, 0)$	$(s^5_2, 0)$	$(s^5_1, 0)$	$(s^5_4, 0)$	$(s^5_4, 0)$	$(s^5_0, 0)$	$(s^5_2, 0)$	$(s^5_2, 0)$
I_{23}	$(s^5_3, 0)$	$(s^5_2, 0)$	$(s^5_2, 0)$	$(s^5_2, 0)$	$(s^5_1, 0)$	$(s^5_4, 0)$	$(s^5_0, 0)$	$(s^5_4, 0)$	$(s^5_0, 0)$	$(s^5_4, 0)$
I_{24}	$(s^5_3, 0)$	$(s^5_1, 0)$	$(s^5_2, 0)$	$(s^5_2, 0)$	$(s^5_1, 0)$	$(s^5_4, 0)$	$(s^5_4, 0)$	$(s^5_4, 0)$	$(s^5_4, 0)$	$(s^5_2, 0)$
I_{32}	$(s^5_1, 0)$	$(s^5_1, 0)$	$(s^5_2, 0)$	$(s^5_2, 0)$	$(s^5_1, 0)$	$(s^5_4, 0)$	$(s^5_4, 0)$	$(s^5_4, 0)$	$(s^5_2, 0)$	$(s^5_0, 0)$
I_{33}	$(s^5_2, 0)$	$(s^5_2, 0)$	$(s^5_2, 0)$	$(s^5_2, 0)$	$(s^5_1, 0)$	$(s^5_4, 0)$	$(s^5_0, 0)$	$(s^5_4, 0)$	$(s^5_0, 0)$	$(s^5_0, 0)$
I_{34}	$(s^5_2, 0)$	$(s^5_1, 0)$	$(s^5_1, 0)$	$(s^5_2, 0)$	$(s^5_0, 0)$	$(s^5_4, 0)$	$(s^5_0, 0)$	$(s^5_4, 0)$	$(s^5_2, 0)$	$(s^5_0, 0)$
I_{42}	$(s^5_2, 0)$	$(s^5_1, 0)$	$(s^5_2, 0)$	$(s^5_2, 0)$	$(s^5_1, 0)$	$(s^5_4, 0)$	$(s^5_0, 0)$	$(s^5_4, 0)$	$(s^5_2, 0)$	$(s^5_0, 0)$
I_{43}	$(s^5_2, 0)$	$(s^5_2, 0)$	$(s^5_2, 0)$	$(s^5_2, 0)$	$(s^5_1, 0)$	$(s^5_2, 0)$	$(s^5_0, 0)$	$(s^5_4, 0)$	$(s^5_0, 0)$	$(s^5_0, 0)$
I_{44}	$(s^5_1, 0)$	$(s^5_1, 0)$	$(s^5_1, 0)$	$(s^5_1, 0)$	$(s^5_0, 0)$	$(s^5_4, 0)$	$(s^5_0, 0)$	$(s^5_4, 0)$	$(s^5_0, 0)$	$(s^5_0, 0)$
I_{51}	$(s^5_0, 0)$	$(s^5_1, 0)$	$(s^5_0, 0)$	$(s^5_0, 0)$	$(s^5_0, 0)$	$(s^5_0, 0)$	$(s^5_0, 0)$	$(s^5_0, 0)$	$(s^5_0, 0)$	$(s^5_0, 0)$
I_{52}	$(s^5_2, 0)$	$(s^5_1, 0)$	$(s^5_1, 0)$	$(s^5_1, 0)$	$(s^5_0, 0)$	$(s^5_4, 0)$	$(s^5_0, 0)$	$(s^5_4, 0)$	$(s^5_0, 0)$	$(s^5_0, 0)$
I_{53}	$(s^5_3, 0)$	$(s^5_2, 0)$	$(s^5_2, 0)$	$(s^5_2, 0)$	$(s^5_1, 0)$	$(s^5_4, 0)$	$(s^5_4, 0)$	$(s^5_4, 0)$	$(s^5_2, 0)$	$(s^5_2, 0)$
I_{54}	$(s^5_2, 0)$	$(s^5_2, 0)$	$(s^5_2, 0)$	$(s^5_2, 0)$	$(s^5_1, 0)$	$(s^5_4, 0)$	$(s^5_4, 0)$	$(s^5_4, 0)$	$(s^5_2, 0)$	$(s^5_0, 0)$
I_{61}	$(s^5_1, 0)$	$(s^5_0, 0)$	$(s^5_1, 0)$	$(s^5_0, 0)$	$(s^5_0, 0)$	$(s^5_4, 0)$	$(s^5_0, 0)$	$(s^5_4, 0)$	$(s^5_0, 0)$	$(s^5_0, 0)$
I_{62}	$(s^5_2, 0)$	$(s^5_1, 0)$	$(s^5_0, 0)$	$(s^5_0, 0)$	$(s^5_0, 0)$	$(s^5_2, 0)$	$(s^5_0, 0)$	$(s^5_4, 0)$	$(s^5_0, 0)$	$(s^5_0, 0)$
I_{63}	$(s^5_3, 0)$	$(s^5_2, 0)$	$(s^5_2, 0)$	$(s^5_2, 0)$	$(s^5_1, 0)$	$(s^5_4, 0)$	$(s^5_4, 0)$	$(s^5_4, 0)$	$(s^5_2, 0)$	$(s^5_0, 0)$
I_{64}	$(s^5_2, 0)$	$(s^5_2, 0)$	$(s^5_2, 0)$	$(s^5_2, 0)$	$(s^5_1, 0)$	$(s^5_4, 0)$	$(s^5_4, 0)$	$(s^5_4, 0)$	$(s^5_2, 0)$	$(s^5_0, 0)$
I_{71}	$(s^5_2, 0)$	$(s^5_0, 0)$	$(s^5_0, 0)$	$(s^5_1, 0)$	$(s^5_0, 0)$	$(s^5_4, 0)$	$(s^5_0, 0)$	$(s^5_4, 0)$	$(s^5_0, 0)$	$(s^5_2, 0)$
I_{72}	$(s^5_2, 0)$	$(s^5_2, 0)$	$(s^5_2, 0)$	$(s^5_2, 0)$	$(s^5_1, 0)$	$(s^5_4, 0)$	$(s^5_4, 0)$	$(s^5_4, 0)$	$(s^5_4, 0)$	$(s^5_4, 0)$
I_{73}	$(s^5_1, 0)$	$(s^5_2, 0)$	$(s^5_1, 0)$	$(s^5_2, 0)$	$(s^5_1, 0)$	$(s^5_4, 0)$	$(s^5_4, 0)$	$(s^5_4, 0)$	$(s^5_4, 0)$	$(s^5_4, 0)$
I_{74}	$(s^5_2, 0)$	$(s^5_2, 0)$	$(s^5_2, 0)$	$(s^5_2, 0)$	$(s^5_1, 0)$	$(s^5_4, 0)$	$(s^5_4, 0)$	$(s^5_4, 0)$	$(s^5_4, 0)$	$(s^5_4, 0)$

shall choose the linguistic term set S^5, since it has the maximum granularity and can express more degree of uncertainty and also because it fulfils Miller's observation [17] regarding the capability of distinction levels of human beings. Therefore, applying the transformation function from Eq. (4), the unified preference values expressed by means of 2-tuple linguistic values are obtained (see Table 3).

3.3.4 Computing Impacts

Applying the 2TWM aggregation operator from Eq. (3) with the weighting vector W^c, the collective value of each impact is computed. Afterwards, the global impact of factors, actions and the project is computed applying the 2TAM aggregation operator in Eq. (2). All results are shown in Table 4.

Table 4 Impact matrix for Moa Port

	a_1	a_2	a_3	a_4	Impact
f_1	–	–	$(M, -0.46)$	$(A, 0.15)$	$(B, 0.17)$
f_2	$(B, -0.46)$	$(M, 0.08)$	$(M, 0.31)$	$(A, -0.38)$	$(M, -0.12)$
f_3	–	$(M, -0.15)$	$(M, -0.23)$	$(M, -0.38)$	$(B, 0.31)$
f_4	–	$(M, -0.23)$	$(M, -0.38)$	$(B, 0.15)$	$(B, 0.14)$
f_5	$(MB, 0.15)$	$(B, 0.38)$	$(A, -0.38)$	$(M, 0.23)$	$(M, -0.40)$
f_6	$(B, -0.08)$	$(B, 0.08)$	$(M, 0.46)$	$(M, 0.23)$	$(M, -0.23)$
f_7	$(B, 0.31)$	$(A, -0.31)$	$(M, 0.38)$	$(M, -0.31)$	$(M, 0.26)$
Impact	$(MB, 0.42)$	$(M, -0.45)$	$(M, 0.10)$	$(M, 0.24)$	$(M, -0.16)$

3.4 Results and Discussion

By using the comparison operation of the 2-tuple linguistic model reviewed in Sect. 2.3, impacts, actions and factors are ranked according to the values computed in the previous step. The lower the better, then the more affected factors and the more aggressive actions have higher impact values. Finally, the rankings of impacts, factors and actions are as follows:

1. Impacts' ranking: $I_{14} \succ I_{72} \succ I_{74} \succ I_{24} \succ I_{53} \succ I_{63} \succ I_{73} \succ I_{23} \succ I_{54} \succ I_{64} \succ I_{22} \succ I_{32} \succ I_{33} \succ I_{42} \succ I_{43} \succ I_{34} \succ I_{13} \succ I_{52} \succ I_{71} \succ I_{44} \succ I_{62} \succ I_{61} \succ I_{21} \succ I_{51}$.
2. Factors' ranking: $f_7 \succ f_2 \succ f_6 \succ f_5 \succ f_3 \succ f_1 \succ f_4$.
3. Actions' ranking: $a_4 \succ a_3 \succ a_2 \succ a_1$.

Rankings suggest on the one hand that the most affected environmental factors are health and hygiene (f_7), and surface/groundwater (f_2). On the other hand, waste accumulation due to creation of industrials drains is the more aggressive action (a_4), followed by the emissions (a_3), related with the throwing to the ground hydrocarbons, liberation to the midway air of gases, noises and materials in particles, contamination of sources for dragging of sediments, hydrocarbons and chemical substances. Previous rankings could support decisions establishing correction measures for handling the more aggressive actions and the highest impacts values in order to reduce such a global impact of the project, which would lead to mitigation of loss of natural resources and human welfare.

The linguistic value for the global impact of the project means that the overall effects caused by the operation of the Moa Port can be expressed in both linguistic term sets of the LH by means of a retranslation process accomplished by $TF_t^{t'}$ without loss of information (see Fig. 3):

$$TF_5^3((s_2^5, -0.16)) = \Delta\left(\frac{\Delta^{-1}(s_2^5, -0.16) \cdot (3-1)}{5-1}\right) = (s_1^3, -0.08)$$

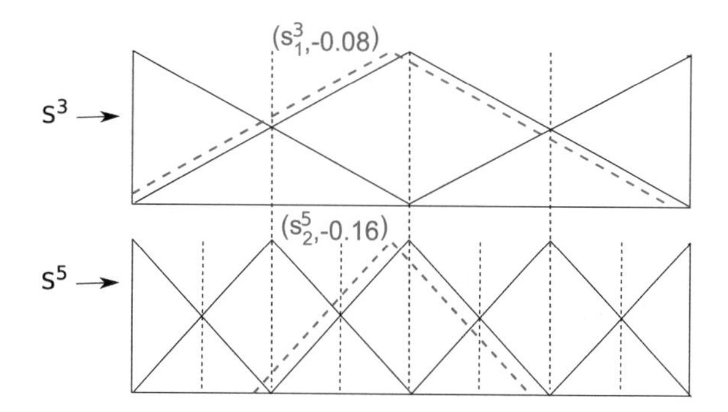

Fig. 3 Global impact of the Moa Port project expressed on both linguistic term set of the LH

4 Conclusions

In this chapter we have proposed an alternative EIA of Moa Port based on the 2-tuple linguistic model. The problem was modelled by means of multiple criteria which can be assessed using different linguistic term sets from a LH. It operates without loss of information and provides easily understandable linguistic impact values that help stakeholders to interpret EIA in a more natural and logic way.

This approach provides three main benefits when developing EIA of maritime port projects: (i) a flexible evaluation framework by enabling the use of multigranular linguistic information; (ii) improvement of the interpretability and therefore less time training to understand the conventional scales used to assess criteria and the meaning of results; and (iii) a enhanced generalisation due to many aggregation operators that can be applied to model other situations and calculate other complex indicators. This could increase the effectiveness of the EIA process while promotes environmental protection and sustainability.

It could be worthwhile to undertake further research on topics such as how multi-granular linguistic term sets are defined, how the information about impacts is gathered and how the application of different aggregation operators affects final results.

Acknowledgements This work is partially supported by the Spanish National research project TIN2015-66524-P, Spanish Ministry of Economy and Finance Postdoctoral Training (FPDI-2013-18193) and ERDF.

References

1. Blanco, A., Delgado, M., Martín, J., Polo, M.: AIEIA: software for fuzzy environmental impact assessment. Expert Syst. Appl. **36**(5), 9135–9149 (2009)
2. Conesa, V.: Guía metodológica para la evaluación del impacto ambiental. Mundi-Prensa, (1997)
3. Degani, R., Bortolan, G.: The problem of linguistic approximation in clinical decision making. Int. J. Approx. Reason. **2**, 143–162 (1988)
4. Delgado, M., Verdegay, J., Vila, M.: On aggregation operations of linguistic labels. Int. J. Intell. Syst. **8**(3), 351–370 (1993)
5. Duarte, O., Requena, I., Rosario, Y.: Fuzzy techniques for environmental-impact assessment in the mineral deposit of Punta Gorda (Moa, Cuba). Environ. Technol. **28**(6), 659–669 (2007)

6. Guilarte, A., Díaz, A., Nápoles, J., Fernández, O., Abalos, A., Pérez, R.: Valoración de impacto ambiental en el Puerto Moa-Holguín. *Revista Colombiana de Biotecnología*, **XVII**(2), 129–139 (2015)
7. Herrera, F., Martínez, L.: A 2-tuple fuzzy linguistic representation model for computing with words. IEEE Trans. Fuzzy Syst. **8**(6), 746–752 (2000)
8. Herrera, F., Martínez, L.: A model based on linguistic 2-tuples for dealing with multigranular hierarchical linguistic contexts in multi-expert decision-making. IEEE Trans. Syst. Man Cybern. Part B: Cybern. **31**(2), 227–234 (2001)
9. Kahraman, C., Onar, S.C., Oztaysi, B.: Fuzzy multicriteria decision-making: a literature review. Int. J. Comput. Intell. Syst. **8**(4), 637–666 (2015)
10. Kiker, G., Bridges, T., Varghese, A., Seager, T., Linkov, I.: Application of multicriteria decision analysis in environmental decision making. Integr. Environ. Assess. Manag. **1**(2), 95–108 (2005)
11. Liu, H., Cai, J., Martínez, L.: The importance weighted continuous generalized ordered weighted averaging operator and its application to group decision making. Knowl.-based Syst. **48**(1), 24–36 (2013)
12. Martínez, L., Herrera, F.: An overview on the 2-tuple linguistic model for computing with words in decision making: extensions, applications and challenges. Inf. Sci. **207**(1), 1–18 (2012)
13. Martínez, L., Liu, J., Yang, J.-B.: A fuzzy model for design evaluation based on multiple criteria analysis in engineering systems. Int. J. Uncertain. Fuzziness Knowl.-Based Syst. **14**(3), 317–336 (2006)
14. Martínez, L., Rodríguez, R., Herrera, F.: The 2-tuple Linguistic Model. Computing with Words in Decision Making, 1st edn. Springer, Berlin (2015)
15. Martínez, L., Ruan, D., Herrera, F.: Computing with words in decision support systems: an overview on models and applications. Int. J. Comput. Intell. Syst. **3**(4), 382–395 (2010)
16. Mendel, J., Zadeh, L., Trillas, E., Yager, R., Lawry, J., Hagras, H., Guadarrama, S.: What computing with words means to me [discussion forum]. Comput. Intell. Mag. IEEE **5**(1), 20–26 (2010)
17. Miller, G.: The magical number seven plus or minus two: some limits on our capacity of processing information. Psychol. Rev. **63**, 81–97 (1956)
18. Peche, R., Rodríguez, E.: Environmental impact assessment by means of a procedure based on fuzzy logic: a practical application. Environ. Impact Assess. Rev. **31**(2), 87–96 (2011)
19. Rodríguez, R., Labella, Á., Martínez, L.: An overview on fuzzy modelling of complex linguistic preferences in decision making. Int. J. Comput. Intell. Syst. **9**, 81–94 (2016)
20. Rodríguez, R., Martínez, L.: An analysis of symbolic linguistic computing models in decision making. Int. J. Gen. Syst. **42**(1), 121–136 (2013)
21. Shepard, R.: Quantifying Environmental Impact Assessments Using Fuzzy Logic. Springer Series on Environmental Management. Springer, Berlin (2005)
22. Wang, J., Hao, J.: A new version of 2-tuple fuzzy linguistic representation model for computing with words. IEEE Trans. Fuzzy Syst. **14**(3), 435–445 (2006)
23. Xu, Z.: A method based on linguistic aggregation operators for group decision making with linguistic preference relations. Inf. Sci. **166**(1–4), 19–30 (2004)
24. Yager, R.: An approach to ordinal decision making. Int. J. Approx. Reason. **2**, 237–261 (1995)
25. Yager, R.: On the retranslation process in Zadeh's paradigm of computing with words. IEEE Trans. Syst. Man Cybern. Part B: Cybern. **34**(2), 1184–1195 (2004)
26. Zadeh, L.: The concept of a linguistic variable and its application to approximate reasoning-I. Inf. Sci. **8**(3), 199–249 (1975)
27. Zadeh, L.: The concept of a linguistic variable and its application to approximate reasoning-II. Inf. Sci. **8**(4), 301–357 (1975)
28. Zadeh, L.: The concept of a linguistic variable and its application to approximate reasoning-III. Inf. Sci. **9**(1), 43–80 (1975)
29. Zulueta, Y., Rodríguez, D., Bello, R., Martínez, L.: A linguistic fusion approach for heterogeneous environmental impact significance assessment. Appl. Math. Model. **40**(2), 1402–1417 (2016)

Printed in the United States
By Bookmasters